Thin Film Optical Coatings for Effective Solar Energy Utilization: APCVD Spectrally Selective Surfaces and Energy Control Coatings

Thin Film Optical Coatings for Effective Solar Energy Utilization: APCVD Spectrally Selective Surfaces and Energy Control Coatings

K. A. Gesheva
Editor

Nova Science Publishers, Inc.
New York

Copyright © 2007 by Nova Science Publishers, Inc.

All rights reserved. No part of this book may be reproduced, stored in a retrieval system or transmitted in any form or by any means: electronic, electrostatic, magnetic, tape, mechanical photocopying, recording or otherwise without the written permission of the Publisher.

For permission to use material from this book please contact us:
Telephone 631-231-7269; Fax 631-231-8175
Web Site: http://www.novapublishers.com

NOTICE TO THE READER

The Publisher has taken reasonable care in the preparation of this book, but makes no expressed or implied warranty of any kind and assumes no responsibility for any errors or omissions. No liability is assumed for incidental or consequential damages in connection with or arising out of information contained in this book. The Publisher shall not be liable for any special, consequential, or exemplary damages resulting, in whole or in part, from the readers' use of, or reliance upon, this material.

Independent verification should be sought for any data, advice or recommendations contained in this book. In addition, no responsibility is assumed by the publisher for any injury and/or damage to persons or property arising from any methods, products, instructions, ideas or otherwise contained in this publication.

This publication is designed to provide accurate and authoritative information with regard to the subject matter covered herein. It is sold with the clear understanding that the Publisher is not engaged in rendering legal or any other professional services. If legal or any other expert assistance is required, the services of a competent person should be sought. FROM A DECLARATION OF PARTICIPANTS JOINTLY ADOPTED BY A COMMITTEE OF THE AMERICAN BAR ASSOCIATION AND A COMMITTEE OF PUBLISHERS.

LIBRARY OF CONGRESS CATALOGING-IN-PUBLICATION DATA

Gesheva, K. A.
 Thin film optical coatings for effective solar energy utilization / K.A. Gesheva.
   p. cm.
 Includes bibliographical references and index.
 ISBN-13: 978-1-60021-643-5 (hardcover : alk. paper)
 ISBN-10: 1-60021-643-9 (hardcover : alk. paper)
 1. Solar cells--Materials. 2. Thin films 650--Optical properties 3. Chemical vapor deposition. 4. Optical coatings. I. Title.
TK2960.G47 2007
621.31'244--dc22
                                2007010122

Published by Nova Science Publishers, Inc. ✤ *New York*

CONTENTS

Preface		vii
Chapter 1	APCVD Spectrally Selective Surfaces *K. A. Gesheva and E. E. Chain*	1
Chapter 1A	Optical Properties of Cermet Materials APCVD Spectrally Selective Surfaces – Black Molybdenum Solar Absorbers *K. A. Gesheva and E. E. Chain*	5
Chapter 1B	Chemical Vapor Deposition (CVD) Technology *K. A. Gesheva and E. E. Chain*	21
Chapter 2	Energy Control Coatings Electrochromism, Materials and Applications *K. A. Gesheva*	97
Chapter 2A	Technology Development and Properties of APCVD WO_3 Electrochromic Thin Films *D. S. Gogova and K. A. Gesheva*	127
Chapter 2B	CVD Molybdenum Oxide Thin Films as Electrochromic Material *T. M. Ivanova and K. A. Gesheva*	165
Chapter 2C	Investigation of CVD Mixed Oxide Films Based on Molybdenum and Tungsten *T. M. Ivanova and K. A. Gesheva*	211
Chapter 2D	Investigation of CVD Obtained Thin Films of Chromium Oxide *T. M. Ivanova and K. A. Gesheva*	247
General Conclusions		269
Appendix I		271
Appendix II		273
Abbreviations, Acronyms and Symbols		283
Index		287

PREFACE

This book has a general purpose to gather the research results on different thin film optical coatings fabricated by atmospheric pressure Chemical Vapor Deposition (APCVD) in its classical versions – chloride CVD process, utilizing dioxydichloride as precursor and carbonyl CVD process, with metal hexacarbonyls as precursors.

This APCVD technology started in Optical Sciences Center (OSC) of Arizona University, where the Editor of this book, Dr. K. A. Gesheva, in 1979/80 had a long term stay working on APCVD chloride process for black molybdenum spectrally selective surfaces. All the details about this complete research are presented in a separate chapter. The results on the chloride black molybdenum (OCBM) are compared with the results from the research done on carbonyl black molybdenum (CBM), developed by Dr. Elizabeth Chain (UTD Dallas Professor, later a long-term employ of Motorolla Co., Phoenix, Arizona), a PhD student at that time of Professor B.O. Seraphin, the leader of Solar Energy Research Group. The results are organized in a way that the advantages of each of the black molybdenum coatings are emphasized through a comparison. Bound theory is adopted to predict the films optical performance reflected by the figure of merit as ratio of solar absorptance and thermal emittance and to estimate its values for the real solar absorber coatings.

The research on molybdenum spectrally selective surfaces has been done under advising of Professor B.O. Seraphin. He was head in Optical Sciences Center at Arizona University of Solar Energy Group with collaborators Dr.D. D. Allred, Dr. M. Jacobson, the PhD students G. Carver and D. Booth, the up going PhD student Elizabeth Chain, the starting his PhD program P. Hillmann, and in collaboration with Professor K. Seshan from the Department of Metallurgy at Arizona University.

Continuation of this work took place upon the arrival of Dr.K.A.Gesheva back in the newly developed Central Laboratory of Solar Energy and New Energy Sources at Bulgarian Academy of Sciences, where she made a "copy" of the APCVD reactor, and continued working on transition metal oxide optical coatings. Dr. Elizabeth Chain's postdoctoral stay at the Academy of Sciences in Sofia, in the Central Laboratory of Solar Energy and New Energy Sources was spent working together employing the newly built up APCVD reactor and started deposition of graded index profile of black metal solar absorbers.

Through some Master of Science programs, followed by two PhD programs of Dr. D.Gogova (1999) and Dr. T. Ivanova (2003), the work progressed to the recent stage on electrochromic optical coatings, a topic on which the main part of the book is devoted.

Through the years, based on this technology, a variety of coatings were developed in response of industry requests as f.i. the one of ISCAR Co. in Israel, a worldwide cutting tools producer. APCVD tungsten coatings as diffusion barriers for the Co containing tools substrates were intended to stop the cobalt diffusion during deposition of overlaying protective hard metal coatings. Here, valuable was the help of Dr.G.Beshkov and Dr.E.Vlakhov, the two from the next-door Institute of Solid State Physics. Metallic APCVD-molybdenum and tungsten films as possible thin film materials with low resistivity for metallization were fabricated. Another type of coating for metallization was developed, WSi_2, which was a product of combined technologies, APCVD and Rapid Thermal Annealing (RTA). Developed also were coatings as In_2O_3:Sn employing spray pyrolysis as a technology, which was applied on larger area glass substrates. These low emissive coatings could be valuable optical coatings for low emissive glass technology. I herein would like to say a few words in honor of my collaborator Oleg Harizanov, who left us too early just at the moment of his readymade undefended PhD dissertation, devoted to sol-gel metal oxide coatings on glass. With Oleg we have had a numerous discussions on optical coatings; I have been missing him since very much.

Some of the publications not related to the main topic are given in Appendix II.

The presented research was following through the years the large scale research literature on optical coatings for solar energy utilization, and thus has reflected it in many publications, references, and citations. We hope this book, which organizes our research results to be a good source for young researchers in handling a complete research on a newly developed thin film material. Since the book relates to thin films produced only with APCVD technology, this would promote the technology as a promising technology of practical importance.

Besides the films already mentioned, the CVD technology of transition metal oxide films finds another appreciation in the area of multilayer optical systems, called X-ray mirrors. In Appendix II are given some aspects of the technology and application of X-ray mirrors, as taken from a work done in collaboration with Bielefeld University, Faculty of Physics (Dr.Frank Hamelmann).

I herein would like to thank my young research collaborator Dr.Tatyana Ivanova (my former PhD student - 2003) who was helping me all around with the book and to Dr. Daniela Gogova (my PhD student – 1999, last several years working abroad on CVD GaN, at the moment she is in Institute for Crystal Growth, Max-Born-Str. 2, D-12489 Berlin, Germany). The two are coauthors in book chapters. I here would like to express my special thanks to Dr. Anna Szekeres from the next door Institute of Solid State Physics, who is always for me a nice, very polite colleague with whom we continue having enjoyable moments in science of material research. Her basic contribution with Spectral Ellipsometry method led to better understanding the optoelectronic properties of our thin film materials. Special thanks are due to G. Popkirov, Assoc.Professor, Dr. from the Central Laboratory of Solar Energy and New Energy Sources, at the moment Docent at Kill University – Germany) who developed for us experimental set-up for electrochromic measurements.

I write this book in honor of my Professor in Material research, Professor B.O. Seraphin.

I devote the book to my family - Dimitur and Evgenia, and therein - Don, Lora Alexandra (11) and Daniel Dimitur (10).

I would like to thank Frank Columbus, Nova Science Publishers President, for inviting me to write this book.

K.A.Gesheva
Sofia, April 2007
Central Laboratory of Solar Energy and New Energy Sources,
Bulgarian Academy of Sciences, Sofia, Bulgaria

Chapter 1

APCVD SPECTRALLY SELECTIVE SURFACES

K. A. Gesheva and E. E. Chain
Central Laboratory of Solar Energy and New Energy Sources,
Bulgarian Academy of Sciences, Sofia, Bulgaria

INTRODUCTION

The impact of thin films on technology lies in their ability to control the flow of photons and electrons precisely with a very small amount of material. The thin films are used as barriers to protect bulk materials from corrosion and abrasion, and are also the basis for the ongoing revolution in electronics and optics. Thin films have become an indispensable, central component of optical technology.

Materials in thin film form rarely duplicate the bulk phase. The structure of thin films differs from that of chemically similar bulk material. Thin film deposition technique far different from the preparation technique of bulk material often generate unfamiliar structures, such as columns, whiskers, and unusual crystalline phases. Thin film composition also diverges from that of nominally similar bulk material. Depending on deposition process, development of vacancies that entrap impurities may be encouraged. Moreover, the proximity of each layer internal volume to adjacent layers promotes inter diffusion, affecting a large fraction of the layer. Similar proximity of the uppermost layers to the atmosphere, coupled with their porous structure, permits gaseous adsorption. These impurities may stabilize unusual crystalline phases.

Many dielectrics, semiconductors and metals can be prepared in thin film form as well as bulk. Thin film technology allows us to produce films with properties that differ from bulk, and to study how these properties are interrelated. Modifications in the composition and microstructure of a thin film induce corresponding variations of most physical properties, a fact that often makes interpretations in terms of bulk material parameters difficult. However, these modifications also permit variations in the properties of a thin film that allow for flexibility in the design for a desired application. This is particularly true for the optical properties of a thin film, which can be modified within wide limits through variations of microstructure and composition.

It is the relation between structure and composition of thin films as determined by their fabrication parameters, on the one hand, and their reflectance, absorptance and transmittance in the spectral regions of solar relevance on the other, which established the scope of the research presented in this book.

Cermet films, i.e. films composed of metal particles in either a dielectric or a conductive host matrix, have been receiving increased attention (Granquist, 1981; Lampert and Washburn, 1979; Ignatiev, O'Neill and Zagac, 1979) sparked in part by their possible

application in spectrally selective surfaces. The major interest lies in the ready synthesis of these materials with optical properties very unlike those of either the metal particles or the matrix. These properties can be optimized within certain limits for a particular application.

For cermet materials, the relevant physical parameters which affect the optical properties are the dielectric functions of the two constituent materials, their volume fractions in the film, and structural factors. Structural factors include the sizes and shapes of grains and their orientation, surface texture, and the film topology, i.e. how the metal particles are dispersed in the film. Detailed structural information is usually the most difficult parameter to obtain in cermet analysis.

In the book, films fabricated basically by chemical vapor deposition will be described.

It is the atmospheric pressure CVD processes employing precursors as dioxydichloride of molybdenum and hexacarbonyl of molybdenum, and the corresponding tungsten salts. The detailed technology and the corresponding results from spectrally selective optical characterization will be given basically for molybdenum, since the case of tungsten did not add especially interesting differences.

Exposing the results will follow systematical way, closing the cycle of relation between technological parameters of CVD deposition process and films structure and composition from one side and the resulting optical properties, from the other.

The method employed is the chemical vapor deposition process proceeding at atmospheric pressure.

CVD is a method of thin film fabrication especially well suited to a study of the relation among film composition, structure and optical properties. CVD processes are fairly complex, with a wide range of adjustable parameters, permitting precise and wide-ranging control over film properties. In this study black molybdenum films have been prepared by two different CVD techniques. The first involved carbonyl process, the second chloride process. By thermal decomposition of molybdenum hexacarbonyl in presence of oxygen, followed by post-deposition anneal in a hydrogen-bearing atmosphere. We refer to these films as to carbonyl black molybdenum (CBM). The second deposition method involves the pyrolitic hydrogen reduction of molybdenum dioxydichloride, and we refer to these films as OCBM (oxychloride black molybdenum).

Black molybdenum films were prepared over a range of process parameters, and their compositional, structural and optical properties were characterized.

The composition and structure of a thin film material are dependent on the process parameters. Deposition of a thin film results in a granular structure, the details of which are dependent on the rate and temperature of the deposition (Movchan and Demchishin, 1969; Blocher, 1974; Tornton, 1974; Bunshah, 1977). For CBM films, the deposition was followed by anneal in a reducing atmosphere, leading to renucleation and grain growth.

Through systematic variation of the process parameters we prepared black molybdenum over a range of structure and composition. To correlate these variations to the resulting optical properties of black molybdenum, the effective dielectric function corresponding to these two-phase cermet materials must be determined. In the past such theoretical calculations have been hampered by insufficient knowledge of microstructure. By specifying the resultant effective dielectric function for black molybdenum films, the optical behavior of these films can be predicted. On this basis we explain the reflectances of the various black molybdenum films in terms of both their structure and composition.

It is the model of the optical properties of black molybdenum which provides the link from film optical properties to fabrication parameters.

The state of the art in material research is employed to establish a correlation between the process parameters and the optical, compositional and structural properties of the resulting black molybdenum films. We identified the various basic optical mechanisms operating in the system by systematically varying the process parameters that strengthen the influence of one at the expense of the others. In the discussion of these results, we establish a link between the properties of carbonyl black molybdenum and oxychloiride

black molybdenum films. Although the films perform similarly, the reaction mechanisms during fabrication are very different, leading to different coatings in terms of structure, composition and optical properties.

By using these results, we can understand how thin film surfaces and interfaces interact with radiation, and apply this understanding to the tailoring of existing materials to meet new optical application. By controlling the composition of black molybdenum films by adjusting preparation parameters, we may modify the structural and thereby the optical properties of the deposited film. This approach to tailoring should use the interaction of film properties and preparation parameters to map the behavior of the material under different conditions.

An important example of tailoring thin film optical properties is provided by the problem of conversion of sunlight into heat at high temperature. Solar radiation is available only at very low energy density as compared to conventional sources of thermal energy. To use solar radiation for heating, cooling, industrial processing, or the generation of electricity by thermodynamic cycles, the dilute solar flux must be interpreted, redirected or concentrated and converted into heat. The materials that accomplish these functions must meet three requirements. First, they must have the optical properties necessary for efficient conversion of the solar radiation. Second, they must have an extended service life at the temperature and under the environmental conditions of operation. And third, large areas of them must be fabricated economically. The cost of materials and their processing makes thin films the most-effective solution, since most optical interactions occur within $1\mu m$ of the front surface. Durability at elevated temperatures is essential and sets solar thin-film technology aside from conventional applications.

In most commercial coatings developed for photo thermal solar conversion, fabrication is not always reproducible and the product is limited in its range of reliable applications. The technology of one of the most promising of these coatings, black chrome, serves as an example. Considerable efforts are under way to lift this technology from its empirical stage, in order to establish the correlation between the optical performance and the process parameters of the fabrication (Ignatiev, O'Neill, and Zajac, 1979; Lampert, 1979; Smith and Ignatiev, 1980). The technological problems of thin film deposition and application are due to a lack of understanding of the basic physical properties, and the inability to explain coating failure. The empirical nature of thin film science is removed when we can systematically relate the properties of films to the conditions of their preparation, as has been done for black molybdenum.

Black molybdenum is a good example of a material that can be produce in a variety of thin film forms and whose structural, compositional, and optical properties are sensitive to the preparation conditions. The high fabrication temperatures inherent in the CVD process may help to prescreen these surfaces against failure at the high operating temperatures necessary for efficient photo thermal conversion. Highly efficient black molybdenum coatings have been developed for use in photo thermal conversion, capable of operating at higher temperatures than any other available coatings of comparable optical behavior.

Our extensive work on molybdenum based cermets led to interest in other refractory metals. Tungsten has similar chemical properties to molybdenum and shares the same column of the periodic table. Its optical properties, as well as its refractory nature, make it a candidate for use in photothermal application as well. The versatility of CVD has led to the development of black tungsten as an extension of the work done on black molybdenum. Films of black tungsten have been prepared by deposition from tungsten hexacarbonyl in the presence of oxygen followed by annealing in a hydrogen-bearing atmosphere, in a manner similar to the preparation of CBM films. In analogy to black molybdenum films they display optical properties that make them desirable for use as photo thermal converter coatings.

Chapter 1A

OPTICAL PROPERTIES OF CERMET MATERIALS APCVD SPECTRALLY SELECTIVE SURFACES – BLACK MOLYBDENUM SOLAR ABSORBERS

K. A. Gesheva and E. E. Chain

*Central Laboratory of Solar Energy and New Energy Sources,
Bulgarian Academy of Sciences, Sofia, Bulgaria*

Inhomogeneties on the length scale 10-1000 Å significantly affect the optical properties of even macroscopically homogeneous thin films (Rouard and Meesen, 1977). When small metallic particles are dispersed in a conducting or dielectric matrix, the optical properties of such a cermet will differ from those of either component. Cermets are two-phase metal-ceramic mixtures; the cermet black molybdenum is a metal-metal oxide composite material.

In this chapter we discuss the optical properties of heterogeneous materials. Their structure is assumed to be macroscopically homogeneous, such that the inhomogeneous cannot be resolved with an optical microscope. But the difference regions are assumed to be large enough so that the separate phases are not mixed on an atomic scale, but rather consist of regions large enough to possess their own dielectric identity. We will now discuss the optical properties of the effective medium so formed, following the approach of Aspnes (1982).

THE EFFECTIVE DIELECTRIC FUNCTION FOR A HETEROGENEOUS MATERIAL

The optical properties of a material depend on its dielectric function, which is complex and also wavelength dependent. The complex dielectric function

$$\varepsilon = \varepsilon_1 + i\varepsilon_2 \tag{1}$$

is defined by
$$D = \varepsilon \vec{E} = \vec{E} + 4\pi \vec{P} \qquad (2)$$

(Jacson, 1975), where D is the macroscopic displacement field, E is the macroscopic electric field, and P is the polarization, or the macroscopic dipole moment per unit volume. ε is a measure of how easily a system can be polarized – it represents the response of the system to the applied field $\vec{E}$.

To arrive at the effective dielectric function for the composite medium, one must first obtain accurate spectra of the dielectric function for each constituent, and then find a reasonable way to average over the spatial variations of the heterogeneous material.

The optical properties of a granular material can be characterized by an effective dielectric function ε, being an average over the dielectric functions of the components, provided that the size of the in homogeneities is much smaller than the wavelength of the radiation. The averaging procedure is far from trivial; several prescriptions exist in the literature. Quantitative models for the properties of heterogeneous materials have been proposed since Faraday in the early 19th century (Landauer, 1978). In the years since, numerous theories of varying complexity have been presented for this averaging procedure.

Inhomogeneities (such as embedded particles) can affect the optical properties of a thin film in two ways (Aspnes, 1982a). First, the dielectric response of an individual homogeneous region, or grain, depends on its composition. Second, screening charges develop at grain boundaries, and are dependent on the size and shape of the grain. This will lead to differences between the local field and the macroscopic applied field of the light wave, and thus influence the film macroscopic dielectric response. The macroscopic dielectric response of a cermet is therefore intimately connected to its composition and microstructure.

Effective Medium Theories

The interaction of a cermet with electromagnetic radiation can, under certain conditions, be equated to the interaction of an equivalent effective medium which is homogeneous. This leads to the following definition of an "effective medium": The behavior of the cermet medium when subjected to electromagnetic radiation should be the same as that of homogeneous material having the effective dielectric function.

In discussing effective medium theories, we shall use the quasistatic (infinite-wavelenght) approximation which is valid if the particle size is small compared to the characteristic wavelength λ. The grain sizes should not exceed $\approx 0.1 \lambda$ (Stroud and Pan, 1978; Aspnes, 1982a). The particles must still be large enough to possess their own dielectric identity. Because the metallic particles in cermets are normally very small, on the order of 100A (Lamb, Wood and Ashcroft 1980), and because cermets are fairly homogeneous on the scale of the relevant light wavelength, effective medium theories are applicable to cermets (Granquist, 1981).

Two effective medium theories are widely accepted, those ascribed to Garnett (1904 and 1906) – known as the Maxwel Garnett theory, and to Bruggeman (1935) – known as the Bruggeman theory or effective-medium approximation. They correspond to two interesting topologies for composite materials - one of isolated inclusions and the other an aggregate or random-mixture topology.

If we refer to the two phases existing in the film as a and b, and if their volumes are large enough to possess their own dielectric identities, then the following general equation applies to this effective medium (Aspnes, 1982a):

$$\frac{\varepsilon - \varepsilon_h}{\varepsilon + 2\varepsilon_h} = f_a \frac{\varepsilon_a - \varepsilon_h}{\varepsilon_a + 2\varepsilon_h} + f_b \frac{\varepsilon_b - \varepsilon_h}{\varepsilon_b + 2\varepsilon_h} \qquad (3)$$

where h represents a host medium into which both phases a and b are embedded. ε_h, ε_a, and ε_b are the dielectric functions of these three media. f_b and f_b denote the "filling factors", i.e. the volume fractions occupied by phases a and b, respectively. ε is the effective dielectric function of the composite media.

Maxwell Garnet

The topology suggested by Maxwell Garnett (Garnett, 1904 and 1906) has inclusions assumed to be spherical. The model takes the particle interaction into account by considering the average effect of the polarization using the Lorentz correction to the local field: the effective electric field E_{eff} for a group of polarizable entities subject to an external field E is given by

$$\vec{E} \quad \vec{E} = \vec{E} + 4\pi/3 \, \vec{P} \qquad (4)$$

The size of the particles is irrelevant as long as it is much smaller than the wavelength and the skin depth for the incident radiation. Each particle is considered to be isolated so the field outside the spheres is an average unaffected by other particles. In this case the effective complex dielectric function is given by Equation (3), with $\varepsilon_h = \varepsilon_0$ (and $f_a + f_b = 1$)

$$\frac{\varepsilon - \varepsilon_b}{\varepsilon + 2\varepsilon_b} = f_a \frac{\varepsilon_a - \varepsilon_b}{\varepsilon_a + 2\varepsilon_b} \qquad (5)$$

Equation 5, and the alternate expression for phase a as the host material, are the Maxwell Garnet effective medium expressions. The theory assumes that the distances between the particles are so large that they act as independent scatterers.

Bruggeman

For the second cermet topology, the aggregate or random-mixture topology is the spatial structure associated with an assembly of microcrystals: a polycrystal. Bruggeman (1935) utilized this concept to develop a very different effective medium theory from the Maxwell Garnet theory. When f_a and f_b are comparable, the roles of host and inclusion are unclear. So one may consider instead that the effective medium is the host into which both a and b are placed. The probability of a particle being a or b is given by its volume fraction f_a or f_b.

In the Brugeman theory a typical element is regarded of the two-phase material which is embedded in an effective medium, whose properties are to be determined self-consistently. To achieve this, one solves for the local field around the element and imposes the condition that the fluctuations of this local field around its effective value should average to zero (Granquist and Hunderi, 1978). The self consistency requirement is then sufficient to specify an expression for the effective dielectric function according to:

$$0 = f_a \frac{\varepsilon_a - \varepsilon}{\varepsilon_a + 2\varepsilon} + f_b \frac{\varepsilon_b - \varepsilon}{\varepsilon_b + 2\varepsilon} \qquad (6)$$

which is equation (3) with $\varepsilon_h = \varepsilon$. This is the Bruggeman expression.

We note that the Bruggeman result is symmetric in the two components while the Maxwell Garnett result is not. In terms of the surface charges and field discontinuities established in a topology of isolated inclusions, it should matter whether islands of medium a are completely surrounded by a sea of medium b or vice versa. In a polycrystal, in which a given microcrystal may never be completely surrounded by a different constituent, the terms " host" or "inclusion" are hardly appropriate (Lamb, Wood and Ashcroft, 1980).

Other Effective Medium Theories

Equations (5) and (6) both assume spherical particles, and thus are often too simple to be directly applicable to experimental samples, which may be characterized by size dependent dielectric functions, non-spherical particles shapes, preferred orientation of elongated grains, and strong aggregation effects (Granquist, 1981). Both theories inadequately describe real systems of particulate selective absorbers which consist of complicated dispersions of metallic or semiconductor particles in dielectric or conductive matrices. Granquist describes elaborations of the above theories to incorporate size distributions, ellipsoidal grains, cube-shaped particles, non-homogeneous particles, and pronounced aggregation of spherical particles. All of these can be incorporated in the two theories to some extent, for low fill fractions.

The Maxwell Garnett model is limited to small filling factors only (Granqvist and Hunderi, 1978). In order to allow wider application the Maxwell Garnet model was made self- consistent by Granquist and Hunderi. Wood and Ashcroft (1977) have extended the Bruggeman theory to describe a "correlated three component system" where the metal particles are coated with a thin layer of dielectric. A recent theory by Ping Sheng (1980) is a symmetrical generalization of the two Maxwell Garnett expressions, in which the relative occurrence of the two kinds of unit cells is obtained from statistical arguments. This list of possible effective medium theories is by no means complete, but serves as an illustration of the range available.

The many effective medium theories give different predictions of the effective dielectric function for a specific fill fraction of metal and oxide in $Co-Al_2O_3$ films (Niklasson and Granqvist, 1981). It is believed that each predicted value of ε corresponds to a definite microstructure (Bergman, 1980). The importance of the composite microstructure in determining the optical response of heterogeneous materials has been established for two gold-dielectric composite systems exhibiting distinctly different topologies (Gibson, Craighead and Buhram, 1982). It thus appears that each effective medium theory would hold for some specific arrangement of regions of phase a and phase b – although not necessarily the arrangement underlying the derivation of the particular effective medium theory. Thus an effective medium approach is expected to be worthwhile only after the microstructure of the medium has been defined.

It is worthy to discuss how absolute limits can be placed on the dielectric function of a composite medium, even if nothing is known about the volume fractions of mictrostructure of its constituents. It can then be shown, as more information is obtained, it can be used to limit further these absolute bounds on ε. Reversing the interpretation, we may also expect to derive information on the microstructure of a given cermet by finding agreement of its observed optical properties with one or the other of the theoretical models.

BOUNDS ON THE EFFECTIVE DIELECTRIC FUNCTION

In order to specify the effective dielectric function of an inhomogeneous material, knowledge is required not only of the compositions of the different phases, their volume fraction and individual dielectric functions, but also of their detailed geometry. In practice the volume fractions and microstructure are often not known; indeed the microstructure can never be exactly known to the detail of location of all voids, extra inclusions, etc. Even so, it is possible to make some simple statements about the types of structural configurations possible, and to arrive at bounds on ε (Aspnes, 1981 and 1982a). A "bound" here implies restriction of ε to a region of the complex ε plane. In the discussion that follows, extensive reference will be made to figure 1, which, as an illustrative example, plots ε_1 vs. ε_2 at $\lambda = 0.5$ μm, for a typical $Mo-MoO_2$ cermet composed of 60 % Mo and 40 % MoO_2.

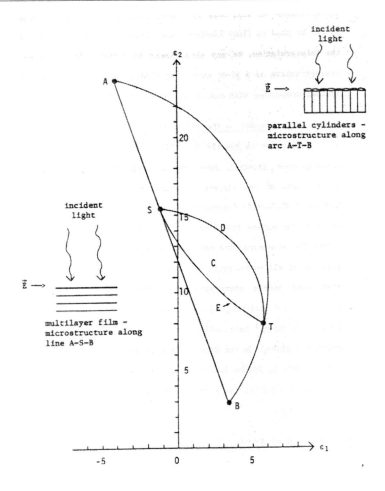

Figure 1. Limits on the allowed range of ε at $\lambda = 0.5\ \mu$ m for a cermet whose composition is 60% Mo and 40% MoO_2.

Wiener Absolute Bounds to ε

Referring to figure 1 at $0.5\ \mu$ m the dielectric functions of single-crystal Mo (Kirillova, Nomerovannaya and Noskov, 1971) and of single-crystal MoO_2 (Chase, 1974; Dissanayake and Chase, 1978) are plotted as one point each in the complex ε plane. The least restrictive bounds on ε for the cermet consist of saying that the effective ε must lie between these two points. ε can immediately be bound more stringently, however, by taking into consideration the screening charges that tend to develop at grain boundaries. These charges cause the local electric field to differ from the macroscopic applied field. The Wiener absolute bounds to ε (Wiener, 1912) are derived by considering maximum and minimum screening effects which are realized for two very simple geometries.

If light is normally incident on a multilayer stack, the applied field E of the light wave is parallel to the stack surface, as shown in figure 1.

All of the stack internal boundaries are thus parallel to the applied field and there is minimum screening. The effective dielectric function is given in this case by:

$$\varepsilon = f_a \varepsilon_a + f_b \varepsilon_b \qquad (7)$$

where a and b are the constituent materials. Equation (7) is equivalent to averaging capacitors connected in parallel. This is indicated by the straight line A-S-B at the left of figure 1, corresponding to the multilayer microstructure.

If light is instead normally incident on a thin film composed of columnar grains oriented perpendicular to the substrate, the film internal boundaries are perpendicular to the applied field of the light wave, giving rise to a maximum screening effect. The effective dielectric function for this case is given by an averaging procedure equivalent to adding capacitors in series:

$$\frac{1}{\varepsilon} = \frac{f_a}{\varepsilon_a} + \frac{f_b}{\varepsilon_b} \qquad (8)$$

Equation (8), for complex ε, ε_a, and ε_b, define a circular arc joining the ε's for Mo and MoO_2 in the complex ε plane. This arc, A-T-B, is indicated on the right side of figure 1, representing columnar grain structure.

The bounds developed so far are absolute; they are true for any microstructure or fractional composition, so long as the microstructural dimensions remain small compared to the wavelength of light.

More Stringent Bounds Based on General Macroscopic Attributes

If the average volume fractions of materials a and b are known, equations (7) and (8) each have a one-point solution; these are labeled in figure 1 as S and T, respectively. More stringent limits to ε, based on this compositional knowledge, have been derived. Hashin and Shtrikman (1962) used a variational approach to show that the two Maxwell Garnett expressions, Equation (5) with $\varepsilon_h = \varepsilon_a$ and $\varepsilon_h = \varepsilon_b$ provide limits to ε if the fractional compositions are known. They showed that for macroscopically uniform and isotropic systems with real ε_a and ε_b these two Maxwell Garnett equations actually form a pair of bounds for ε, irrespective of the topology of the system. Their work was generalized to complex ε by Bergman and Milton (Bergman, 1980 and 1981; Milton, 1980 and 1981). The Hashin-Shtrikman bounds define a region of one composition, in which the structure is allowed to vary. These bounds, given by the two Maxwell Garnett expressions, are arcs in the complex ε plane which intersect at points S and T to bound the area C. The upper arc on figure 1 curve D, is defined by points S, T and A, the ε of MoO_2. The lower arc, curve E, is defined by points S, T and B the ε of Mo. The region C represents a region of constant composition over which the microstructure varies; each value of in this region represents a different microstructure but the same volume fractions of Mo and MoO_2. To obtain points S and T, and the region C, the volume fractions of 60 % Mo and 40 % MoO_2 were supplied to the theory.

Along both curves, D and E, that bound the region C, the cermet must be composed entirely of singly-coated spheroids, as shown in figure 2.

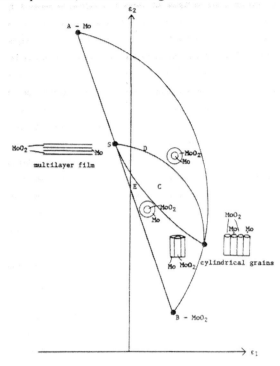

Figure 2. The dependence of ε on film microstructure over region C with composition constant.

Along curve D the grains are MoO_2-coated Mo ellipsoids, and along curve E the particles are Mo-coated MoO_2 ellipsoids. The two spheroidal surfaces of a single-coated grain must be confocal, and the volume fractions, f_a and f_b must be the same in all the coated grains, but the volume of a grain is arbitrary. Space can be packed completely in this fashion (Bergman, 1980). The composite spheroids all have the same orientation and are densely packed since they fill all space, and this requires a specific but not unique distribution of particle sizes (Milton, 1980).

At point S the film must be composed of ellipsoids oriented parallel to the substrate. They are long and thin, so much so that it is difficult to tell which material is the core and which is the coating. Moving along curve D from S toward point T, the ellipsoids become increasingly shorter and wider and deeper, while always maintaining constant volume fractions of the two components, until halfway between S and T where the structure is one of MoO_2-coated Mo spheres. A further progression along curve D toward T results in ellipsoids oriented perpendicular to the substrate, and at T they are again so long and thin that it is difficult to designate either material as the coating or the core. The lower arc, curve E, represents a cermet with the same structural variations encountered in proceeding from S to T, but where the roles of Mo and MoO_2 are reversed.

A path leading from curve E, which corresponds to Mo-coated MoO_2 ellipsoids, to the curve D which corresponds to MoO_2-coated Mo ellipsoids, describes a medium of densely packed, parallel doubly-coated composite ellipsoids (Milton, 1980). Each composite cylinder consists of a cylindrical shell of one component sandwiched between a cylindrical core and an outer cylindrical shell of the other component, as in figure 2. The volume fractions of each component must still remain constant, while the radius of each shell of the grain varies along such a path. In this way the entire region C can be mapped, with differences in the dielectric response due exclusively to the film microstructure.

The limit theorems show that if ε_a and ε_b are similar, then ε is determined almost entirely by composition and is essentially independent of microstructure. If ε_a and ε_b are substantially different, then microstructure is the dominant variable (Aspnes, 1981). Niklasson and Granqvist (1981) showed for a Co-Al$_2$O$_3$ cermet that all of the predictions from all effective medium theories lie within the bounds predicted by these theorems.

THE REFLECTANCE OF A CERMET MATERIAL

The reflectance of light normally incident on a surface in air (n = 1) is given by

$$R = (n-1)^2 + \frac{k^2}{(n+1)^2 + k^2} \qquad (9)$$

n is the refractive index (the real part of the refractive index - it determines the phase change suffered by the wave on traversing the film) and k is the extinction coefficient (determines the fall in intensity due to absorption). Both n and k are calculated from the complex dielectric function as follows:

$$\varepsilon_1 = n^2 - k^2, \qquad \varepsilon_2 = 2nk \qquad (10)$$

A reflectance can be calculated corresponding to any value in the complex ε plane by using Equations (9) and (10). Using this procedure, the theory of bounds on ε has been used to calculate corresponding bounds on the reflectance R of Mo-MoO$_2$ composite films. Because ε varies with wavelength for both Mo and MoO$_2$ both the size and the location of the region C in the complex ε plane will be wavelength-dependent, leading to wavelength-dependent predictions for R.

To apply this effective medium theory to black molybdenum, the microstructure must be defined. To facilitate comprehension of the importance of structural factors and the theory of bounds to the determination of the optical processes operating in black molybdenum, we will briefly describe their preparation. In the carbonyl process, films of nearly stoichiometric MoO$_2$ are deposited. These are annealed in hydrogen to partially reduce the MoO$_2$ to Mo and so produce the cermet structure; hydrogen enters the film along the grain boundaries and reduces a MoO$_2$ grain to Mo from the outside toward the center. We call the films so prepared CBM (Carbonyl Black Molybdenum). In contrast, the cermet structure is attained in one step in the dioxydichloride process, in which grains of Mo and grains of MoO$_2$ are codeposited. We refer to these films as OCBM (Oxychloride Black Molybdenum).

Refer again to figure 1 and 2, CBM films consist of columnar Mo coated MoO_2 grains. Thus the CBM structure places it along curve E, and near to point T (it lacks in figure 2). OCBM films consist of individual columnar grains of Mo or MoO_2. So the OCBM structure places it at point T on this same figure. A comparison of CBM and OCBM films of equal volume fractions, of Mo and MoO_2 should yield different optical behavior for the two films based on differences in microstructure.

Figures of Merit for Real Spectrally Selective Surfaces

We will first apply the theory of bounds to ε to an ideal Mo: MoO_2 cermet and predict the reflectance as a function of wavelength. We adopt this procedure to determine whether such a material displays spectral selectivity. Spectral selectivity denotes an optical property such as reflectance that varies significantly with wavelength. Spectral selectivity permits the spectral filtering or selection of radiative interaction. For the conversion of solar radiation into heat, spectral selectivity exploits the wavelength separation of the solar input from the reradiative loss. 98.5 % of the solar insolation is found at wavelengths below $2\,\mu$m (Hahn and Seraphin, 1978) while, for temperatures below 500°C, 98 % of the thermal infrared radiation occurs at wavelengths greater than $2\,\mu$m (Agnihotri and Gupta, 1981). A selective absorber efficiently captures solar energy in the visible and the near infrared spectral regions while radiating poorly in the thermal infrared. Ideally, it would have a step-function spectral profile as illustrated in figure 3. The optimum location of the step depends on the operating temperature of the spectrally selective surface; for 500°C, the step should be at approximately $2\,\mu$m.

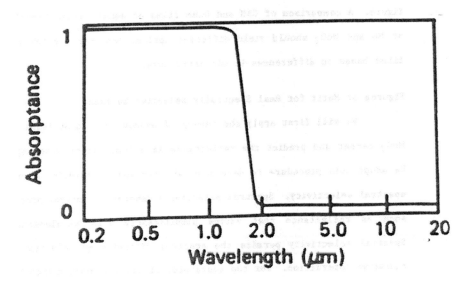

Figure 3. Spectral profile of the absorptance for an ideal photothermal converter surface.

Because the solar flux is fairly dilute, on the order of 1 kW/m^2, the incoming solar radiation must be concentrated to attain these operating temperatures (Seraphin and Meinel,

1976). This has the effect, shown in figure 4, of multiplying the solar input by a concentrating factor. Accordingly, a selective coating will absorb and retain a high amount of energy while a non-selective surface, such as an ordinary black paint, will lose much of its absorbed energy by reradiation.

Thus far we have described the ideal step-function optical behavior leading to maximum spectral selectivity. However, neither complete solar absorption nor perfect thermal suppression is possible, whatever the absorption profile. Unavoidable overlap of the emission bands of sun and converter will cause long wavelength solar photons to be rejected, while energy will leak through the part of the thermal emission spectrum tailing into the absorbing region. In practice, the solar absorber should possess the maximum possible absorptance in the solar spectrum while maintaining a minimum infrared emittance. Evaluation of the spectral profile of a real surface leads to the solar absorptance (equations 11 and 12 below).

$$a_n = \frac{\int_0^\infty [1 - R(\lambda)] \Phi(\lambda) d\lambda}{\int_0^\infty \Phi(\lambda) d\lambda} \qquad (11)$$

$$e(T) = \frac{\int_0^\infty [1 - R(\lambda)] e_{bb}(\lambda, T) d\lambda}{\int_0^\infty e_{bb}(\lambda, T) d\lambda} \qquad (12)$$

In Equation (11) $\Phi(\lambda)$ is the spectral solar irradiance in W/m^2. The hemispherical absorptance $\alpha(\lambda)$ of a given surface is defined as the fraction absorbed from a solar flux $\Phi(\lambda).d(\lambda)$ incident hemispherically on the surface. Since this varies strongly with location and time, the solar absorptance depends not only on the absorptance characteristic of the surface, but on the nature of the solar spectrum. Equation (ll) is transformed into (12) by replacing the hemispherical absorptance by the hemispherical emittance $e(\lambda)$, define the fraction of energy emitted by a real surface into a hemisphereas compared with that of an ideal blackbody, $e_{bb}(\lambda, T)d(\lambda)$, at the same temperature T. Integration over all wavelengths gives the total hemispherical emittance, or thermal emittance $e(\lambda)$.

In both integrals, the temperature dependence of both spectral functions $\alpha(\lambda)$ and $e(\lambda)$ (Trotter and Sievers, 1981) has been ignored. In contrast to the solar absorptance, the thermal emittance is generally regarded as a property of the radiating surface since the blackbody radiation function is completely specified once the surface temperature is given. The emittance as compared to a 500°C blackbody, e_{500oC}, shall be given simply as e.

The optical constants of the converter surface determine the solar absorptance a and thermal emittance e, which are the optical properties of interest for a spectrally selective surface. These figures of merit ultimately determine the fraction of the incident energy drawn from a converter at a given temperature, and in a given configuration.

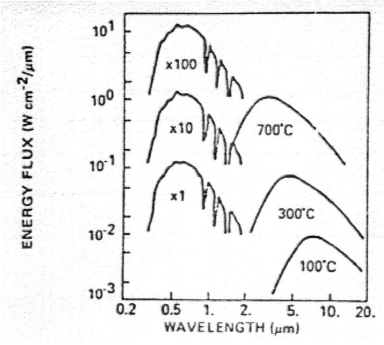

Figure 4. Optical profile of the energy flux of solar input for three concentrations, and of reradiative loss at three converter temperatures.

The Spectral Selectivity of Mo-MoO$_2$ Cermet

We have used compositional information for black molybdenum to apply the more restrictive Hashin-Shtrikman bounds to ε as opposed to using just the Wiener absolute bounds. The Hashin-Shtrikman bounds, corresponding to region C in figures 1 and 2, are obtained for a fixed, known composition, with variations in ε due to variations in film microstructure. One important question is whether, on the basis of the Hashin-Shtrikman bounds to ε, spectral selectivity may be expected in any films of the Mo-MoO$_2$ cermet system. To obtain a spectrally selective reflectance profile, the theory must predict significantly different optical behavior in the visible than in the infrared. That is, the region bounded in the complex ε plane at 0.5 μm must be different from that bounded at 5 μm. As long as these two regions do not overlap, different optical behavior can be expected at the two wavelengths. The theory for bounds to ε must predict a dependence of reflectance on wavelength that approximates the ideal step-function of figure 3.

Using the reflectances calculated from the theory of bounds data, the solar absorptance and thermal emittance have been calculated for the Mo - MoO$_2$ cermet according to the methods described in the preceding section. The two figures of merit, a and e, have been calculated for films with four different volume fractions of the constituent phases, and for 6 points equally spaced along the upper and lower curves (D and E) joining points S and T. The four different volume fractions were chosen for comparison with results on both CBM and

OCBM films. The results are displayed in figures 5 through 8, with a and e both given for each point. On comparing the four figures, we see first that the area mapped out for a film in the complex ε plane according to the theory of bounds is dependent on the volume fractions of Mo and MoO_2. Of course, the prediction for a film of stoichiometric MoO_2 would be one point. The bounded area then increases as the percentage of metallic molybdenum in the film increases. But as the volume fraction of molybdenum approaches unity, the bounded area again decreases. In other words, ε can be known more precisely for films that vary only slightly from stoichiometry. Conversely, for a film whose volume fractions of Mo and MoO_2 are known to be comparable, rather less restrictive bounds on ε emerge. Thus, at low or high fill factors the cermet's optical properties are largely composition-dependent, while at intermediate fill fractions, the microstructure dominates.

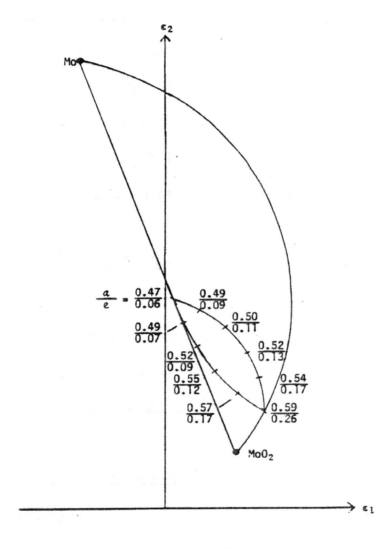

Figure 5. Figures of merit a and e calculated from the theory of bounds to ε for a cermet composed of 40% Mo and 60% MoO_2

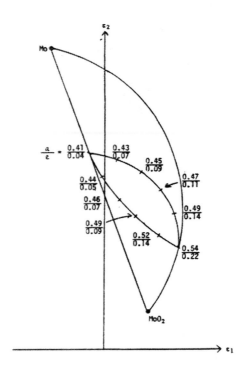

Figure 6. Figure of merit a and e calculated from the theory of bounds to ε for a cermet composed of 60% Mo and 40% MoO_2

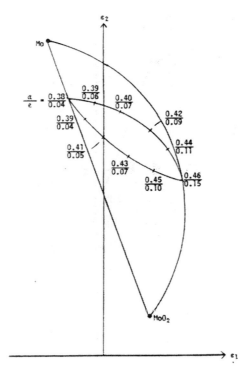

Figure 7. Figure of merit a and e calculated from the theory of bounds to ε for a cermet composed of 80 % Mo and 20 % MoO_2

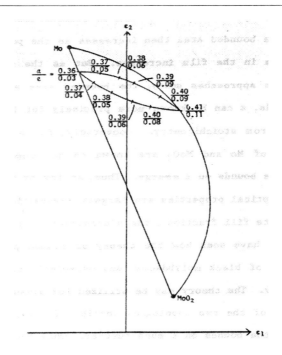

Figure 8. Figure of merit a and e calculated from the theory of bounds to ε for a cermet composed of 90 % no and 10 % MoO$_2$.

We have seen how the theory of bounds predicts the optical properties of black molybdenum, and especially its degree of spectral selectivity. The theory may be utilized for given composition only, but knowledge of the two topologies possible for black molybdenum films restricts the bounds on ε much further. Only through the specification of both composition and microstructure can ε be completely specified.

In addition to the optical effects predicted by the theory of bounds on ε, a combination of effects operate to generate the optical properties of black molybdenum. The structural factors to be considered include the film topology, surface texture, and the presence of a possible amorphous phase. We will further discuss the implications of the bounds derived for ε for real black molybdenum films.

Chapter 1B

CHEMICAL VAPOR DEPOSITION (CVD) TECHNOLOGY

K. A. Gesheva and E. E. Chain

Central Laboratory of Solar Energy and New Energy Sources,
Bulgarian Academy of Sciences, Sofia, Bulgaria

Since 1980, when first applied in the new lamp industry, chemical vapor deposition has been employed in a diverse group of technologies (Jacobson, 1982). At present, CVD plays vital role in microelectronics, wear- and radiation-resistant coatings, fiber optics, and the purification and fabrication of exotic materials, from ultra-low expansion glasses to high purity refractory metals. CVD has four major advantages over most other thin film deposition techniques. First, the process allows tight control over gas stream flow rate and composition which leads to predictable and repeatable film composition and graded structures, if desired. Second, the thermal activation of the reaction establishes thermal equilibrium at the site of film deposition, producing tight, highly coordinated structures. Third, the throwing power of CVD is excellent, allowing for the coating of less accessible surfaces such as the inside of tubes. Last, the techniques is especially well suited to the deposition of refractory materials which is difficult by other techniques.

In Chemical Vapor Deposition, the thin film materials are deposited through a thermally activated chemical reaction. A mixture of gases, a chemical vapor, is transported to a reaction chamber. If a substrate is placed in this chamber and heated, and other parameters are properly chosen, the desired chemical reaction will be activated. The gas either breaks up at the surface of the hot substrate or reacts with other gas constituents, leaving behind the desired solid material. The reactants in a CVD system are usually diluted by an inert carrier gas. The carrier gas forms the bulk of the gas stream, preventing premature decomposition by diluting the reactants, while entraining the reactant gases to the substrate and establishing the flow necessary for uniform deposition.

Significantly CVD and PVD (physical vapor deposition) differ on a more fundamental level. In PVD, it is the absence of equilibrium between film and vapor that provides the drive for the condensation of the evaporant, and hence, the growth of the film. Since there is also a lack of equlibrium between film and substrate, the growth of thin films produced by PVD may be considered a two-stage non equilibrium process in which thermal equilibrium with substrate is achieved only after deposition. On the other hand, CVD is an equlibrium process. The attainment of thermal equilibrium between the components of the gas phase and the substrate at the reaction temperature induce the chemical reaction to produce the film material. Thus, the film is constantly in thermal equilibrium with the substrate as it grows.

CVD relies on the thermal energy of the substrate to activate the desired reactions. In PVD, the kinetic energy of the evaporated or sputtered material delivers it to the substrate; in electrodeposition, for instance, the electrical conductivity of the substrate and source electrodes permit the flow of ions onto the growing film. Therefore, while the stability of substrates in CVD does not depend on electrical conductivity, their temperature dependent properties must be considered. Problems can occur when a high Ts takes the substrate through a structural phase transformation, such as the ferritic/austenitic change of stainless steel at $767^\circ C$.

New materials require completely different CVD procedures. To some extent, the simplicity of the CVD system is balanced by the greater complexity of the CVD process. The wide range of adjustable parameters available to the producer, however, permits precise and wide-ranging control over film properties. CVD thus offers an invaluable tool for unraveling the effects of composition and structure on optical properties of materials.

Gas composition and purity, substrate temperature, flow-rate, and reactant composition can all be monitored and maintained at precise levels in CVD. This leads to films of structural uniformity, chemical purity, hence predictable optical properties. At the same time, the high temperatures and thermal equilibrium characteristics of CVD tend to drive the layer toward greater overall order, reducing energy and stress levels. Such films are less subject to sudden changes in structure or penetration by contaminants during use.

Commonly considered disadvantages of CVD include restrictions on substrates to those stable at high temperatures, stresses in the film during the cool down period following deposition, and inter diffusion among films and the substrate at high temperatures. Other problems with CVD include obtaining thickness uniformity, and the poisonous, air sensitive, and/or corrosive nature of certain CVD reactants and end products. Proper control over substrate heating and gas flows, plus precautions concerning gas containment in a CVD system can circumvent these problems.

To make far-reaching improvements in the quality of thin films, profound changes in deposition technique must be considered. CVD may be vehicle for such a radical departure. At beginning of technologies development few have applied CVD for optical coatings. Nevertheless, the potential of CVD has been clearly demonstrated even at an early stage of development and today even in field of optical coatings achievements are on the scale. Appendix I shows some of the most recent achievements in the area of CVD multilayer optical systems, the so-called X-ray mirrors. Until now only PVD was considered for deposition of optical multilayer systems, but recently very high achievements in the area of X-ray mirrors prove for this side of the CVD technology. Some aspects of the technology and physics, and some practical application aspects are given in this Appendix.

The CVD process promises the production of films with higher purity and denser structure. The improved optical and mechanical properties that result should lead to better performance in areas where conventional coatings are limited. These areas include durability under high radiant fluxes, resistance to environmental damage, and lower optical scattering. All are critical parameters in several areas.

WHY MOLYBDENUM FOR SPECTRALLY SELECTIVE SURFACES

The transition metals and their oxides have unique physical properties (high melting temperatures, for Mo - 2622°C, very small thermal expansion coefficient - $5.35 \times 10^{-6}/°C$ for 20°C). Besides, these materials posses' electronic properties defining low visible reflectance, a basic requirement for spectral selectivity. The Fermi - level intersects the d-band, which is overlapped by the s-band. Intersection of the d-band by the Fermi-level means that the d-electrons participate in defining the optical properties in the visible and the infrared regions, and in the ultraviolet as is in case of silver, for which the Fermi-level lies around 3 eV above the d-band. On figure 9 the overlapping of the two bands with the Fermi-level position is presented.

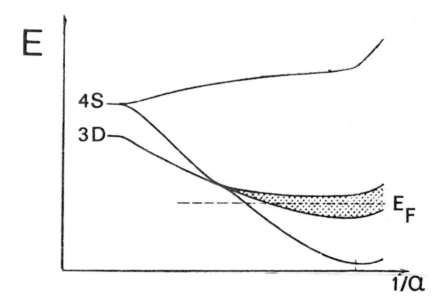

Figure 9. Transition metal s- and d-band energy overlapping and Fermi-level position.

Because of the interactions between the s- and d-electrons, the mobility of the s-electrons is decreased, due to scattered by the lattice phonons s-electrons which fall in the unoccupied cites in the d-band. In result, the reflectance in the visible is lowered. On figure 10 the reflectance spectra of molybdenum, compared with silver spectra and of ideal spectrally selective surface are presented.

This specific property of transition metals could be successfully used for developing of spectrally selective surfaces with low visible reflectance. The idea was how to further decrease the visible reflectance without decreasing the infrared one. Transition metal based composite structure, consisting of metal oxide matrix in which pure metallic particles are embedded is the found answer of this question. CVD technology has been successfully applied for fabrication and tailoring of molybdenum based cermet materials with decreased reflectance in the visible keeping unchanged the high infrared reflectance of pure molybdenum.

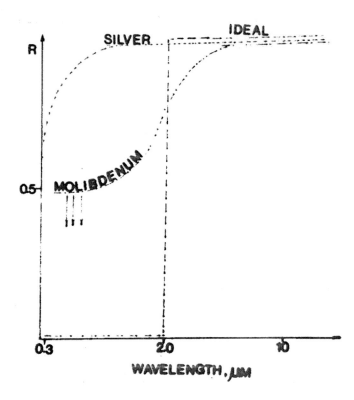

Figure 10. Reflectance Spectra of Silver and Molybdenum as compared with ideal spectrally selective surface profile.

Thin film technology is an art to be conducted very carefully. Physical vapor deposition is a powerful technique usually leading to inclusion in the growing film of considerable amount of impurities like O, N, C. They often are situated at the grain boundaries and can substantially influence the film properties. They may also be incorporated in the crystal lattice of the metal forming impurities band, if the concentration is high enough. The grain boundaries with impurities incorporated inside lead to increased resistance for the electron carriers under an external electrical field. This resistivity is a function of two parameters: grain boundaries and the coefficient of reflection of the electrons from the grain boundaries (Mayadas A.F. and Shatzkes, Phys.Rev. B1, 1382 (1970);Mayadas a.F., M.Shatzkes and J.F.Janak, Appl. Phys.Lett., 14 345, 1969). When there is an increased impurities concentration, this may change the film structure by stabilizing of crystal lattice in a new, abnormal, different from the volume sample crystal lattice. Also if a film is deposited on a metallic substrate because of the difference in the thermal expansion coefficient, in the stage of cooling the hot substrate, deformation in the film may appear leading to different inter atomic distances in the crystal lattice. This may change the periodicity of the internal field from one point to the next and this could also be an additional mechanism of scattering. The film surfaces are also disconnections in the potential of the crystal lattice leading to enhanced electron scattering. This effect is strong when the free electron path is comparable with the film thickness. After deposition, the exposing of the film to the environment gas absorption is possible leading to high impurities concentration in this part of the film. So, in the metal films defects are generated and can be built up in the crystal lattice during the film growth, cooling or exposing to environment conditions. These defects may appear in molybdenum films and the observed difference in infrared reflection of a film and a volume sample are related to the presence of such defects in the film (Carver G. – "Optical Properties of CVD Molybdenum Films"-Theses, Tucson, AZ, 1980). Which are the defects causing less than 1-2 % change in the reflection, and which cause bigger than 1-2 % changes in the reflection? Seto D.K., V.Y. Doo and S.Dash, Proceeding of the Third International Conference on CVD, F.A. Glaski, ed., Salt Lake City, 1972 observes column small grained structure of CVD molybdenum films, a big number vacations (the density of the obtained films is at least 95%) and elastic deformated crystal lattice. The author points that the presence of three types of defects increased the electrical resistivity of the films with less than 2 $\mu\Omega$.cm above the value for the volume sample, which is 5.7 $\mu\Omega$.cm. Chemical Vapor Deposition process in case of Mo deposition on substrate with equal to its thermal expansion coefficient can result in a film with high density and lack of defects, such as vacations and elastic deformations. The density, which could be reached, is 99.99 % (Holzl R.A., Techniques of Metals Research, v.1, R.F. Bunshah, ed. (Interscience N.Y. 1968). So, the influence of these two kinds of structural defects on the CVD optical materials could be ignored. Maisel (125dis Maissel, Thin Film Technology, Handbook, 1977. When the grain size in the film is equal or smaller than the free electron path of electrons in molybdenum, a possible quantum size effect may exists which could influence the dielectric function and in result the reflectance (Lushnikov A.A., B.B. Maksimenko and A.Simonov, SSP, 20, 292, 1978.

Another modification in the Mo film composition as compared with volume sample is accumulation of defects at the grain boundaries area. Oikawa and Tsuchiya (127 Oikawa H., and T.Tsuchiya, J.Vac.Sci.Technol. 15, 1117 (1978) have shown that the grain boundaries with accumulated impurities increase the resistivity of evaporated molybdenum films, while Nestell and Christy (129dis Nestell J.E., Jr., R.W. Christy, J.Vac. Sci.Technology 15 (2), 1978 connect the very low reflectance of deposited in vacuum Mo films by the impurities accumulation during the film deposition. If the concentration is very high, a second phase is formed, MoO_x reaching a fraction of 20-30 % of the film volume. The influence of the second phase on the optical properties is considered by the effective medium theory.

Here, down below we present detail data for the technology and investigation for APCVD chloride black molybdenum, one of the first successful high-temperature spectrally selective surfaces, which sustain a long-term 500°C testing in a fore vacuum environment for a long time (the test was interrupted after 2000 hours). If expose to air, these solar absorber coatings sustain a test at 350°C for 1500 hours. After this period deterioration appears leading to increasing in the thermal emission.

APCVD TECHNOLOGY OF OXYCHLORIDE AND CARBONYL BLACK MOLYBDENUM

There are different methods for deposition of refractory metals and their compounds. But there is no chemical reaction for deposition of thin films from refractory metals such as Mo, W, Nb, Ta from only one electrolyte (Holzl R.A., Techniques of Metals Research, v.1, R.F. Bunshah, ed. (Interscience N.Y. 1968). Not all of the metals can be deposited electrochemically, because during deposition besides the basic reaction, secondary ones appear (for instance H_2 producing). The transition metals can be deposited electrochemically, if the solutions used contain besides these metals also other metals. For instance, W and Mo are deposited, by alloying with Fe, Ni, Co for the purposes of corrosion sustaining coatings. If the electrochemical coating contains besides oxides also fine particles from these alloys, such a coating has spectral selectivity of its optical properties. In this way the famous absorbing coatings Cromol2 and Cromol 3 with solar absorptance 0.95 are obtained (Smith G.B., A.Ignatiev, SEM, v.4 2(1981). These absorbers are more stable compare to the black chromium and the black cobalt, the stability of which does not exceed $300^{\circ}C$. Disadvantage of the electrochemical plating of Cromol 2 and Chromol 3 are the high current density, achieving $3000A/m^2$, even $4000A/m^2$ while it is $1200A/m^2$ for plating of black chromium. The high current density is a general disadvantage of the electrochemical plating.

The carbonyl process occurs at low temperatures (200 to $400^{\circ}C$) while the dioxydichloride process operates above $500^{\circ}C$. The starting material in the carbonyl process is molybdenum haxacarbonyl - $Mo(CO)_6$ used as the source material. A separate oxygen flow is usually added to the system to permit the fabrication of CBM films. Black tungsten was deposited in the carbonyl system through use of the starting material $W(CO)_6$. In the dioxidichloride system, pyrolitic hydrogen reduction of $Mo_2O_2Cl_2$ was used to produce black molybdenum with oxygen entering as a part of the reactant molecule rather than through a separate oxygen flow. In both systems, variation of substrate temperature and gas flow-rates allowed for some adjustment of composition and crystal structure.

Both the carbonyl and the dioxydichloride processes utilized starting materials which are solids at room temperature, necessitating heating to temperature T_1 to achieve a sufficiently high chemical vapor pressure. Decomposition temperature T_3, must be chosen high enough to achieve the desired reaction with sufficient deposition rate. The temperature of lines heating T_2 should be equal or higher than the vapor source temperature (T_1) to avoid either premature reaction or recondensation of the source material within plumbing.

Dioxydichloride Chemical Vapor Deposition Process

On figure 11 APCVD experimental set-up is shown. To generate a cermet material it is necessary to have in the final film structure an oxide matrix in which pure metallic particles are embedded. Usually, as in the case of carbonyl process, the oxygen needed to form the basic oxide material matrix is intentionally introduced in the chemical reaction. In the dioxydichloride process, the oxygen is part of the precursor molecule and by pyrolitic decomposition, accompanied by hydrogen reduction, in one-step process black molybdenum with composite, cermet structure could be produced. OCBM films are deposited by the following reactions:

$$MoO_2Cl_2 + H_2 \longrightarrow MoO_2 + 2HCl \tag{13}$$

$$MoO_2 + 2H_2 \longrightarrow Mo + 2H_2O \tag{14}$$

The reaction goes to completion at temperatures greater than $500^{\circ}C$.

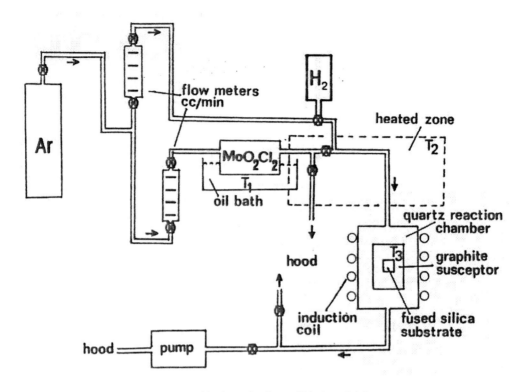

Figure 11. Experimental CVD set-up used in the technology of black molybdenum.

The extent to which it is followed by the further reduction of MoO_2 to molybdenum depends on the temperature and the concentration of the hydrogen in the gas phase during deposition. The source was maintained at 90°C while the substrate temperature was varied between 550°C and 710°C. Two gas mixtures – 10% hydrogen and 8.5 % hydrogen in argon – were used to compare depositions through different substrate temperatures for two values of hydrogen concentration in the reactant gas phase. These MoO_2Cl_2 : H_2 ratios are 0.015:1 and 0.022:1. For the 10% hydrogen mixture, the flow rate of the source carrier gas was 1300 cm^3/min of which 33% went through the Mo_2Cl_2 source. For the lower H_2 content, the corresponding flow rate was 1750 cm^3/min, of which 27 % went to the source line. For the two values of hydrogen concentration in the reactant gas phase the substrate temperature was varied over the range 550 to 710°C. The variations of these two parameters change film composition, which in turn influence their solar absorptance and thermal emittance (Gesheva, Chain, Seraphin, 1981).

On figure 12 dependence of film composition determined by microprobe analysis, as a function of deposition temperature for two values of hydrogen concentrations in the reactant phase is shown. For low temperatures, the composition of films is close to stoihiometric MoO_2.

BLACK MOLYBDENUM FILMS CHARACTERIZATION

Auger Electron Spectroscopy and Microprobe Analysis for Chemical Composition Determination

The first method realized by SEMQ-ARL [Scanning Electron Microprobe Quantum] is connected with using an electron beam falling on sample surface, in result atoms start radiating different types of particles and radiation. The technique under consideration uses x-ray radiation, which comes out of atoms in result of the electron beam action. This radiation is analyzed by a suitable analyzer. It is possible to determine the identity and also the absolute amount of elements in the film, as compared with standard sample.

The quantitative analysis of one element in a certain sample depends on the accuracy of measurement the X-ray radiation coming from the atoms of this element. The absolute measuring of the radiation is difficult because each wavelength is diffracted by the crystal and detected by the detector with different efficiency.

The method of X-ray microanalysis (used to study the chemical composition change with the substrate temperature and for the two hydrogen concentration, see Fig. 13) measures the relative intensity of one and the same X-ray radiation coming from the film and from the standard sample. The concentration of a certain chemical element of a film can be determined by comparing the intensity of the radiation coming from the element with intensity of radiation coming out of the same element from a standard sample, the chemical composition of which is known. The wavelength range in which this quant meter works is 1Å-95.2Å. The minimum quantity to be measured is $10^{-14} - 10^{-16}$ g, or 0.005% to 0.05%. The energy of the electron beam is 5KeV. Three, four elements were under interest in this study, Mo, O, C, and Si. The first two elements are basic in the coating composition, while C and Si come out of the substrates and the vacuum part of CVD equipment.

The second method, used for chemical composition determination and depth distribution of the chemical elements through the thickness is Auger Electron Spectroscopy.

The Auger process can be presented considering the ionization process of one isolated atom under electron bombardment. When one electron with enough energy ionizes a certain level (K-level), the electron vacancy is filled by another electron coming from above level (L_1-level). The energy released from this process as characteristic X-ray radiation or by or it can be release as Auger electron by transferring to another electron from higher energy level (L_2-level). The Auger electrons appear as small peaks in the spectrum of distribution function N(E) in dependence of electron kinetic energy. These maxima become better expressed by electron differentiation of this function, by which the phone signal due to elastically scattered Auger or back scattered initial electrons is separated. In this way the spectrum of electron energy distribution is obtained as dependence of N[E]/E on kinetic energy (Joshi A., L.E. Davis and P.W. Palmberg, Methods of Surface Analysis, A.W. Czanderna,ed.(Elsevier, Scientific Publishing Company, N.Y., 1979). The principle of the Auger Spectroscopy consists in the following: electron beam with energy 2-3 KeV bombards the sample surface, which radiates Auger electrons from a very thin layer (5-20 Å) at the surface of the sample. Since the energy of the radiated Auger-electrons depends on the type of atom, from which they originate, these energies could be used to determine the film chemical composition. The value of the atomic concentration of certain element expressed in percents can be determined by the following formula:

$$At.\%X = \frac{I_x/S_x}{\sum I_i/S_i} \qquad (15)$$

where I_i is the Auger signal for the ith element, S_i is the factor of sensitivity of the equipment for this element. This factor gives estimation for the following: because of the loss of energy of the Auger electrons after their extraction from the atoms, only the atoms, situated at a few atomic distances from the surface can be identified in conditions of high vacuum. The sensitivity of the equipment for a certain element depends on that in what extend the electrons decrease their energy through the film thickness when they try to leave out the film. The sensitivity factors for Mo are 0.28, for O-0.40 and for the C-0.14 (Palmberg P.W., G.E. Riach, R.E. Weber and N.C. Mac Donald, Handbook of Auger Electron Spectroscopy (Pgisical Electronics Industries, Edena, Minnesotta, 1972).

Besides determination of the chemical elements concentration, Auger analysis may be used for thickness distribution of these elements. For this purpose the film is sputtered in Ag atmosphere in vacuum of 5.10^{-5} torr with a certain speed (120 Å/min), and at a certain thickness levels the Auger spectrum is measured (as peak to peak intensity in the differential spectrum) in dependence on the sputtering time. Using the formula given above the concentration of the elements in the film composition is calculated as dependent on the sputtering time.

As substrate temperature increases the molybdenum content increases at the expense of the oxygen content. Near 630°C the concentrations of molybdenum and oxygen are equal and for higher temperatures the films are rich in molybdenum. At about 690°C the molybdenum concentration reaches a maximum. More hydrogen in the gas stream reduces the fraction of oxygen in the film at high substrate temperatures but not in the low-temperature region.

Results of Auger analysis depth profiling are shown on figure 12.

The composition is uniform with depth, except for small deviations near the interfaces. Thus, explanation of the optical properties of OCBM films must be based on a material uniform with depth.

Changes in composition in result of variation of the substrate temperature and the hydrogen content of the reactant gas result in changes of the optical properties. Figure 13 shows the change in chemical composition of films deposited at different substrate temperatures and two values of hydrogen concentration as mentioned above.

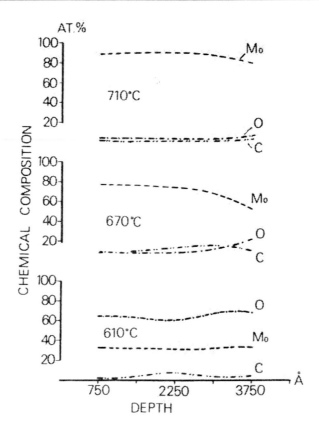

Figure 12. Auger analysis of black molybdenum coatings deposited by APCVD using chloride process at different reaction (substrate) temperatures.

Experimental Results from Chemical Composition Study

The investigation of the chemical composition was performed employing X-ray by the Microprobe analysis microanalysis. Dependence of the concentration in atomic percents of molybdenum and oxygen on the substrate temperature for the two amounts of hydrogen used in technology (presented on figure as $MoO_2Cl_2/H_2 = 0.22$ and respectively $MoO_2Cl_2/H_2 = 0.015$) is presented on figure 13. As is seen from the figure, at low substrate temperatures, and at the lower hydrogen content, films rich of O are deposited, in which the Mo content is low. At high substrate temperatures, the situation is opposite. Films are rich of Mo and contain less oxygen. In order to study how the hydrogen concentration influences the film composition, analysis was made for coatings obtained at higher amount of hydrogen. As the figure show at higher substrate temperatures (higher than 630°C) and high hydrogen concentration defined with the ratio $MoO_2Cl_2/H_2 = 0.015$ leads to higher amount of Mo in the coatings. This can be explained as follows: the high substrate temperatures assure high speed of pyrolisis of the precursor molecules, the hydrogen reduction leads to high amount of MoO_2, which at higher available amount of hydrogen goes further to Mo. This is not true for the lower substrate temperatures. Obviously, the kinetics of MoO_2Cl_2 in this range is such that increasing of hydrogen content influences the presence of oxygen in the growing film.

Chemical Vapor Deposition (CVD) Technology

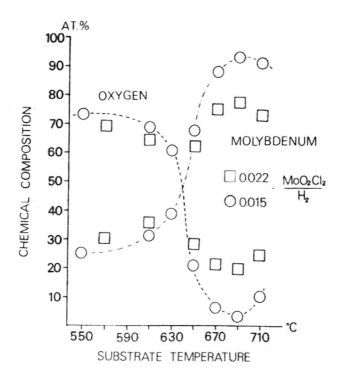

Figure 13. Chemical composition changes in dependence of substrate temperature and hydrogen concentrations.

As seen from the figure 13 and figure 17 (given below, presenting the change in solar absorptance and thermal emittance values in dependence on substrate temperature and the two values of the H_2 concentration), we can fabricate either films rich of oxygen which have higher solar absorptance and comparatively higher thermal emittance, or films rich of molybdenum, which have higher reflectance in the visible resulting in lower solar absorptance and respectively lower thermal emittance. This is controlled by the substrate temperature and the hydrogen concentration in the reaction during the film growth (Gesheva K.A., E.E.Chain, and B.O.Seraphin, *Solar Energy Materials* 3, 415, 1980).

Structural Characterization of Black Molybdenum Coatings

Two methods were employed for structural characterization: X-ray diffraction (XRD) and transmission electron microscopy (TEM). X-ray diffractometer-General Electric XRD diffractometer was used. Hitachi0 microscope was used for the electron-microscopic studies. The voltage of the electron beam was 150 KeV. The magnification of the microscope is 1000-20000 times.

The samples for TEM were prepared by the following method: cooper or nickel sets with diameter approximately 2 mm were first covered by a liquid plastic membrane. After drying over it a carbon film was deposited. On the film a Si_3N_4 coating was deposited in order for the substrate defects to be avoided. The results obtained by the two methods were compared.

Besides these methods, scanning electron microscopy (SEM) was used for the film morphology characterization.

The X-ray diffraction data, obtained from X-ray spectrum (dependence of diffracted X-ray intensity of $\lambda = 1.54$ Å/Kα of cooper/ on diffraction angle 2θ (the angle between the direction of monochromatic radiation and beam passed through crystal plane with indexes hkl). The diffracted rays from crystal planes at a distance d between them result in increased intensity beam giving a maximum in the diffraction spectrum, only if the Bragg, s condition is satisfied: $\lambda = 2d\sin\theta$. It is seen from the table that 2θ values at which maximums are observed coincide well with the powder sample 2θ data taken from diffraction analysis table data for monoclinic MoO_2 (Powder Diffraction data for Metals and Alloys, Joint Committee on Powder Diffraction Standards (JCPDS) Swarthmore, PA 19081).

Table 1. X-ray diffraction results for oxychloride CVD black molybdenum coating, deposited at substrate temperature of 610°C; H_2 flow-rates of 140 cm^3/min or ratio MoO_2Cl_2 = 0.022

Material	hkl	Expected Results			Experimentally Observed Results		
		I/Io	d/Å/	2θ	I/Io	d/Å/	2θ
MoO_2	110 111	100	3.41	26.1	100	3.42	26
MoO_2	020 111	85	2.42	37.12	28	2.42	37.2
MoO_2	222 112 022	80	1.54	53.74	24	1.70	53.74
MoO_2	313	35	1.52	60.80	7	1.52	60.80
MoO_2	131 202	50	1.39	66.92	3	1.39	66.92
MoO_2	133 404	30	1.20	79.64	3	1.20	79.64

As the results presented in the next table (table 2) show, at low substrate temperature and low hydrogen content in the reaction, pyrolitic decomposition of the precursor MoO_2Cl_2 to MoO_2 takes place without further partial reduction of it to pure Mo. Thus, at the lowest substrate temperature used, $T_s = 590°C$, and low hydrogen content, only MoO_2 is produced in the growing film. The same is at slightly higher substrate temperature, $T_s = 610°C$, when in the structure of the grown film all the observed diffraction lines are MoO_2 ones, as seen from table 2.

Table 2. X-ray diffraction results for oxychloride CVD black molybdenum coating, deposited at substrate temperature of 630°C; H_2 flow-rates of 140 cm³/min or ratio $MoO_2Cl_2 = 0.022$

Material	hkl	Expected Results			Experimentally Observed Results		
		I/Io	d/Å/	2θ	I/Io	d/Å/	2θ
MoO_2	001	19	4.78	18.54	25	4.8	18.4
MoO_2	110	100	3.41	26.1	8	3.4	25.9
MoO_2	020	85	2.42	37.12	100	2.42	37.2
Mo	110	100	2.225	40.5	88*	2.23	40.4
MoO_2	310	25	1.54	60.2	9	1.56	59.0
MoO_2	221	25	1.46	63.54	5	1.45	64
MoO_2	113	25	1.287	72.74	10	1.28	74

* Intensity of Mo line is compared with the intensity of the strongest MoO_2 line in the spectrum.

As it can be seen from the table, in the diffraction spectrum of the coating obtained at T = 630°C there is one Mo line (the strongest intensity line of space centered cubic lattice. The rest lines are characteristic for monoclinic MoO_2. Observation of Mo line is in accordance with the results for the chemical composition of the coatings. At this temperature the molybdenum content in the film increases (appr.39%) probably due to a partial reduction of MoO_2 to pure Mo.

As it is seen from the table, at substrate temperature of 670°C, the film grows with crystalline centered cubic lattice. At this temperature the quantity of the oxygen decreases, it is only 7%, as the chemical composition results show. The data for the d-spacings of this modification, as well as the theoretical data for the powder sample are taken from the tables for standard powder samples [(Powder Diffraction data for Metals and Alloys, Joint Committee on Powder Diffraction Standards (JCPDS) Swarthmore, PA 19081).

In the X-ray diffraction data, MoO_2 line also observed, but with a very small intensity. This is in agreement with the investigation of the chemical composition of the coating. The O concentration is very low, 3.4% (see the graph for Auger analysis).

Table 3. Data for the X-ray diffraction of black Mo obtained at T = 670°C; H_2 flow - rates of 140 cm³/min or ratio $MoO_2Cl_2 = 0.022$

Material	hkl	Expected Results			Experimentally Observed Results		
		I/Io	d/Å/	2θ	I/Io	d/Å/	2θ
MoO_2	200	100	2.430	36.9	6⁺	2.430	36.6
Mo	110	100	2.225	40.5	100	2.225	40.5
Mo	200	21	1.570	58.6	6	1.570	58.9
Mo	211	39	1.280	73.7	20	1.280	73.9
Mo	220	11	1.110	87.6	4	1.110	88.0
Mo	310	17	0.990	101.4	2	0.990	101.9
Mo	222	7	0.910	116.0	3	0.910	116.4
Mo	321	26	0.840	132.6	6	0.840	133.2

**Table 4. Data for the X-ray diffraction of black Mo obtained at T = 690°C;
H$_2$ flow-rates of 140 cm^3/min or ratio MoO$_2$Cl$_2$ = 0.022**

Material	hkl	Expected Results			Experimentally Observed Results		
		I/Io	d/Å/	2θ	I/Io	d/Å/	2θ
MoO$_2$	200	50	2.500	36.9	1.4$^+$	2.500	36.5
Mo	110	100	2.225	40.5	100	2.225	40.5
Mo	200	21	1.570	58.6	3	1.570	58.8
Mo	211	39	1.280	73.8	15	1.280	73.9
Mo	220	11	1.110	87.6	3	1.110	87.8
Mo	310	17	0.990	101.4	1.5	0.990	102.5

**Table 5. Data for the X-ray diffraction of black Mo obtained at T = 710°C;
H$_2$ flow-rates of 140 cm^3/min or ratio MoO$_2$Cl$_2$ = 0.022**

Material	hkl	Expected Results			Experimentally Observed Results		
		I/Io	d/Å/	2θ	I/Io	d/Å/	2θ
MoO$_2$	200	50	2.430	36.9	3.4$^+$	2.460	36.5
Mo	110	100	2.225	40.5	100	2.220	40.6
Mo	200	21	1.570	58.6	7	1.570	58.8
Mo	211	39	1.280	73.7	28	1.280	73.9
Mo	220	11	1.110	87.6	4	1.110	88.0
Mo	310	17	0.990	101.4	3	0.990	101.8
Mo	222	7	0.910	116.0	4	0.910	116.4
Mo	321	26	0.840	132.6	18	0.840	133.2

The results presented in the above tables are for black molybdenum coatings obtained at H$_2$ flow-rates of 140cm^3/min (ratio MoO$_2$Cl$_2$ = 0.022). Similar XRD spectra measurements are made also for coatings deposited at the same temperatures and high hydrogen content, MoO$_2$Cl$_2$/H$_2$ = 0.015.

Films grown in the temperature range 550-630°C were oriented along both the [010] and the [100] directions, while only the [100] orientation was observed for films grown at 650 - 710°C.

JCPDS table data confirm the monoclinic MoO$_2$ and bcc Mo lines.

Transmission Electron Microscopy of Oxychloride Black Molybdenum

The samples for TEM were prepared by the following method: cooper or nickel grid sets with diameter approximately 2 mm were first covered by a liquid plastic membrane. After drying over it a carbon film was deposited. Over the film surface a Si$_3$N$_4$ coating was deposited to avoid surface defects. The results obtained by the two methods were compared. Scanning electron microscopy (SEM) was used for the film morphology characterization.

Throughout the range of two process parameters (substrate temperature and reactive gas flow rate) films of a grain size between 180 and 300 Å were obtained. These values are averaged from many dark micrographs made to highlight the grains.

TEM results are obtained for oxychloride black molybdenum films, deposited at low (590°C) and high (650°C) substrate temperature and low (100 cc/min) and high (775cc/min) flow rates of hydrogen. The results are shown on figures 14, 15 and 16.

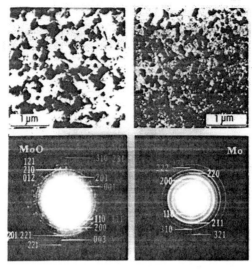

Figure 14. TEM analysis of oxychloride black molybdenum films, films: Dark field micrographs for (001) reflection and (100) reflection and electron diffraction pattern (the line below) of black molybdenum films prepared at: substrate temperature of 590°C and at two flow rates, low - 100 cm^3 min^{-1} and high - 775 cm^3 min^{-1}.

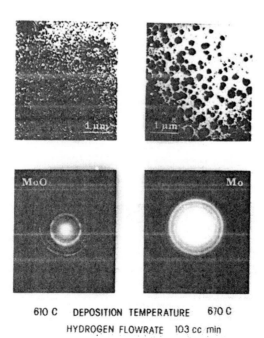

Figure 15. TEM analysis of oxychloride black molybdenum films: Above: dark field micrographs for the (001) reflection and the dark field micrograph for the (100) reflection and electron diffraction patterns (below) of black molybdenum films prepared at: two substrate temperatures of 610°C and 670°C and low flow rate of 103 cm^3 min^{-1}.

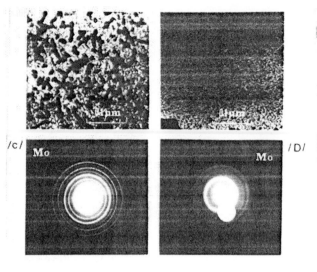

590 C   DEPOSITION TEMPERATURE   650 C
HYDROGEN FLOWRATE 775 cc min

Figure 16. TEM analysis of oxychloride black molybdenum films: dark field micrographs and electron diffraction patterns (below) of chloride black molybdenum films deposited at: low and high substrate temperatures of 590°C and 650°C, and high hydrogen flow rate of 775 cm^3 min^{-1}.

Optical Spectrophotometry Measurements

For calculation of solar absorptance and thermal emittance of black Mo coatings, we measure their reflectance spectra in the visible and near infrared region /0.3-2.4μm/.

The measurements were made with spectrophotometer with integrated sphere, a modified model of Perkin Elmer 21, described in Jacobson M.R. and R.D. Lamoreux, 2nd Ann.Conf. on Absorber Surf.for Solar Receivers, Boulder CO.,1979. Calculating the solar absorptance on the basis of the reflectance, data for the solar spectrum calculated by Thomas and Richmond 3-Thomas A.-NASA/Goddart Space Flight Center and J.C.Richmond, National Burea of Standards, Washington D.C. 20234 were used (European standard for this calculation is usually applied by the researchers, as cited above), solar spectrum for air mass m = 2, α = 1.3 and β = 0.04 is used. Here α and β are Angstrom coefficients for the clearness of the atmosphere. For calculation of solar absorptance the equations (11) is used, and for emittance, respectively equation (12), considering that $a(\lambda) = 1 - R(\lambda)$, where $R(\lambda)$ is the measured hemispherical reflectance measured with spectrophotometer in the spectral range where the sun radiates, 0.3-2.5 μm. R(λ), and the $\Phi(\lambda)$ is the standart solar spectrum for air mass 2. When calculating the thermal emittance, we first measure the reflectance in the Infrared region, 2.5-15μm, and replace the $e(\lambda)$ in formula (12) with $1 - R(\lambda)$.

There are two ways for measuring the hemispherical reflection: the first is when the radiation falls at a definite angle at the sample surface and the reflected in all directions rays are detected; and secondly, when the radiation from the source of radiation falls at different angles over the sample surface (after being reflected from internal wall of a sphere, inside of which the sample is situated) and is detected at a certain angle. The measurements were made employing the second approach. First we measure the reflectance curve of a black surface, the reflectance of which is known (4.5%) and of standard aluminum sample, deposited by thermal evaporation in vacuum. We should note that all the measurements are made with respect to reference sample - a metal piece on the internal wall of the sphere, from where a reference reflected beam originates. The fone signal is eliminated in the following way: First eliminating in monochromator the scattered rays, which are detected by black paint covered part of light chopper, by subtracting it from the reflectance factor of the aluminum sample; secondly by comparing of the known and the measured reflectance of the standard black film, the rest of the noise is calculated and also subtracted. At this stage we have the corrected value reflectance of Al sample. Then we measure our sample, and the data for its reflectance are also corrected by subtracting of fone signal. We divide the reflectance factor of our sample by the one of the Al sample, by which we eliminate the reflectance of the reference signal coming from the metallic surface of the internal wall of the integrated sphere. The last step is multiplication of this ratio by the absolute reflectance value of Al, which is taken from the literature (Bennet H.E., J.M. Bennett and E.Ashley, J.of Opt.Soc.Am, 52, 1245,1962). This dependence on data of such Al sample would be eliminated if the absolute reflectometer of Bennett is used, with which the reflectance could be measured with accuracy of 0.1% 136. Further, reflectance is measured in spectral range 0.4-2.7 µm at 240 points. The reflectance data corresponding to 100 coordinates, determined from Thomas and Richmond Thomas A.-NASA/Goddart Space Flight Center and J,C. Richmond, National Bureau of Standards, Washington D.C. 20234 are summarized and the obtained result is divided by coordinates number, so the value of the solar absorptance is obtained. The whole measuring and calculation process is conducted by a program.

Reflectance dependence on angle θ/ reflected at angle θ radiation by the sample / is studied for different values of this angle/ with special knob the sample position can be changed/, so the calculated solar absorptance for different angles is one and the same in the range of the accuracy /approximately 1%/. The measurements are made at $\theta = 15°$ towards the horizontal axis.

Reflectance spectra are measured at room temperature, thus dependence on temperature is neglected. In order to estimate the expected change in reflectance by temperature we will take into account the results of Carver and Seraphin, Appl. Phys.Letters, 34, 279 (1979). The authors measure the reflectance of a solar absorber /amorphous silicon, deposited on molybdenum reflector film/ at high temperatures, by heating up the sample from room to 500°C in the range 0.4-2.7µm and at room temperature (20°C). The calculated values for solar absorptance at these two temperatures are correspondingly 0.71 and 0.76. Maybe some structural changes lead to decreasing of the reflectance in the visible and the infrared, which means increase of solar absorptance and the thermal emittance.

Measurement of Infrared Reflectance in 2.5 - 15 µm Range and Calculation of Thermal Emittance of Black Molybdenum Coatings

For determination of thermal emittance of a coating it is necessary to measure the infrared reflectance. The reflectance measurements in this range are performed on Perkin-Elmer 137B. This spectrophotometer gives a possibility for near normal reflectance in the range 2.5-15µm.

The equipment is two-ray one, the relative reflectance is determined as compared to evaporated in vacuum standard aluminum film. The absolute values of Al sample data are taken from the work of Bennett et al., spectra were processed with digitizer, and the following formula for thermal emittance was used:

$$e_n = \frac{\int_0^\infty [1 - R(\lambda, T, \Theta = 7o)] e_{BB}(\lambda, T = 500\ oC)\, d\lambda}{\int_0^\infty e_{BB}(\lambda, T = 500\ oC)\, d\lambda} \qquad (17)$$

Instead of hemispherical reflectance we have used R measured at 7°. This might insert a certain uncertainty in determination of the emittance: as it is known at a certain angle of incidence on a metal or dielectric surface /the Bruster angle) a full polarization appears. Then this angle is determined from the equation

$$tg\theta = n \qquad (18)$$

n is the refraction coefficient of the metal or the dielectric. For different materials the Bruster angle has different values. At an angle equal to the Bruster angle the reflectance passes through a minimum, where the averaged value of the reflectance of the wavelengths polarized parallel and perpendicularly to the direction of the incidence is smaller than at normal incidence.(Drummeter L.F. and Hass G. Physics of Thin Films, 2, 305, 1964). When this effect is obvious the hemispheric emittance is (1.25-1.3) e_n for metals with smooth, clean surfaces with high reflectance in the infrared. The normal emittance can be even smaller than hemispherical when the surface of coating is rough. As our SEM studies showed, the films surfaces are smooth, so difference is not expected between the hemispherical emittance and the one for normal incidence reflectance.

When determining the thermal emittance, another assumption is also taken into account, namely the reflectance does not depend on temperature. In this case the emittance, calculated from the expression /17/ will change with temperature only through the temperature dependence of the black body spectrum. With temperature increasing, the spectrum of the black body moves towards the shorter wavelengths, where the reflectance of the coating starts dropping down. Since the black molybdenum is expected to work at 500°C, the thermal emittance is calculated using the data for the blackbody spectrum for 500°C at the body surface. For estimation of the fact how much the reflectance would change we again will use the results of Carver and Seraphin, Appl. Phys.Letters, 34, 279 (1979). Heating the sample from 20°C up to 500°C they measure the reflectance of the two layer absorber of amorphous Si and Mo as a reflector. For the thermal emittance at room temperature and at 500°C they obtain 0.06 and 0.07 correspondingly. As it could be seen the expected changes in the reflectance at elevated temperatures are not very big.

Besides the described spectrophotometers Perkin-Elmer 450 was also used which was supplied with addition part for transmittance measurements in the range 0.165-2.7μm. This measurement was used intensively at the very beginning of technology optimization when the smallest time duration of deposition had to be determined for each value of the process parameters, substrate T and hydrogen concentration.

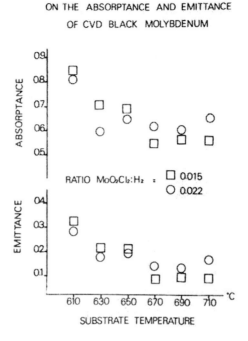

Figure 17. Solar absorptance and thermal emittance of chloride black molybdenum films deposited at different substrate temperatures (610-710°C) and for two values of hydrogen concentrations.

Both the absorptance and the emittance decrease as the substrate temperature increased - solar absorptance of up to 0.85 can be obtained at low substrate temperatures at the expence of a high emittance of 0.33, while for high temperatures the extremes are reversed. Here emittance values as low as 0.07 are accompanied by absorptance values of only 0.55.

By systematically varying the two principle process parameters – the substrate temperature and the hydrogen concentration (H_2), OCBM films with different properties were deposited. In table 6 we summarize the results arranged for low and high values of the substrate temperatures and the hydrogen concentration. Inside each of the four blocks the three major properties are given for films prepared under the corresponding conditions. The first line lists the solar absorptance a and the thermal emittance e, the second provides the chemical composition with respect to the atomic concentration of Mo, O, and C as determined my microprobe analysis, and the third relates the structure of the films, as determined by TEM or XRD.

Table 6. Solar absorptance a, **thermal emittance** e, **chemical composition and structure of oxychloride black molybdenum films: films, deposited at high/low substrate temperature, and high/low hydrogen flow-rate**

H_2 flow rate	Low substrate temperature 590°C	High substrate temperature 650°C
Low-100 cm³/min	Film A: A = 0.74; e = 0.28 Mo - 45at. % O - 55at. % C – 0 at % TEM, MoO_2 XRD, MoO_2	Film C: a = 0.75; e = 0.28 Mo – 50at. % O – 45 at. % C – 5 at. % TEM, Mo +AF XRD, Mo
High – 775 cm³/min	Film B: A = 0.43; e = 0.09 Mo-93.4% O – 4.6 at. % C – 2 at. % TEM, Mo XRD: Mo	Film D: A = 0.60; e = 0.20 Mo = 75 at. % O = 17 at. % C = 8 at. % TEM, Mo+AF; XRD - Mo

Film A of table 6 was deposited at low temperature, 590° C, and at low hydrogen flow rate, 100 cm³/min, resulting in a film with a 300 A grain size and with the monoclinic structure of MoO_2, as shown in figure 4.4 (a) and (b). The monoclinic structure was identified by d – spacings, the obtained values for which were in exact agreement with the expected theoretical JCPDS data for powdered sample of MoO_2. Microprobe analysis shows molybdenum concentration that is 18 at. % in excess of that expected for stoichiometric MoO_2, although no bcc molybdenum lines were identified in either TEM or XRD. The complete reduction of the MoO_2Cl_2 molecule according to Equation (13) and (14) requires the presence of an ample amount of hydrogen, corresponding to Case B in table 6, which is a combination of high hydrogen flow rate - 775 cm³/min and low substrate temperature. Complete reduction to Mo is favored under these conditions, as the films show molybdenum concentrations of about 94%. Pure Mo films were not formed in the range of temperatures and hydrogen partial pressure used for deposition. Figure 16 presents a dark field micrographs for the (100) reflection and electron diffraction patterns of the film, deposited at low substrate temperature of 590 C and for low and high hydrogen concentrations - cases A and B film. The ratios of TEM d - spacings agree with those of bcc molybdenum.

The influence of temperature in determining the extent of reduction can be seen in cases C and D with a higher substrate temperature of 650°C. The reaction kinetics is affected since increasing the deposition temperature from 590 to 650°C increases the film growth rate from 1.5 to 4 A/sec. In cases C and D, both XRD and TEM measurements indicate bcc metallic molybdenum as the only crystalline phase present. Microprobe analysis, however, shows considerable amount of oxygen in both types of films, probably because the higher growth rate results in trapping of the liberated oxygen.

Case C represents a film deposited at a low hydrogen flow rate and high substrate temperature. This film contains a considerable amount of oxygen and some carbon as determined from microprobe analysis. While the absorptance of this film is higher than that of molybdenum, TEM and XRD data show that Mo is the only crystalline phase present. Therefore, an amorphous Mo oxщcarbide phase may have been formed.

Films deposited at high substrate temperature and high hydrogen flow rate, case D, again contain oxygen and carbon although Mo is the only observed crystalline phase. The higher growth rate due to increased substrate temperature again may be responsible for the incorporation of excess oxygen into growing film in Case C. Note that the amount of excess oxygen decreases as the hydrogen flow rate increases at this temperature, which may indicate that the trapping of oxygen is less effective in the presence of increased amounts of hydrogen.

Graded Index Profile Black Mo

The CVD process of pyrolitic decomposition of MoO_2Cl_2 leads to obtaining black molybdenum coatings with different structure, composition and optical properties by controlling the two process parameters - substrate temperature and hydrogen concentration. If the parameters are fixed, we deposit homogeneous coatings. The black Mo properties change in a large range - from highly reflective (poor of O composition) to highly absorptive (rich of O composition). If during the deposition, the substrate temperature is changed or the hydrogen flow rate, we obtain films for which the structure, composition and properties change through the film thickness. In this way, graded index profile coatings could be obtained.

The CVD process offers elegant way for gradual or sharp change in composition and from here the optical profile of the film properties. The film nature could be changed according to in advanced required optical profile, similar to the properties of multilayer system. For instance films of SiO_xN_y are obtained to serve as antireflective coatings for Si (Donadieu A. and Seraphin B.O., J.Opt.Soc. Am. 68, 292 (1978). In these films the composition changes through the film thickness starting with Si_3N_4, followed by SiO_xN_y film and at the surface finishes with SiO_2. Such antireflecting coating "kills" the reflection much better than if the system was multilayered with separately deposited sublayers.

Similarly the graded index profile conception could be applied to a two layer system consisting of reflecting and a film absorbing in the solar spectrum and transparent in the Infrared. The last requirement is necessary in order for the reflecting film to determine the low thermal emittance. This configuration can be achieved with the CVD technology of black molybdenum by changing the chemical composition through the film thickness keeping one – layer (one stage process) character of the coating. The film deposition starts at process parameters leading to film rich of Mo defining high infrared reflection and gradually reaching the coating surface forming a film with high solar absorptance.

Thickness Optimization

Practically the graded index profile coating was deposited by hydrogen reduction of MoO_2Cl_2. The thickness of the absorbing film should be such that the coating could absorb the whole solar radiation. The film below should also have optimal thickness. The emmitance of a coating depends on its thickness. Always should be found the thickness at which the coating for the first moment becomes nontransparent in the solar spectrum. If the film thickness arises further, the absorption keeps the value but the emittance continues increasing.

In order to find the conditions at which the film should start the growing several runs were made at substrate temperature 690°C and hydrogen flow rate 775 cm$^{3/}$min, conditions leading to highly reflecting in the infrared optical coatings. After deposition of coatings with different thickness the reflectance in the range 0.3-15 μm was measured and the solar absorptance and thermal emittance coefficients were calculated. The results are presented in table 7.

Table 7. Dependence of solar absorptance and thermal emittance on film thickness

Sample number	Thickness, Å	Solar absorptance, %	Thermal emittance/500°C/
1	760	43	7.8
2	860	45	7.45
3	1040	45	11.9
4	1250	45	12

As seen from table 7 at thickness of 860Å the solar absorptance reaches its maximal value and further thickness increase leads to increasing only in thermal emittance without contributing to the overall coating efficiency. We note that the thickness was measured with profilometer in 6 points of the surface and the accuracy was 10Å.

Further we determined the optimal thickness of black Mo coating. A set of samples on quartz substrates with in advanced deposited highly reflective Mo films were used and films with different thickness at substrate temperature of 610°C and hydrogen flow rate of 100 cm^3/min were deposited. For all of the samples the reflectance in 0.4-15μm was measured and the solar absorptance and thermal emittance were calculated. The results are shown in table 8.

As seen from table 8 the solar absorptance of almost all samples has a value of 0.74, and the thermal emittance changes in a large scale from 0.31 to 0.10. The ratio a/e$_{500oC}$ changes with the thickness of absorber layer from 2.5 for thickness of 1850 Å to 7.3 for thickness of 450 Å. We should note that reflectance spectra of all films with thickness below 1000Å posses interference minimum at λ = 0.55μm, which shows presence of interference in thin film system. The graphic dependence figure of merit a/e$_{500oC}$ vs. film thickness/time of deposition/ is shown on figure 18. As seen from figure 18 the ratio a/e increases rapidly at film thickness 450Å. At much smaller thickness /time of deposition 2 minutes/ there is a certain increase in thermal emittance and respectively smaller value of figure of merit.

Table 8. Data from thickness optimization with respect to solar absorptance and thermal emittance (for 500°C black body temperature) of oxychloride black Mo coatings

sample	Deposition time, min	Film thickness, Å	Solar absorptance,%	Thermal emittance,%	Figure of merit,a/e$_{500oC}$
A83	15	1850	74	30	2.5
A84	10	1030	74	31.2	2.4
A85	8	990	72	28.4	2.5
A86	5	830	74	28.3	2.6
A87	3.5	530	73	21.6	3.4
A88	3	510	74	13.7	5.4
A89	2.5	450	74	10	7.3
A90	2	...	75	14.8	5.1

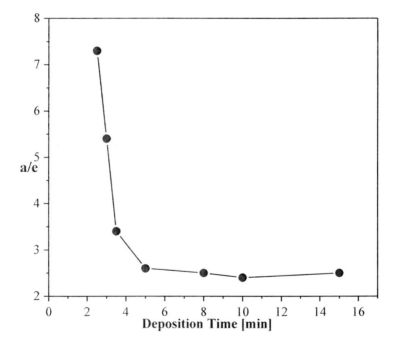

Figure 18. Dependence of the figure of merit of black molybdenum coatings on the film thickness (deposition time).

So, in order to fabricate an optimal black molybdenum coating, the following technological steps should be followed:

On a metal substrate we deposit first a film at substrate temperature 690°C, H_2 flow rate 775cm³/min, deposition time 5 minutes, corresponding to a thickness of 860Å

At the beginning of the 6th minute, the process parameters change to: T = 610°C and H_2- 100cm³/min, time duration 2.5 minutes.

We should emphasize that the ratio a/e = 7.3 is calculated for 500°C - thermal radiation data.

Most of the literature data for this ratio values are for blackbody spectrum for 20°C. This is not correct, since the spectrally selective surfaces are directed to work at elevated temperatures.

In order to illustrate the differences in the ratio values for different temperatures, we present the calculated thermal emittance values for 500°C, 300°C and 100°C. For this evaluation, samples with optimized characteristics were used. The results are presented in the next table 9.

Table 9. Dependence of the figure of merit on black body temperature

sample	Absorptance %	$e_{500°C}$ %	$e_{300°C}$ %	$e_{100°C}$ %	$a/e_{500°C}$ %	$a/e_{300°C}$ %	$a/e_{100°C}$ %
	74	13.7	8.2	5.3	5.4	9.0	13.9
	74	10.2	6.9	4.7	7.3	10.8	15.7
	75	14.8	9.9	7.1	5.1	7.6	10.5

As it is seen from table 9 the ratio changes considerably in the temperature range 300°C-500°C. Calculated for 100°C this ratio differs very strongly from its value for 500°C. On figure 19 the graphic dependence of this ratio versus temperature is presented and the data for 20°C are also included.

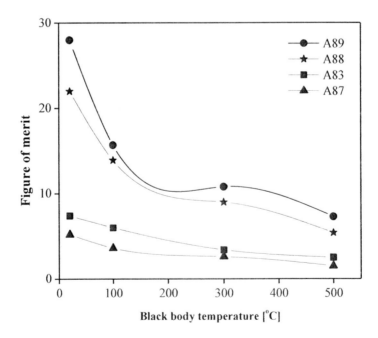

Figure 19. Dependence of OCBM figure of merit on black body temperature.

As seen from figure 19 the value of the figure of merit changes considerably in black body temperature range 20-500°C.

Another possibility for grading of film composition through the thickness is by changing the process parameters during the film growth in a way that the film composition is almost pure Mo phase at the beginning of the film growth, going through two-phase material

CVD Silicon Nitride Protective and Antireflecting Coating for Black Molybdenum Solar Absorbers

All coatings foreseen to work at elevated temperatures have to be protected against oxidizing processes, appearing at high temperatures. Molybdenum needs such a protective coating since at 540°C its oxidation increases rapidly (Harwood J.J., Materials and Methods 44, 84-89 (1956). At this temperature MoO_2 transforms in the volatile MoO_3 which is the final product of molybdenum oxidation and thus a protective coating can not be formed. The Black molybdenum is expected to work at 500°C and one thin protective coating with proper thickness in order to play the role of antireflective coating would prolong considerably the life-time of the films at the working conditions. The interest towards Si_3N_4 is based on its very high diffusion barrier ability toward different ions.

Table 10. Optimization of the thickness of CVD Si_3N_4 antireflection coating deposited on chloride black molybdenum films

No Si_3N_4		A78	B78	A79	B79	A80	B80	A81	B81
absorptance		0.778	0.77	0.757	0.760	0.755	0.743	0.729	0.736
$e_{500°C}$		0.316	0.316	0.330	0.321	0.340	0.331	0.356	0.356
$e_{300°C}$		0.249	0.249	0.261	0.252	0.274	0.265	0.293	0.293
$e_{100°C}$		0.187	0.187	0.197	0.187	0.207	0.202	0.224	0.224
λ_{min}		4800	6000	7000	8100	9000	9970	1.25µ	1.5µ
D(Å) Monitor		600	750	875	1010	1125	1250	1560	1875
With Si_3N_4		0.88	0.925	0.923	0.914	0.925	0.894	0.849	0.846
deposited on the film		0.373	0.377	0.368	0.35	0.358	0.354	0.369	0.380
Si_3N_4 deposition	T, °C	600°C	600°C	600°C	600°C	600°C	600°C	600°C	600°C
	H_2 l/m	4	4	4	4	4	4	4	4
	SiH_4 cc/m	60							
	N_2H_4 cc/m	2.08,							
	Time minutes	3.75	3.33	4.50	4.75	5.25	5.67	7.50	9.00
MoO_2Cl_2 deposition	T°C	610°C	610°C	610°C	610°C	610°C	610°C	610°C	610°C
	H_2 l/m	5.5	5.5	5.5	5.5	5.5	5.5	5.5	5.5
	Time, minutes	25	25	25	25	25	25	25	25

At the beginning on black molybdenum surfaces Si_3N_4 films with thickness of 750 Å. were deposited Optimization of the thickness was made in the following way: A set of black molybdenum samples was obtained at substrate temperature of 610°C, H_2 - 100cm^3/min and for one and the same deposition time duration. The reflectance spectra of all samples were measured in the range 0.4-15µ and the solar absorptance and thermal emittance for 100, 300 and 500°C was calculated. Over the coatings, by technology Si_3N_4 films were deposited with thickness from 600Å to 1870 Å. With monitor during the deposition of these films the reflection of a monochromatic light was detected and a film thickness was reached when the reflection has a minimum. After covering by antireflection coating the absorptance and emittance were again determined. The data from this study are presented in table 10.

The Si_3N_4 coating is deposited when sample is heated at 600°C, and silane (10%) and hydrazine (2%) is introduced in the CVD reactor by He as a carrier gas, with flow rates correspondingly 60 cm^3/min, 375 cm^3/min and 4 l/min. The deposition time is changed for the 8 samples as follows: 2.08 min, 3.33 min, 4.5 min, 4.75 min, 5.25 min, 5.67 min, 7.5 min and 9 min. As it could be estimated from the table Si_3N_4 coating with thickness of 1125 Å is the best antireflecting coating. With this the solar absorptance of black Mo coating increases from 0.75 to 0.93 (18%), while the thermal emittance increases with only 1%. As a final result, with Si_3N_4 antireflecting coating of thickness 1125 Å on optimal coating, the following optimal results were obtained: a = 0.84 and e_{500oC} = 0.12 or as = 0.91 and e_{500oC} = 0.16.

Carbonyl Black Molybdenum Spectrally Selective Surfaces (CBM)

The experimental set-up for the carbonyl black molybdenum is shown on figure 20.

Deposition of black molybdenum thin films by the pyrolysis of molybdenum hexacarbonyl Mo $(CO)_6$ proceeds at atmospheric pressure in the presence of oxygen. The primary reaction is:

$$Mo(CO)_6 + O_2 \rightarrow MoO_2 + 6CO \tag{19}$$

A substrate temperature of 300° C allows for the best combination of deposition rate and film adherence. The flow rate of the argon gas that carries the carbonyl source material to the reactor is 900 cm^3/min. Of the total argon gas flow, 220 cm^3/min is directed through the $Mo(CO)_6$ powder, which is maintain at 70°C. The rest of the argon gas flow is used to dilute the reactant flow. Oxygen is bled into the carrier gas stream at a rate less than 3 cm^3, 0.4% of the argon flow. Deposition rates ranged from 4 to 5 Å/ second on one inch square fused silica substrates, leading to opaque films with thicknesses between 5400 and 5900 Å.

Post deposition anneal of the films was carried out in a reducing atmosphere at temperatures between 700°C and 1000°C. Variations of gas composition during deposition, and post-deposition anneal parameters, produced samples of varying properties.

Changes in the oxygen partial pressure in the reactant gas mixture result in corresponding changes in the compositional, structural and optical properties of the resulting films. The variation of the oxygen flow rate with respect to the constant flow of carbonyl vapor during deposition results in varying amounts of oxygen incorporated in the growing films, as evidenced by microprobe analysis. This causes a and e to vary, as shown in figure 21.

An increase in the oxygen content of the film correspond to a change in composition from MoO_xC_y (with x+y = ½) (Carver, 1981) to MoO_2 with corresponding increase in a. At the same time, the structure changes from fcc to monoclinic. Films prepared with oxygen flow rate greater than 3cm^3/min delaminate during post-deposition anneal, indicating poor adhesion to the quartz substrates.

Chemical Vapor Deposition (CVD) Technology

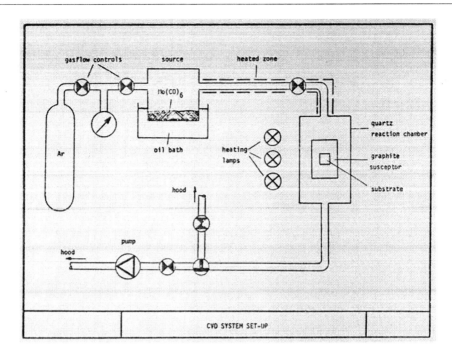

Figure 20. Experimental set-up for deposition of carbonyl black molybdenum.

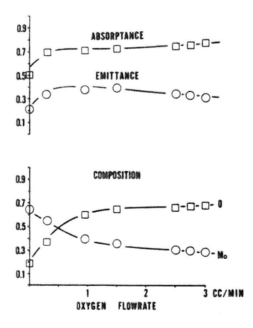

Figure 21. Solar absorptance, thermal emittance and chemical composition of as-deposited CBM films in dependence on oxygen flow rate.

As a result of this systematic variation of reactant ratios we arrived at an optimum operating point of $3 cm^3/min$ oxygen flow rate during deposition.

The observed x-ray and electron diffraction data for as-deposited CBM films are shown in table 11.

Table 11. The observed X-Ray and electron diffraction lines for as-deposited CBM

hkl	X-Ray	TEM
110/11$\bar{1}$	*	*
020/111	*	*
210	*	*
31$\bar{1}$/220	*	
22$\bar{2}$/112/022	*	*
11$\bar{3}$		
13$\bar{1}$	*	*
31$\bar{3}$		
131	*	*
33$\bar{1}$/422		
040/33$\bar{2}$	*	*

* - indicates that line is present.

When the d spacings are calculated for each diffraction line and compared with the JCPDS data, the best match for the most intense lines is obtained with monoclinic MoO_2. However, the 001, 201, 200, 202, and nine other possible MoO_2 reflections are missing. These systematic absences can be explained on the basis of the zone law (Cullity, 1978) which states that diffraction will not be observed from crystal planes perpendicular to the crystal orientation. Thus, for a film with [uvw] orientation, reflections hkl with properly hu + kv + lw = 0 will not be observed. Using the zone law, these films were identified as MoO_2 with a [010] orientation. Lines with intensity less than 30% of the most intense MoO_2 line were not observed in the electron diffraction pattern.

Stress in thin films tends to broaden x-ray peaks and change the measured d-spacings. Because film grain size (300 A) as calculated from XRD via the Scherrer formula (Cullity,1978,p.102), namely

$$T = \frac{0.9\lambda}{B\cos\theta} \tag{19}$$

where λ is the x-ray diffraction wavelength, B the full angular width of the diffraction peak at half maximum intensity, and θ the angular peak location, agrees with that observed in TEM, the x-ray lines do not appear broaden. The as-deposited CBM films are believed to be stress-free, [010] oriented, polycrystalline MoO_2 with a 300 Å grain size.

A comparison between high-resolution bright- and dark-field micrographs shows (not presented) the presence of islands in the as-deposited films which do not change contrast going from bright to dark field. The possibility that these islands are due to substantial thickness variation through surface roughness was eliminated because the films are too smooth to produce significantly no uniform thicknesses. It is likely that these islands consist of a noncrystalline or amorphous phase. There is on the order of 2 at.% carbon in these films and the possibility of an Mo_xC_y amorphous phase cannot be ruled out as this phase is small and it scattered much less than crystalline MoO_2, an amorphous ring pattern may exist and yet not appear under TEM analysis.

Post-deposition anneals have been used by Carver (1979) to raise the infrared reflectance of CVD-Mo films, obtained by the carbonyl CVD process. Post-deposition anneal of CBM films in a mixture of flowing He (90%) and H_2 can remove most of the oxygen and convert the MoO_2 to Mo. Anneal proceeds according to the following reaction:

$$MoO_2 + xH_2 \rightarrow \frac{x}{2}Mo + (1-\frac{x}{2})MoO_2 + xH_2O \qquad (20)$$

at a temperature T > 700°C. Controlling both x and T produces different ratios of molybdenum to molybdenum dioxide in the post-anneal film.

XRD analysis on partially annealed films reveals a mixture of Mo and MoO_2. Since the structure of MoO_2 is fine grained, and because the nascent Mo particles are even smaller, it is difficult to determine the precise distribution of Mo in MoO_2 during early stages of the reduction process. On the basis of the TEM micrograph and the model based on diffusion data (Seshan, K., P.D. Hillman, K.A.Gesheva, E.E.Chain, and B.O.Seraphin, 1981) the Mo is expected to form on multiple sites along the MoO_2 grain boundaries during the anneal process to produce a complex mixture of Mo and MoO_2. The reduction of MoO_2 should proceed as hydrogen diffuses (or permeates) through the MoO_2 film by following the grain boundaries, producing islands of Mo along this path.

As the anneal progresses the reflectance in the infrared rises faster than that in the visible.

There is first 5 minutes of anneal at 770°C while e drops substantially, indicating an increased infrared reflectance. After 5 minutes of anneal at 770°C the solar absorptance also drops, because the visible reflectance begins to increase.

The film optical properties depend on both the temperature and duration of anneal. As the anneal progresses, oxygen is removed from the film, as shown by microprobe analyses. This causes a and e to depart from values of nearly stoichiometric MoO_2, as shown in table 12 and figure 24.

Table 12. The results of the anneal progression as depicted in figure 24

Anneal Parameters	Mo, volume fraction	MoO_2, volume fraction	Absorptance, a	Emittance, e
No anneal	0.0	1.0	0.77	0.31
2 min., 770°C	0.4	0.6	0.74	0.14
5 min., 770°C	0.6	0.4	0.74	0.08
6 min., 770°C	0.8	0.2	0.59	0.06
8 min., 770°C	0.9	0.1	0.47	0.06
complete anneal	1.0	0.0	0.37	0.03

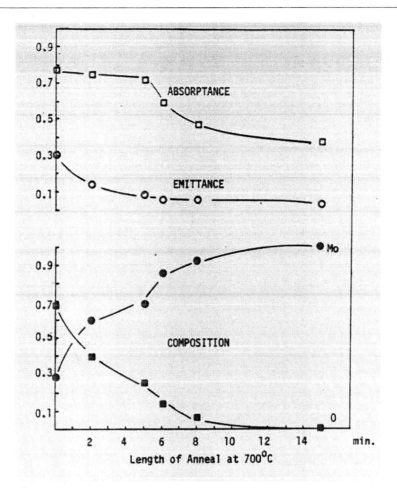

Figure 22. Change in the composition and the corresponding solar absorptance and thermal emittance in result of intensive annealing process of carbonyl black molybdenum films.

As the anneal progresses, the reduction of oxygen content induces the growth of bcc molybdenum at the expense of the monoclinic MoO_2; both Mo and MoO_2 coexist. After complete anneal (Carver and Seraphin, 1979b) at 1000°C for 10 minutes, the bcc crystal structure and the high reflectance of CVD molybdenum are attained.

The volume fraction of Mo and MoO_2 are based on the microprobe compositional data. Absorptance and emittance data are also given for each film.

Figure 23 plots reflectance versus wavelength for carbonyl black molybdenum films at the six stages of anneal. These reflectance traces are on "distorted-λ" plots.

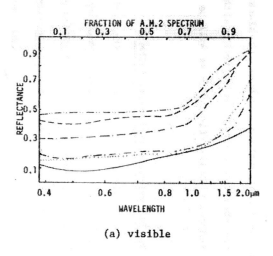

(a) visible

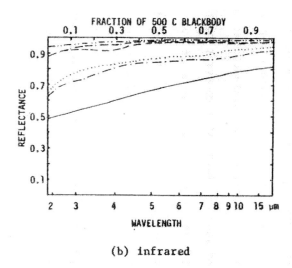

(b) infrared

Figure 23. Reflectance versus wavelength for carbonyl black molybdenum at the six stages of annealing presented in table 12. These reflectance traces are on "distorted-λ" plots.

On figure 24 are shown the Auger Electron Spectroscopy sputter profiles for as-deposited carbonyl black molybdenum and for the partially annealed film. Both profiles show that composition is uniform with film thickness, except for small deviations near the surface and the substrate.

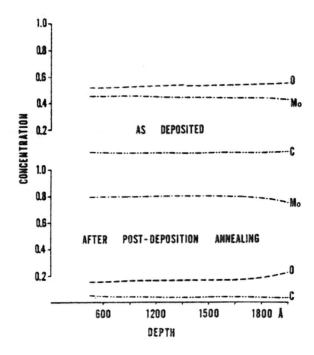

Figure 24. Auger Electron Spectroscopy sputter profiles for as-deposited carbonyl black molybdenum and for the partially annealed film.

We conclude that during film growth oxygen is incorporated into the films at a constant rate, and during the reduction is removed uniformly from all parts of the films. Any interpretation of the optical results must thus be based on a material that is uniform with dept, rather than on effects due to a graded-composition profile.

DISCUSSION OF OPTICAL PROPERTIES OF BLACK MOLYBDENUM FILMS

Black molybdenum, whether deposited from the carbonyl followed by anneal (carbonyl black molybdenum), or directly from the dioxydichloride (oxychloride black molybdenum) is a composite material consisting of MoO_2 particles with smaller Mo inclusions. The properties of Black molybdenum can be related to film composition and structures.

The optical properties of a Mo - MoO_2 cermet can be interpreted in terms of effective dielectric function ε which depends on three parameters: (1) The dielectric functions of two constituent phases (Mo and MoO_2) (2) the volumetric fractions of each phase, and (3) structural factors.

Since the different microstructures of the two types of black molybdenum films are known from the analysis, the observed differences in their optical properties can confirm the applicability to this group of materials of the theory of a bounded dielectric function. We will perform the comparison and explain black molybdenum optical properties in terms of that theory.

COMPARISON OF THE PHYSICAL PROPERTIES OF CARBONYL AND OXYCHLORIDE FILMS

In spite of these different techniques, films prepared in either way were similar in grain size, composition, and optical properties, though different in topology and surface roughness. Black molybdenum, whether CBM or OCBM, is a spectrally selective cermet material composed of Mo particles dispersed on MoO_2. Both processes lead to films with columnar grains similar in size. Throughout the range of the dioxydichloride process parameters, OCBM grain sizes between 180 Å and 300Å were obtained. CBM films are deposited with MoO_2 grain size of 300Å, and molybdenum grains formed during anneal are even smaller. Because the grain sizes are all of the order of 0.1 or less of the film wavelength where λ is in the range of 0.3 – 15 μm, they are small enough for effective medium and bounded ε theories (Stroud and Pan, 1978; Aspnes, 1982b).

Changing the relevant CBM and OCBM process parameters permits a range of composition and microstructure. However, even for films with equal volume fraction of Mo and MoO_2, two major structural differences, other than grain size, can be taken into account in explaining the observed reflectance differences between CBM and OCBM films: (1) topology and (2) surface roughness.

The different growth mechanisms at work during deposition and anneal must lead to cermets with distinctly different topologies. Because anneal of CBM films must reduce the MoO_2 grains from the outside inward, columnar Mo-coated MoO_2 grains result. In contrast, the dioxydichloride process is a random mixture of separate Mo and MoO_2 columnar grains.

Figure 27 compares the surface morphology, as revealed by SEM, of CBM and OCBM films. This SEM analysis shows that OCBM films are even smoother than CBM films. Because it has already been determined that the slight surface roughness of CBM films cannot be solely responsible for the observed spectral selectivity, and because this SEM analysis shows that OCBM films are even smoother than CBM films, surface roughness also cannot solely account for the observed spectral selectivity of OCBM films.

Understanding the difference in reflectance between CBM and OCBM films in terms of the theory of bounds to ε will confirm the effects of film topology on the optical properties of black molybdenum. Table 14 compares the reflectances at selected wavelengths, along with the integrated absorptance a and emittance e calculated from reflectance measurements, for five CBM and OCBM film compositions ranging from stoichiometric MoO_2 to a composition of 90% Mo and 10% MoO_2.

One sees that for corresponding CBM and OCBM films there is no significant difference in the solar absorptance a, because the visible reflectances of these films are essentially equal. But emittance values show significant differences between the two film topologies;

In all cases the emittance of an OCBM film exceeds that of the corresponding CBM film. For two films of equal solar absorptance, the film with the lower thermal emittance is the more spectrally selective. For the case of stoichiometric, noncermetic MoO_2 films, there are no significant differences between a and e.

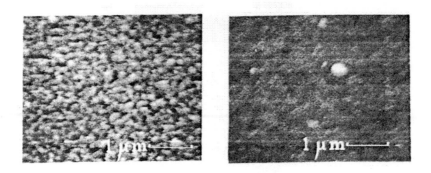

DEPOSITED FROM CARBONYL   FROM DIOXYDICHLORIDE

Figure 25. SEM micrograph comparing the surface texture of a CBM and OCBM film.

Table 13. A comparison of CBM and OCBM films with equal volume fractions of Mo and MoO_2

Sample	Volume fraction, Mo	Volume fraction, MoO_2	Absorptance, a	Emittance, e	R (0.5μm)	R (2.7μm)	R (10μm)
CBM	None	All	0.77	0.31	0.11	0.51	0.80
OCBM	None	All	0.75	0.32	0.16	0.44	0.84
CBM	0.4	0.6	0.74	0.14	0.18	0.71	0.88
OCBM	0.4	0.6	0.70	0.24	0.20	0.65	0.86
CBM	0.6	0.4	0.74	0.08	0.18	0.72	0.90
OCBM	0.6	0.4	0.74	0.20	0.18	0.57	0.79
CBM	0.8	0.2	0.59	0.06	0.31	0.88	0.95
OCBM	0.8	0.2	0.57	0.20	0.33	0.75	0.84
CBM	0.9	0.1	0.47	0.06	0.44	0.90	0.96
OCBM	0.9	0.1	0.50	0.15	0.41	0.85	0.96

INTERPRETATION OF THE RESULTS IN TERMS OF THE BOUNDS ON THE EFFECTIVE DIELECTRIC FUNCTION

The optical properties of a cermet are dependent on several factors, primarily on the optical properties of the constituent phases. Both Mo and MoO_2 display slight spectral selectivity. Both have lower reflectances in the visible than in the infrared (figure 26).

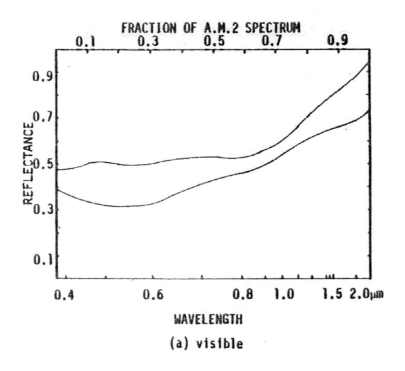

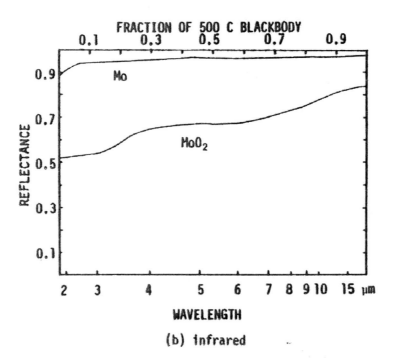

Figure 26. Reflectance vs. wavelength for Mo and MoO_2 single crystals: Mo (Kirillova et al., 1971), and MoO_2 (Chase, 1974).

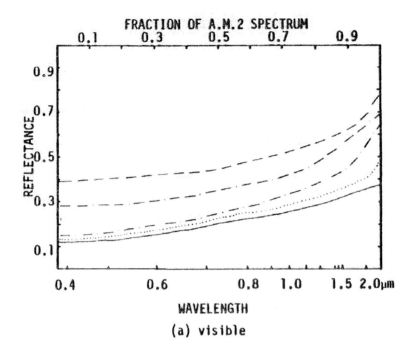

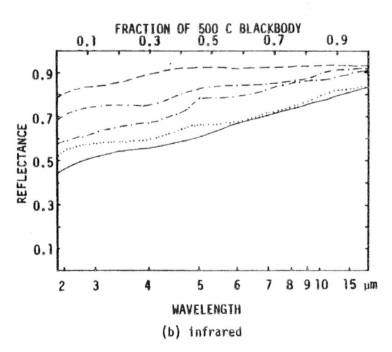

Figure 27. Reflectance versus wavelength for OCBM films deposited with five different Mo:MoO$_2$ ratios: Black line - 0.00:1.00 - pure MoO$_2$, Dash with points - 0.40 : 0.60; Dashpoints line - 0.60 : 0.40; Doubledash with points - 0.80 : 0.20 and Dash line - 0.90 : 0.10.

Molybdenum has a solar absorptance of 0.34 combined with a thermal emittance of 0.03, while for molybdenum dioxide the corresponding values are 0.67 and 0.32 respectively. Combining the low visible reflectance of MoO_2 with the high infrared reflectance of Mo would result in a spectral selectivity greatly superior to that of either pure material. This is the advantage of the cermet structure, as it was illustrated by a and e values given in the above table 13.

The films presented in table 1 are the same five films for which the reflectance curves are plotted on figure 27 for the OCBM films and figure 30 for the CBM.

Molybdenum has a solar absorptance of 0.34 combined with a thermal emittance of 0.03, while for molybdenum dioxide the corresponding values are 0.67 and 0.32 respectively. Combining the low visible reflectance of MoO_2 with the high infrared reflectance of Mo would result in a spectral selectivity greatly superior to that of either pure material. This is the advantage of the cermet structure, as it was illustrated by a and e values given in the above table 13.

The films presented in table 1 are the same five films for which the reflectance curves are plotted on figure 27 for the OCBM films and figure 30 for the CBM.

From the data in table 13 and figure 27 and figure 28 one sees that for corresponding CBM and OCBM films there is no significant difference in the solar absorptance a, because the visible reflectances of these films are essentially equal. But emmitance values show significant differences between the two film topologies; in all cases the emittance of an OCBM film exceeds that of the corresponding CBM film. For two films of equal solar absorptance, the film with the lower thermal emittance is the more spectrally selective. For the case of stoichiometric, noncermetic MoO_2 films, there are no significant differences between a and e.

We talked about how the theory of bounds to ε predicts that spectral selectivity may be expected in films of the $Mo-MoO_2$ cermet system and how this theory predicts that for a given fractional composition, differences in black molybdenum topologies must lead to differences in reflectance. An OCBM film with separate columnar grains of Mo and MoO_2 must have a dielectric reflectance, than a CBM film with columnar Mo-coated MoO_2 grains. Figures 29, 30, 31 and 32 display reflectance versus wavelength, again on distorted-λ plots, over (a) the visible, and (b) the infrared. These figures compare the results for the CBM and OCBM films whose absorptances and emittances are tabulated in table 13. Each figure plots observed reflectances values for the two black molybdenum films at one fractional composition of the two constituent phases Mo and MoO_2, and also compares these to the maximum and minimum reflectance calculated from the theory of bounds.

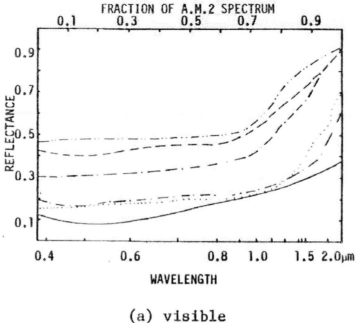

(a) visible

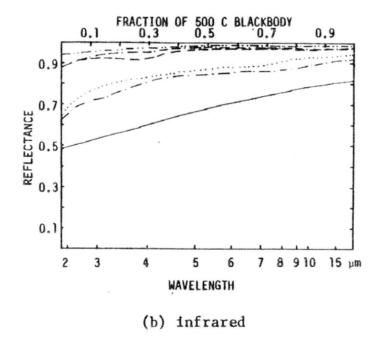

(b) infrared

Figure 28. Reflectance vs. wavelength for CBM films deposited with five different Mo:MoO$_2$ ratios: Black line - 0.00:1.00 - for pure MoO$_2$, Dash with dots line - 0.40:0.60; Dash-dots line - 0.60:0.40; Double-dash with dots line - 0.80:0.20.

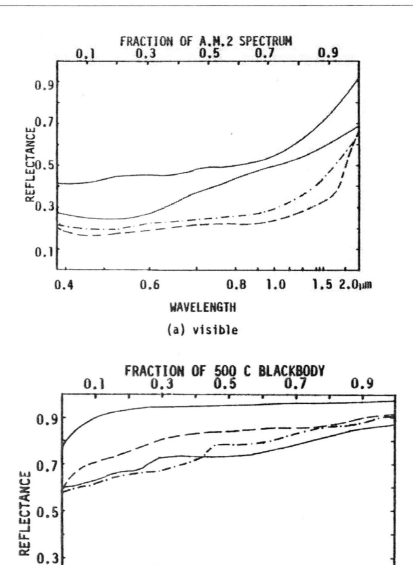

Figure 29. Reflectance vs wavelength for black molybdenum composed of 40% Mo and 60% MoO_2, compared to minimum and maximum reflectance predicted by the theory of bounds to ε : black line – predicted values of reflectance by the theory of bounds. Dash with points line - OCBM, a = 0.70, e = 0.24: Dash line - CBM film; a = 0.74, e = 0.14.

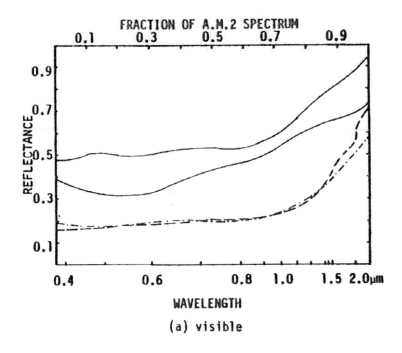

(a) visible

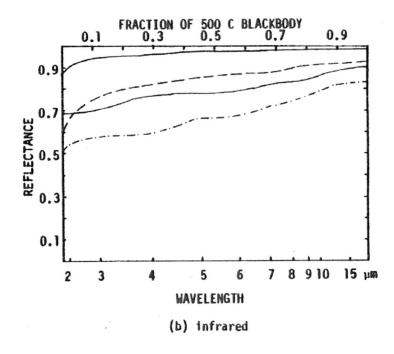

(b) infrared

Figure 30. Reflectance results for OCBM and CBM with fractional composition of 60% Mo and 40% MoO_2. As one sees VIS reflectance are well below the predicted by the theory of bounds. IR reflectance of CBM falls in the predicted range.

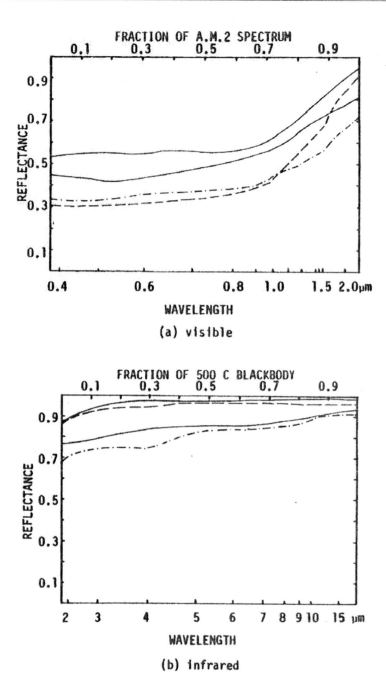

Figure 31. Reflectance results for OCBM and CBM fractional composition of 80% Mo and 20% MoO_2. The IR reflectance of OCBM is lower than the one of CBM and is closed to the predicted value.

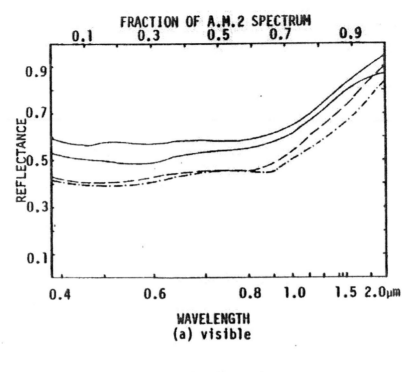

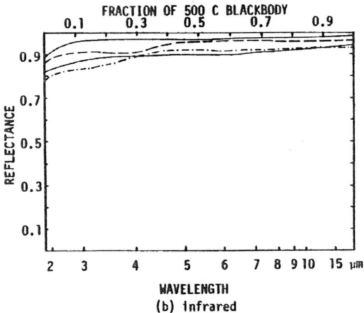

Figure 32. Reflectance data of OCBM and CBM for a fractional composition of 90% Mo and 10% MoO_2. The IR reflectance values as measured and predicted are close – all in the range 0.8-0.9.

These predicted films reflectance are indicated by the solid curves, and the theory may predict on the basis of microstructural transformation, any value within these limits.

Figure 29 shows that over visible wavelengths the CBM reflectance is slightly lower than that for the OCBM, with both lower than reflectances predicted by the theory. In the infrared region CBM reflectance is higher than OCBM reflectance, with the latter nearly coincident with the minimum predicted reflectance.

Figure 30 displays results for a fractional composition that is 60% Mo and 40% MoO_2. In the visible region the reflectances of the CBM and OCBM are essentially the same and below the range of predicted values. In the infrared region the CBM reflectance is considerably higher than that for the OCBM film, and only the CBM reflectance trace falls within the predicted region.

Figure 31 provides results for a fractional composition that is 80% Mo and 20% MoO_2, and figure 32 continues with similar data for a fractional composition that is 90% Mo and 10% MoO_2. In the visible range the reflectances of the CBM and OCBM films are essentially the same, and less than theoretical values. In the infrared region the CBM reflectance is higher than the OCBM reflectance, while the latter nearly coincides with the minimum theoretical reflectance. Though a comparison of these four figures we can make several general statements concerning the reflectances of black molybdenum films:

First, as the fractional composition of metallic molybdenum is increased in both CBM and OCBM films, their reflectances increase over the entire spectral range. The theory of bounds predicts that as the amount of metallic phase increases, the dielectric function will tend toward the dielectric function of Mo and away from the dielectric function of MoO_2, as evidenced in the compositional progression from figures 7, 8, 9 and 10. The calculated reflectances increased in this progression as the amount of Mo is increased, and thus this experimental result agrees with the theory;

Second, the predictions from the bounds theory successfully explain the infrared reflectance of black molybdenum. The OCBM reflectance is always lower than that of the CBM film, confirming the influence of topology. Also, the OCBM reflectance nearly coincides with the minimum predicted reflectance which corresponds to a topology of separated columnar grains of Mo and MoO_2, as is expected for OCBM films;

Third, there is very little difference in the visible reflectances of CBM and OCBM films for a given composition, while the theory for bounds to ε predicts a lower reflectance for OCBM films than for CBM films. The theory does not explain the optical behavior of black molybdenum in this spectral range, as reflectances of both types of black molybdenum fall well below the lowest values predicted. Also, the theory of bounds cannot explain the apparently anomalous behavior of the infrared reflectances of the OCBM film in figure 30, which is below even the minimum predicted reflectance.

We must consider other structural effects to explain these two observations.

Enhancement of Selectivity Due to Surface Texture

While the theory of bounds to ε explains many aspects of the optical behavior of black molybdenum, both the CBM and OCBM films are more spectrally selective than is predicted by the theory. This is because reflectances for both types of black molybdenum fall well below the lowest predicted reflectances. Consideration of surface texture effects will help to explain this discrepancy.

A texturized surface will generally exhibit higher absorptance (i.e. lower visible reflectance) than a smooth surface, if the dimension d of the texture is of the order of the wavelength. This is the case for black molybdenum, as in above shown on figure 25 surface texture with d = 0.1- 0.2 μm for a CBM film, and an even smoother surface for an OCBM film. Surface texturing with a sufficient density of voids leads to a reduction of the effective refractive index, causing the index of the film to be "graded" from that of the incident medium to that of the film.

Surface texturing in a thin film can be modeled as a graded-index multilayered coating covering a perfectly smooth film. The indices of the grading layers are assumed to lie between that of the film, as the bottom layer, and of air, as the top layer. To construct such a model one fits the shape of the surface texture to the variation of the complex refractive index between these two limits.

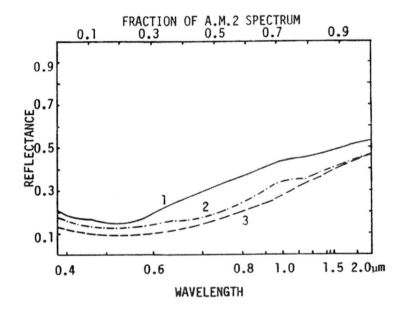

Figure 33. Comparison of the visible reflectance of single-crystal MoO_2 of unknown crystallographic orientation with that of black molybdenum films whose composition is close to that of stoichiometric MoO_2 (1-single-crystal, 2 - OCBM, 3 - CBM).

Figure 33 compares the visible reflectance of single-crystal MoO_2 of unknown crystallographic orientation with that of black molybdenum films whose composition is close to that of stoichiometric MoO_2. Since none of these curves is for a cermet, the theory bounding ε cannot be invoked to explain the reflectances differences among the three curves. Rather, the lower reflectance of both types of black molybdenum films can be explained by the surface structure. Carver et al (1982) found that surface texture effects can increase the absorptance of unannealed CBM films by 0.18 without increasing the thermal emittance. Thus, the visible reflectance of the black molybdenum cermet films may be significantly lowered due to surface texture. On the basis of the observed texture, we may expect the visible reflectance of both types of black molybdenum films to be lower than theoretically predicted. Because the CBM films are rougher than the OCBM films, their visible reflectance will be affected more. Thus, even though the theory of bounds predicts that OCBM reflectance will be lower, the surface texture of the CBM films will drive their visible reflectance as low or lower than that of the OCBM films, resulting in the curves of figures 29, 30, 31 and 32.

Graded-Index Multilayer Approach to Match the Theory with Experiment

Because the theory of bounds predicts black molybdenum infrared reflectance yet fails in the visible, we wanted to learn whether the graded-index approach would yield a lower visible reflectance without changing the infrared prediction.

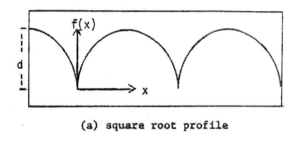

(a) square root profile

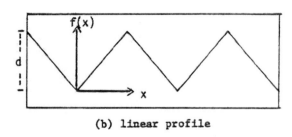

(b) linear profile

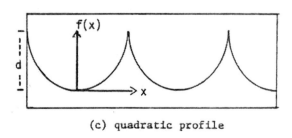

(c) quadratic profile

Figure 34. Highly stylized diagram depicting three possible types of surface texturing and for each plots the shape f(x) of the surface texture as a function of the thickness x in the film for texture of thickness d.

Figure 34 is a highly stylized diagram depicting three possible types of surface texturing and for each plots the shape f(x) of the surface texture as a function of the thickness x in the film for texture of thickness d. By considering the texture shape $f(x)$, it is easy to see why case (a) would be the best approximated as a squire root dependence on thickness [$f(x) = x^{\frac{1}{2}}$], case (b) as a linear dependence on thickness [$f(x) = x$], and case (c) as a quadratic dependence [$f(x) = x^2$] on the thickness of the surface roughness.

Using the graded-index multilayer approach, the thin-film optics computer program OPTF (Macleod, 1982) performed the calculations for a CBM film with a comparison of 60% Mo and 40% MoO_2 to determine the effects of surface texturing with dimensions of 0.15 μm (as deduced from SEM analysis) on both visible and infrared reflectance. OPTF calculates transmittance and reflectance of multilayer coatings. The index of refraction n, extinction coefficient k, and thickness of each layer must be specified.

Reflectance was calculated with OPTF at 0.5 μm and at 10μm for each of the three-model surface textures depicted in figure 34.

All calculations assume a total multilayer thickness of 0.15 μm in agreement with the observed dimensions of the surface texture in black molybdenum. This surface texture was approximated by a 10-layer overcoating. One calculation was performed with 30 layers with an improvement over results obtained for the 10-layer coating; hence for the other calculations 10 layers were termed sufficient to approximate the surface roughness.

The values of reflectance calculated by this method are listed in table 14, along with the experimentally observed reflectance, and that predicted by the theory of bounds to ε.

Table 14. A modeling of the surface roughness of black molybdenum

λ (μm)	Texture shape, f (x)	R (10 layer calculation)	R (30 layer calculation)	R (observed)	R (bounds prediction)
0.5	$x^{1/2}$	0.16	0.16	0.18	0.32
	x	0.10			
	x^2	0.11			
	$x^{1/2}$	0.85		0.81 and 0.96	0.86
	x	0.85			
	x^2	0.85			

The n and k used in this calculation for the black molybdenum bottom layer were derived from the theory of bounds.

At 0.5μm, the observed reflectance was 0.18 while that predicted from the theory of bounds was 0.32. The multilayer graded index calculation reduces the predicted reflectance close to the observed value, independent of texture shape. At 10μm the bounds theory produces a reflectance of 0.86, close to the observed values of 0.81 and 0.96 for the two films. Applying the graded index calculation at this wavelength changes this prediction only slightly, independent of texture. Thus, the discrepancy between the theory of bounds to ε and the experimentally observed visible reflectance of CBM and OCBM films can be largely explained by surface texture effects.

In addition to the structural factors of film topology and surface texture, the reflectance of black molybdenum may also be affected by the presence of an amorphous phase. There is about 1 at. % carbon in CBM films, and up to 8at. % carbon in OCBM films, and the possibility of an MoO_xC_y amorphous phase in these films cannot be ruled out (Gesheva, Seshan and Seraphin, 1981). Such a phase could help to account for the low visible reflectance of both CBM and OCBM films and also the anomalously low infrared reflectance of some OCBM films.

We have shown that the dependence of the optical properties of black molybdenum on those of the constituent Mo and MoO_2, and their volume fractions in the film, can be explained qualitatively by the theory of bounds to ε. This theory can also explain differences between CBM and OCBM reflectance on the basis of their differing topologies. Other structural factors – surface texture and the presence of other compounds and phases in these films – may refine our understanding of the reflectance of black molybdenum.

APPLICATIONS OF BLACK MOLYBDENUM SOLAR ABSORBER COATINGS

The strong dependence of thin film properties on structure and composition allows the tailoring of films to match a given optical application. In particular, we have prepared black molybdenum films through variation of the deposition and anneal conditions with wide ranging optical properties at given wavelengths. We established a link between the process parameters and the compositional, structural, and optical properties of the films. This correlation guides the development of films toward selectivity to meet the requirements of photothermal solar energy conversion.

We are concerned with photothermal conversion at temperatures on the order of $500°C$. At these elevated temperatures the straightforward application of most conventional thin film materials is impossible. In contrast, refractory metals such as molybdenum and tungsten are better suited for these high temperatures. The CVD of refractory materials proceeds without the high source temperatures required in evaporated systems.

The refractory metal tungsten has similar chemical properties to molybdenum and shares the same column of the periodic table. Its optical reflectance, however, differs from molybdenum such that it promises to be even more spectrally selective. We will discuss why we anticipate an improved spectral selectivity for tungsten-based cermet films over what has been achieved for black molybdenum. The versatility of CVD has enabled us to prepare thin films of the tungsten-tungsten oxide system appearing black to the eye. We call these films "black tungsten" (Thomas and Chain, 1983). The development of black tungsten progressed naturally from our work on black molybdenum.

Black Molybdenum Thin Films in Photothermal Solar Energy Conversion

Molybdenum exhibits the moderate spectral selectivity characteristic of refractory transition metals (Hahn and Seraphin, 1978).

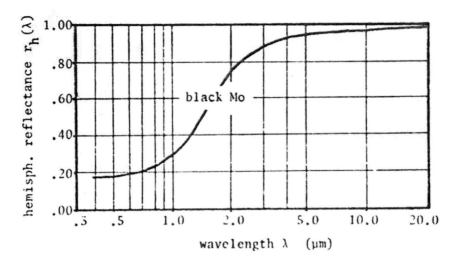

Figure 35. A representative reflectance curve for black molybdenum film.

Bare molybdenum would function as an absorber with $a = 0.37$ and $e = 0.03$, while bare molybdenum dioxide would function as an absorber with $a = 0.67$ and $e = 0.32$. As we have shown, black molybdenum displays a greater selectivity than that of either of these constituent materials.

The term "black molybdenum" is not unique to the selective surface of this study, as at least two other coatings bear a similar name. A black protective coating called "moly black" has been known for over 100 years (Killeffer and Linz, 1952). This coating can be formed as a cathodic deposit on aluminum, cooper, iron or their alloys by electrolysis of alkaline or weakly acidic solutions of molybdates. This electroplated moly black is believed to be Mo_2O_3 or one its hydrates but its composition may vary depending on plating parameters. Recently, a solar selective coating prepared in a similar manner has been patented (Shardein and Lioid, 1979); it has $a = 0.91$ and an $e(150°C)$ of 0.16. Formed by electrolyzing aluminum in an ammonium molybdate solution, it is said to consist of a thin molybdenum and/or molybdenum oxide film on aluminum. CVD black molybdenum is quite different from these two materials, and its properties are particularly well, suited to solar thermal applications.

The refractory nature of the black molybdenum films, as well as the high temperature of their fabrication, recommend their use in applications at temperatures above 300°C. This section introduces applications which might exploit the optical and thermal properties of CVD black molybdenum.

The Requirements of Photothermal Conversion

The purpose of photothermal conversion is to collect the solar insolation falling on a converter surface and convert this radiant energy to heat, while suppressing thermal losses due to reradiation. The basic principle of Photothermal solar energy conversion is illustrated in the energy-flow diagram of figure 36 (Liebert and Hibbard, 1964).

Some optical device focuses the solar flux onto the absorber-converter, concentrating it by the ratio of the concentrator and the absorber. At the absorber surface, the concentrated flux W_1 is divided into at least three components: W_2 is reflected directly, and W_3 represents the radiation in the thermal infrared of the surface at temperature T_1. The reminder W_4 of the solar input is passed on as heat to the next stage, generating Carno work W_5 and waste heat W_6. By proper engineering and choice of operating conditions, additional losses caused by convection and conduction can be made negligible at the converter stage. Obviously, we want to minimize W_2 and W_3, the reflection and reradiation losses, and maximize the useful component W_4, by the use of spectrally selective surface. High selectivity of the converter surface, characterized by a high a and a low e, maximizes energy transfer to the next stage, typically a heat transfer fluid within the absorber-converter element.

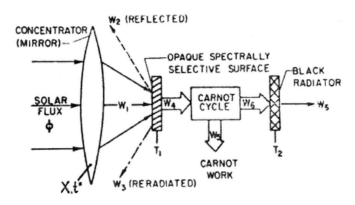

Figure 36. Energy flow in a photothermal solar energy conversion unit.

To be effective, these coatings must endure the high temperatures that they are designed to produce. Specific operating temperatures depend on the collector system, as they are a function of concentration ratio, conduction and convection losses, and optical component efficiencies; they vary from under 100°C for some flat plate collectors to over 1000°C for two-axis tracking parabolic dish concentrators and central receivers. Since the fuel is free, we shall receive returns on the initial monetary investments in direct proportion to the lifespan of the installation. Finally, since solar fuel is rather dilute, large intercepting areas are required to collect it, calling for low cost materials and fabrication. Spectral selectivity, long lifetime at high temperatures, and reasonable cost are, therefore, the basic requirements that the photothermal converter surface must meet. Coatings must be deposited onto metal which efficiently transports heat into a liquid heat transfer medium. These requirements can most easily be met in a coating of simple design – one with as few interfaces as possible (Seraphin, 1982).

CVD Black Molybdenum in Photothermal Converters

Because the optical properties of black molybdenum depend very strongly on the amount of incorporated oxygen, any change in the oxygen content, whether an increase or a decrease, would be understandable. Silicon nitride deposited by CVD has been used to passivate (protect) black molybdenum against any changes in its oxygen content. The thickness of the nitride layer (750°C) was chosen so that its first interface fringe coincides with the solar peak. Thus, a passivating Si_3N_4 layer can be used to antireflective the black molybdenum film over a wide band in the solar emission range. By using the proper nitride thickness a can be enhanced without a substantial increase in e.

Selective, durable coatings must, of course, be applied to suitable substrates. Extremely selective and temperature-stable (at 500°C in vacuum) coatings have been produced on quartz substrates, but for efficient heat transfer to the working fluid in a solar collector, the substrate should be of a metal such as Stainless steel 316, or Incoloy 800. Deposition and anneal of black molybdenum films proceed at temperatures in excess of 700°C, which encourage interactions with the underlaying metal substrate; the stainless steel undergoes transition from austenitic to ferritic state if the temperature exceeds 725°C. Because of this the deposition of the chloride black molybdenum was kept at temperature equal or below 700°C. The same temperatures were assured for the annealing process of carbonyl black molybdenum. Mo deposits tend to diffuse into metallic substrates at these temperatures. This necessitates a barrier layer to minimize diffusion.

OCBM films have been deposited onto both stainless and Incoloy, protected against diffusion by a native oxide barrier layer. By incorporating a native oxide layer formed by oxidizing the substrates for 20 minutes in air at 500°C prior to deposition, we can deposit CBM films on Incoloy 800 having an absorptance of 0.74 with an emittance $e = 0.11$.

CBM has now been successfully deposited and partially annealed on both metals as well. Catastrophic diffusion of the black molybdenum into the substrate is no longer a problem, as has been evidenced by depth profiling. Depth profiles of CBM (figure 6.2a), as well as OCBM (figure 6.2b) films show distinct regions in the coatings. The substrate, oxide barrier, and black molybdenum layer are all separate, with limited, desirable diffusion near the interfaces (Chain, Gesheva, Seraphin, 1981a).

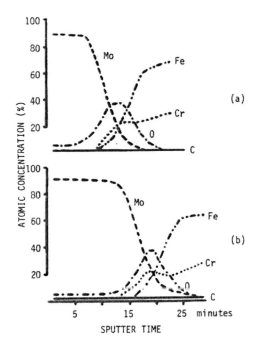

Figure 37. Depth profiling is AES of black molybdenum deposited on oxidized Incoloy alloy 800. The normalized chemical concentrations of Mo, O, Cr and Fe are given. A) CBM; B) OCBM.

Chemical Vapor Deposition (CVD) Technology

The optical performance of passivated black molybdenum films on metal are listed in tables 15 and table 16.

Table 15. Solar absorptance and thermal emittance of passivated CBM films deposited on metal substrates

	absorptance	e (100°C)	e (300°C)	e (500°C)
Best value	0.93	0.09	0.12	0.17
Range of typical values	0.92 - 0.93	0.09-0.10	0.12-0.13	0.17-0.20

Table 16. Solar absorptance and thermal emittance of passivated OCBM films deposited on metal substrates

	absorptance a	emittance e (100°C)	emittance e (300°C)	emittance e (500°C)
Best value	0.92	0.12	0.14	0.17
Range of Typical values	0.86-0.92	0.12-0.24	0.14-0.28	0.17-0.34

Figure 38 displays reflectance plotted against distorted-λ for antireflected black molybdenum samples on Incoloy and Stainless steal 316; curve one is for CBM and curve 2 is for OCBM films.

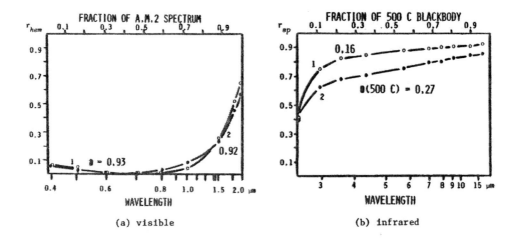

(a) visible (b) infrared

Figure 38. Room temperature reflectance as a function of wavelength for black molybdenum films deposited on oxidized metal substrates: Curve 1 CBM; Curve 2 OCBM.

The solar absorptance and thermal emittance of the operational stacks, coated with 750 Å of CVD-Si_3N_4, as calculated from reflectance data are given next to each curve.

Another requirement of photothermal conversion, reasonable cost of converter fabrication, can be satisfied by the CVD process. From a systems viewpoint, CVD can proceed at atmospheric pressure without needing costly vacuum pumps, and is capable of depositing some refractory materials like molybdenum without the high temperature crucibles needed in evaporating systems. CVD can uniformly coat odd shapes such as long pipes. Operation at one atmosphere permits a sequence of deposition chambers to be lined up-separated by O-ring seals or curtains of flowing inert gas – to deposit the various components of a selective stack in a continuous manner.

EXTENDED LIFETIME-STABILITY TESTS OF BLACK MOLYBDENUM FILMS DEPOSITED ON METAL SUBSTRATES

Black Molybdenum films displaying the spectral selectivity required for efficient photothermal conversion have been successfully deposited on Stainless Steel 316 and Incoloy 800. Anti-reflected samples were tested for durability in a fore pump vacuum of 100 at $500^\circ C$ and in air at temperature up to $400^\circ C$. The coatings were deposited at two different temperatures, on both types of substrates, and tested in both environments, making a total of eight sample sets. 100% of the films deposited at $690^\circ C$ survived tests at $500^\circ C$ in vacuum for over 200 hours. These samples showed an increase of less than 0.014 in thermal emittance, based on room temperature reflectance data weighted for a blackbody temperature of $300^\circ C$, and a decrease of less than 0.006 in solar absorptance. For similar samples in air, 2/3 survived temperatures up to $350^\circ C$ for over 500 hours with a thermal emittance increase of less than 0.040, and a solar absorptance decrease of less than 0.002. Although only 3/5-ths of the samples deposited at $610^\circ C$ survived the vacuum testing at $500^\circ C$, they seemed to withstand air testing at $400^\circ C$ better than samples deposited at $690^\circ C$, 2/5-ths withstood the $400^\circ C$ air test with a thermal emittance increase less than 0.03 and virtually no measurable change in absorptance. Before testing began, the range of absorptance was 0.924 to 0.957 for deposition at 690 C and 0.95 to 0.876 for deposition at $610^\circ C$, while the range of emittance was 0.294 to 0.142 and 0.334 to 0.221 for the different deposition temperature, respectively.

One of the main drawbacks from using CVD to produce photothermal converter coatings was the difficulty of depositing these films onto metal substrates. In the past extremely efficient and temperature stable (at $500^\circ C$ in vacuum) coatings have been produced, but only through deposition onto quartz substrates. For efficient heat transfer to the working fluid in a solar collector, the pipes should be of a high heat conducting material, such as Stainless Steel 316 and Incoloy 800. Although these metals do not have extremely high values of heat conductivity, other factors as cost corrosion resistance must be considered.

These black molybdenum coatings were deposited by hydrogen reduction of MoO_2Cl_2. In this process, the substrate temperature and hydrogen concentration determine the structure and composition of the film, and thereby the optical properties. Empirical correlation and thermodynamic calculations produced two sets of deposition parameters of interest. A deposition temperature of $690^\circ C$ with a low hydrogen flow rate produced films with a high solar absorptance, but a relatively high thermal emittance. Films of both types were produced on both Stainless Steel 316 and Incoloy 800 to be tested

In fore pump vacuum in air. Before testing, the samples were anti-reflected with Si_3N_4 at 600 nm. Since Si_3N_4 was also deposited by CVD, a continuous mass production process is possible because the deposition pressure for both layers is one atmosphere.

The effect of anti-reflection increased the solar absorptance of all samples to about 0.9, but the emittance increased also, the final value being linearly proportional to the initial value. The proportionality constant being 1.11.

During the test, the samples were taken out periodically for measurements of reflectance in the range from 0.4 to 15 microns. The measurements from 0.4 to 2.7 microns were performed on a computer run integrated sphere reflectometer which also calculated the solar absorptance based on air mass 2. The thermal emittance calculations were based on reflectance measurements from 2.5 to 15 microns on a Perkin Elmer spectrophotometer. Because of the numerous measurements, the samples were subjected to thermal circling, a rigorous process that usually destroys most coatings. For the air test, testing started at 250°C and was increased 50°C every 500 hours.

Sample results are plotted in the figures 41, 42, 43, 44, 45, and 46 where the best and worst samples in each test were chosen.

Spectrally selective surfaces fabricated by chemical vapor deposition on stainless steel substrates as well as on Incoloy 800 covered by antireflective coating of Si_3N_4 were tested at conditions close to real ones-fore pump vacuum (0.1torr) at 500°C and in air at 400°C.

The coatings were deposited at two different temperatures on the two types of substrates. Two sets of samples were prepared, each of 8 samples, deposited at 690°C and H_2 -775 cm^3/min and correspondingly 610°C and H_2 – 100 cm^3/min.

The following stack configuration was used in lifetime testing of black molybdenum:

Incoloy 800 or Stainless steel 316 substrates
Native oxide barrier layer, formed by oxidizing the substrates for 20 minutes in air at 550°C
prior to deposition
Black molybdenum (3000 Å) thick
Si_3N_4 (750Å)

A Si_3N_4 overcoat is used to passivate the dioxydichloride black molybdenum against changes in oxygen content during operation. Because there is a good match between the refractive indices of Si_3N_4 and black molybdenum, the thickness of the passivating layer can be adjusted so that antireflection action results at the same time (Carver G. and Chain E. J.de Physique, Colloque C-1, 42, C1-203 (1981)). When dioxydichloride black molybdenum is passivated in this manner, the operational stacks have a solar absorptance of 0.92, and emittances (calculated for 500°C blackbody reflectance curve) are of the order of 0.25 for films deposited on Stainless Steel 316 and Incoloy alloy 800 metal substrates. Emittance calculated at 300°C is 0.22, and e(100 C) = 0.18.

All the samples prepared at temperature of 690°C withstand testing at 500°C in rough vacuum for more than 2000 hours. These samples showed increasing in thermal emittance less than 0.014, determined from the reflectance spectra, measured at room temperature and for black body radiation spectrum data for 300°C. The solar absorptance decreases in result of testing with less than 0.006.

For similar samples, tested in air, 2/5th from them withstand 350°C for more than 500 hours with increasing of thermal emittance with 0.040 and correspondingly decreasing in the absorption with less than 0.002. Although only 3/5ths of the samples obtained at 610°C withstand air test, they seem as more stable at 400°C in air, in comparison with the samples obtained at 690°C. 2/5^{ths} from them withstand air test at 400°C with thermal emittance increasing of 0.03 and practically no change in the solar absorptance. Before starting the test, the samples showed solar absorptance from 0.924 to 0.857 for these deposited at substrate temperature 690°C and correspondingly 0.95-0.88 for the samples obtained at 610°C, and the range of thermal emittance values were 0.28-0.14 and 0.33-0.22 for the two deposition temperatures.

During the testing samples were taken out periodically for reflectance measurements in the range 0.4-15μm. Because of the great amount of measurements during the testing period, they practically undergone thermal cycling-heating/cooling, again heating up to 500°C or to smaller temperature in the case of testing in air. This cycling is a test, which destroy most of the known coatings. In testing in air, the initial temperature was 250°C and it was increased with 50°C each 500 hours. The test results are presented on the figures 39-46: On the figures the best and the worst results are presented to evaluate the repeatability of the technology.

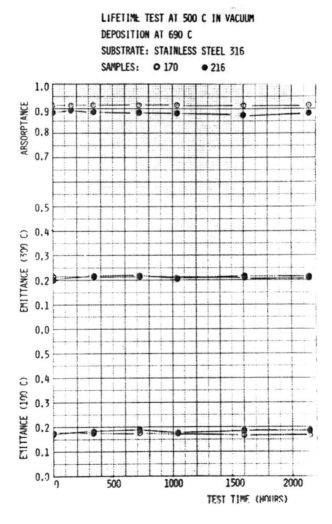

Figure 39. Optical performance results of Oxychloride black molybdenum coatings, deposited at 690°C on Stainless Steel 316 after 2000 hours testing in fore pump vacuum at 500°C.

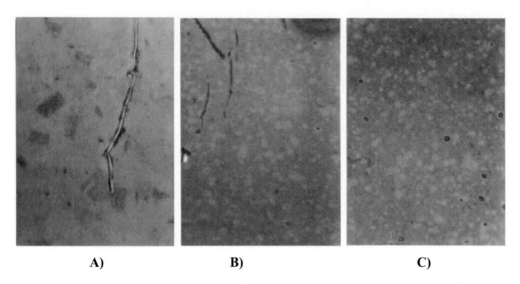

A)　　　　　　　　　　B)　　　　　　　　　　C)

Figure 39-1. SS316_170: the film deposited at 690°C and tested in vacuum for 2000 hours at 500°C. Optical performance results are perfect, no changes are detected in solar absorptance and thermal emittance values (figure 39).

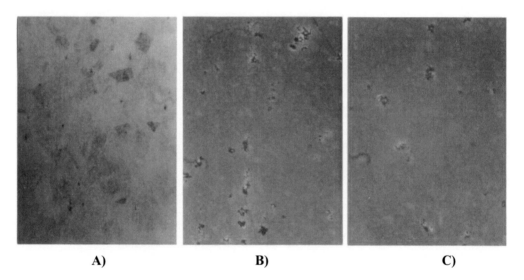

A)　　　　　　　　　　B)　　　　　　　　　　C)

Figure 39-2. SS-316_216, photographs at 5000 X magnification of the surface of: A – oxidized Stainless Steel 316; B – o hours testing (film covered by Si_3N_4 layer); C - after test of OCBM at 500°C for 1600 hours. The sample has survived life-time testing at 500°C for 2000 hours in fore pump vacuum of 100 mtorr.

After 1600 hours, increasing temperature each 500 hours with 50°C, the sample starts failing, emittance increases in different extend for the two tested samples. Samples are chosen as the best and the worst in one and the same run.

Testing was followed by making photographs of: A) – oxidized substrate either stainless Steel 316 or Incoloy 800 substrate; B) photograph after deposition of chloride black molybdenum film and over it a protective Si_3N_4 layer, called as "0 hours test stage"; C) – after either 2000 hours test at 500°C in fore pump vacuum or 1600 hours testing in air at 250-400°C. We note that during testing in air, the temperature was arising each 500 hours with 50°C. Besides, samples were taken out for reflectance measurements, and returning back in the furnace. This cooling and heating is really a harmful stress on the samples. It coincides well with daily heating and nightly cooling the black absorbers in real conditions.

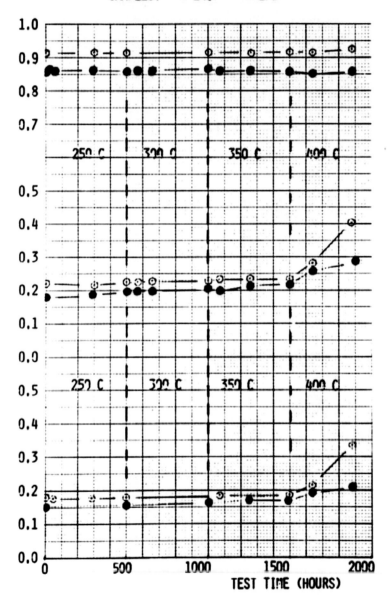

Figure 40. Optical performance results of Oxychloride black molybdenum coatings deposited at 690°C on Stainless Steel 316 after test in air in the temperature range 250-400°C.

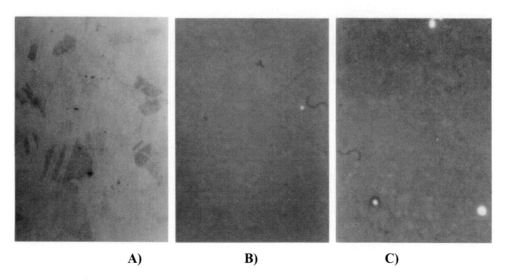

A)   B)   C)

Figure 40-1. SS316 -149 sample photographs.

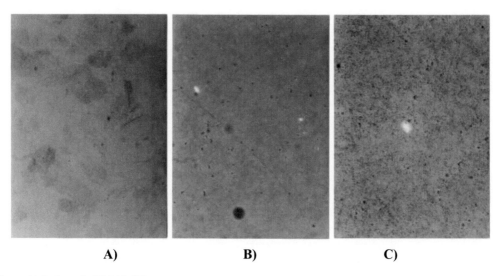

A)   B)   C)

Figure 40-2. Sample SS 316_173.

OCBM films deposited at 690°C on Stainless Steel 316 after testing in air for 1600 hours in the range 250-400°C, 50 hours, shortly after entering the range of 400°C, the film undergoes changes:

As seen the emittance value of sample SS316-149 arises from 0.18 – 0.34 (for 100°C blackbody temperature), and from 0.22 to 0.40 (for 300°C blackbody temperature). No changes in the solar absorptance are detected. For the sample SS 316-173 the emittance arises from 0.15 to 0.2 (100°C) and from 0.2 – 0.28 (300°C), no changes in the solar absorptance are detected. The photographs of the sample at 0 test hours (the middle) show some defects, through which obviously the oxygen enters the film structure and changes the film optical performance.

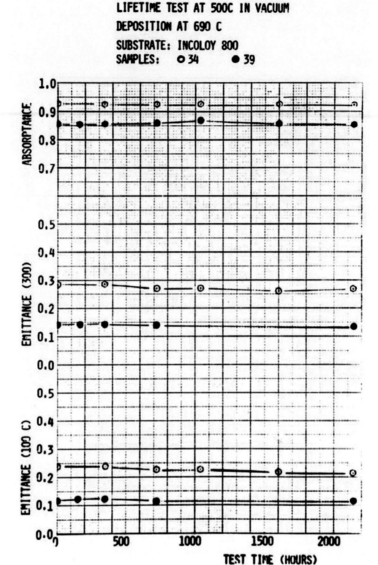

Figure 41. Life time test at 500°C in vacuum for two samples of OCBM films, deposited at 690°C on oxidized Incoloy-800 substrates. Films are covered with Si_3N_4 protective layer. The two films are from one and the same CVD deposition run.

Photographs of the surfaces of oxidized Incoloy 800 substrate – A) and covered with protective Si_3N_4 layer OCBM film deposited at 690°C substrate temperature on Incoloy-800 substrate – B) (no available photograph), and after 2000 hours heating at 500°C in fore-pump vacuum – C). No changes in optical performance are noted, as seen on figure 41. It is interesting that some pinholes defects are noticed in the photographs presented in C). But the lack of oxygen in the vacuum environment assures no change in the film performance. The test results for the solar absorptance and thermal emittance graphs are seen on figure 41.

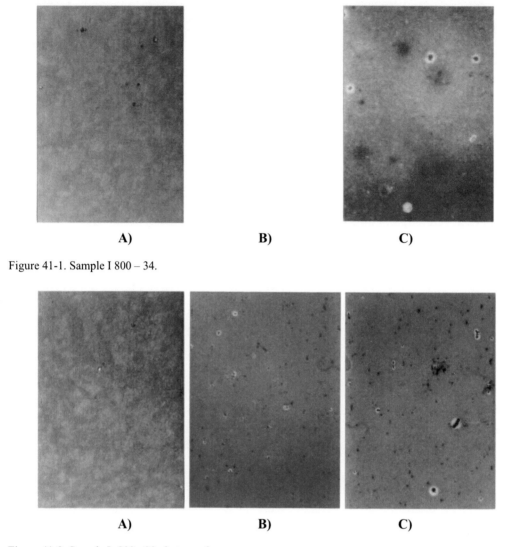

A)   B)   C)

Figure 41-1. Sample I 800 – 34.

A)   B)   C)

Figure 41-2. Sample I -800 - 39 photographs.

The sample withstood 1600 hours heating in air at 250-400°C and entering the range of 400°C, the first 50 hours at that temperature, the emittance, calculated for 100°C and 300°C black body temperature arises rapidly from 0.25 to 0.5 (100°C) and from 0.35 to 0.55 (300°C) (figure 42). Possible reason is pinholes on Si_3N_4 layer.

This sample (number 35) Oxychloride Black molybdenum deposited at 690°C on Incoloy 800 and after 1600 hours testing, in result of increasing the temperature to 400°C, sample fails although no serious physical defects of the surfaces are noticed on photographs. On the graphs presented on figure 42, the emittance (calculated for 100°C and 300°C blackbody temperature) arises slower than that of sample 32, from 0.15 to 0.3, and from 0.20 to 0.40, respectively. Obviously air environment is a harmfull environment at high temperatures, and oxygen enters film structure and changes the sample optical performance.

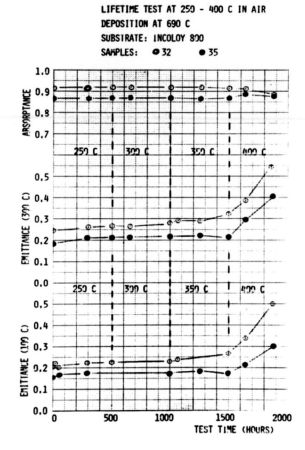

Figure 42. Life time test at 250-400°C in air for two samples of OCBM films, deposited at 690°C on oxidized Incoloy 800 substrates. Films are covered with Si_3N_4 protective layer. The two films are from one and the same CVD deposition run.

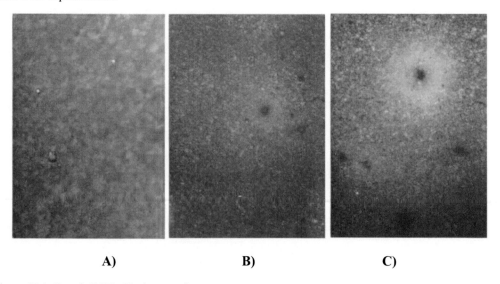

A)   B)   C)

Figure 42-1. Sample I-800- 32 photographs.

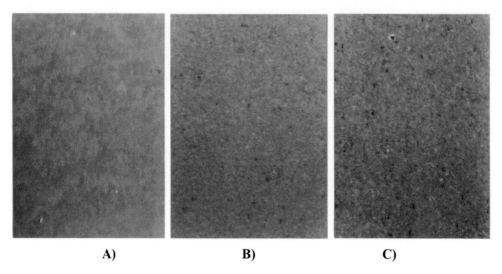

Figure 42-2. Sample I-800-35 photographs.

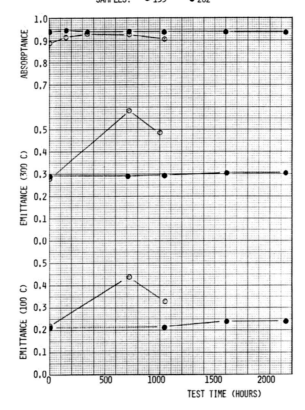

Figure 43. Life time test at 500°C in fore-pump vacuum for two samples of OCBM films, deposited at 610°C on oxidized Stainless Steel 316 substrates. Films are covered with Si_3N_4 protective layer. The two films are from one and the same CVD deposition run.

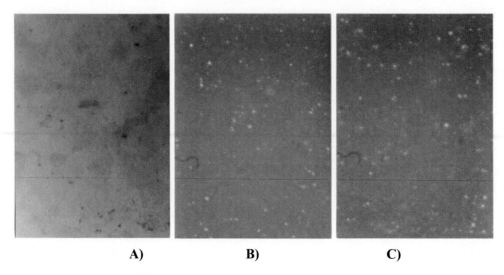

A) B) C)

Figure 43-1. Sample SS 316-133 photographs (5000X).

The sample (number 133) deposited on oxidized Stainless Steel substrate after heating at 500°C for 720 hours in vacuum fails, the emittance (100°C) increases from 0.2 to 0.45, the emittance value calculated at 300°C blackbody temperature – from 0.3 to 0.58. The photograph shows pinholes in the Si_3N_4 layer, the environment is vacuum, no oxygen available, another reason could be taken into account, like sample position in the furnace or its position towards the gas flow during the film growth.

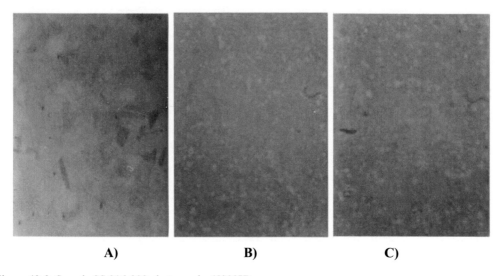

A) B) C)

Figure 43-2. Sample SS 316-202 photographs (5000X).

The sample (number 202) Oxychloride Black molybdenum coating, deposited on oxidized Stainless Steel 316 at 610°C, and annealed at 500°C in vacuum. No change in optical performance is detected after more than 2000 hours of testing. After the long test no changes in the sample surface is noticed.

Chemical Vapor Deposition (CVD) Technology

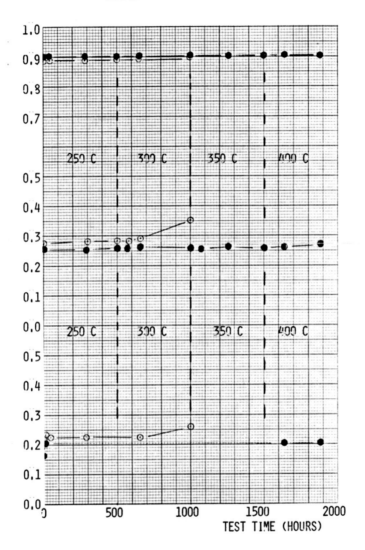

Figure 44. Life time test at 250 – 400°C in air for two samples of OCBM films, deposited at 610°C on oxidized Stainless Steel 316 substrates. Films are covered with Si_3N_4 protective layer. The two films are from one and the same CVD deposition run.

The sample (number 26) is oxychloride black molybdenum, deposited at 610°C on oxydized Incoloy 800 and covered by Si_3N_4 protective layer, The photographs are made on sample surface after annealing in air for 1600 hours. The sample withstands test in air very successful. No defects are seen (see the optical performance on figure 44, related to samples 26 and 139).

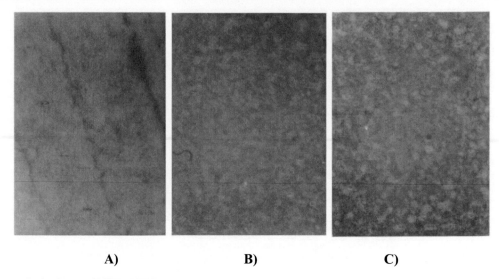

A)   B)   C)

Figure 44-1. Sample SS 316-26 photographs.

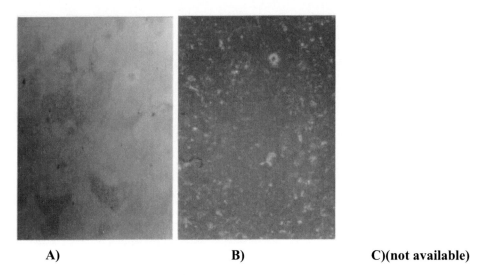

A)   B)   C)(not available)

Figure 44-2. Sample SS 316-139 photographs.

The sample (number 139) is an oxychloride black molybdenum film deposited at 610°C on Stainless steel 316. The sample fails after testing in air for 650 hours shortly after entering the range of 300°C the film starts deteriorating. Possible reason is the defects on the Si_3N_4 layer, where the oxygen in the air environment enters the film structure (see photograph on position B and optical performance behavior of sample number 139 on figure 44).

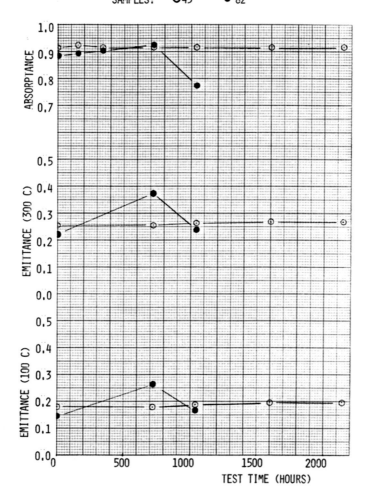

Figure 45. Life time test at 500°C in fore pump vacuum for two samples of OCBM films, deposited at 610°C on oxidized Incoloy 800 substrates. Films are covered with Si_3N_4 protective layer. The two films (numbers 43 and 82) are from one and the same CVD deposition run.

Sample 43 has a very good optical performance of OCBM film, deposited at 610°C on Incoloy 800 and tested in vacuum at 500°C. As seen from figure 45, this film has a perfect optical performance even after 2000 hours test. Sample 82: Film of OCBM deposited at 610°C on Incoloy 800 substrate. After 720 hours test in vacuum at 500°C film undergo changes, the emittance, calculated for 100°C blackbody temperature, arises from 0.15 to 0.27, and from 0.22 to 0.37 (for 300°C blackbody temperature). After 1050 hours the emittance drops down again to 0.17, together with the film absorptance, which decreases from 0.92 to 0.77. The sample surface has some defects, possible reason for the failure of the observed optical performance results.

Figure 45-1. Sample I -800-43 photographs.

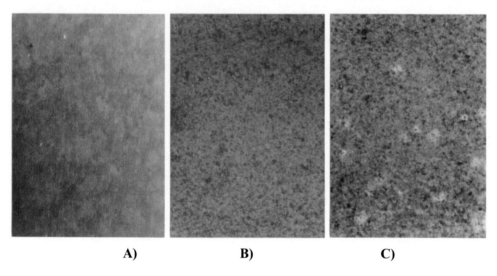

A)   **B)**   **C)**

Figure 45-2. Sample I – 800 - 82 photographs.

On the photograph are seen some defects, leading to the failure. No defects are observed on Si_3N_4 layer surface at 0 hours testing (the photograph in position B).

Chemical Vapor Deposition (CVD) Technology

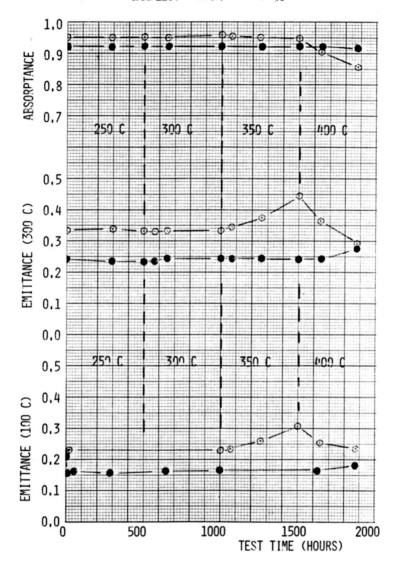

Figure 46. Life time test at 250-400°C in air of numbers 34 and 48 OCBM films, deposited at 610°C on oxidized Incoloy 800 substrates. Films are covered with Si_3N_4 protective layer. The two samples are from one and the same CVD deposition run.

Sample number 34 after 1050 hours heating, entering the range of 350°C starts deteriorating, the emittance arises from 0.23 to 0.31 (100°C) and from 0.34 to 0.45 (300°C). Reaching 1600 hours testing, which coincides with 400°C test temperature, the emittance starts decreasing again together with a decrease in solar absorptance. The surface of the film (Sample 48) has (see figure 46) only slight change in optical performance (the emittance arises from 0.16 to 0.18 (for 100°C blackbody temperature) and from 0.24 to 0.28 (for 300°C blackbody temperature) after testing in air for 1600 hours in temperature range 250 - 400°C.

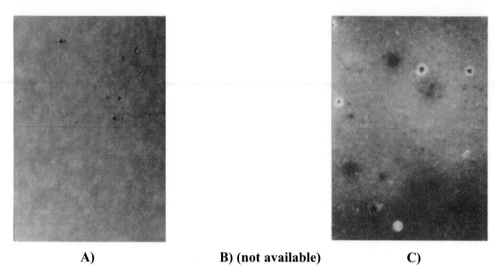

A)   B) (not available)   C)

Figure 46-1. Sample I-800-34 photographs.

The photograph of sample 34 surface in position C) is taken after 1600 hours testing in air. The defects seen on sample surface originate from defects in the substrate and the protective layer (no photograph is available at position B) - the so-called 0 hours test stage – a sample of oxidized substrate, with deposited black Mo on it and with a protective layer of Si_3N_4 over the black Mo film).

Figure 46-2. Sample I-800-48 surface photographs.

For the sample 48, as seen from figure 46-2, no defects on the sample surface were seen neither on the 0 hour test surface (after covering the black molybdenum with Si_3N_4 protective layer, photograph B), nor after 1600 hours testing in air.

From the test results of Black molybdenum solar absorber coatings (OCBM) the following conclusions could be made:

- Films deposited at higher temperatures withstood in vacuum test better than the films deposited at lower deposition temperatures. The test in air lead to different conclusions: the lower deposition temperatures produce films rich of oxygen so heating in air does not lead to significant influence of the air. This is valuable result, since these films show the best solar absorptance.
- Depositions on Incoloy-800 substrate lead to coatings with lower solar absorptance, while deposition on stainless steel leads to films with lower thermal emittance.
- For the two temperatures of deposition, the ranges in which the values of absorptance and emittance change were considerably smaller for the stainless steel substrates.
- The observed difference in the properties of the films, obtained at equal technological conditions we considered as due to difference in the substrate surfaces in result of using different polishing machines.Both deposition temperatures yield films of similar absorptance after antireflection but the low temperature films have a significantly higher thermal emittance.
- In conclusion it can be emphasized that the obtained spectrally selective surfaces sustain temperatures of $500^\circ C$ in rough pump vacuum for a long time (2000 hours) while if tested in air, they sustain shorter periods if temperature is $400^\circ C$. We should note that these solar absorbers are directed for applications in vacuum collectors so the convection losses could be minimal. The stability in air is needed only for shorter periods when some parts should be changed in case of failure.

In conclusion, from the OCBM no failures were observed besides the three films (sample numbers 133, 139 and 82) presented above. One possible explanation could be a combination of reasons due to film position in the furnace, the film position with respect to gas stream in the CVD reactor during the deposition, and no good quality of protective layer surface, through which the oxygen enters in film structure.

Carbonyl Black Molybdenum Testing

The carbonyl black molybdenum were deposited on quartz and Incoloy 800 metal substrates and they remained intact after 1000 hours of lifetime testing in a roughing pump vacuum at $500^\circ C$, while OCBM films deposited on both quartz and metal substrates have shown no changes in their optical properties after 2000 hours in the same environment at $500^\circ C$. After the successful testing in vacuum these samples were tested in air by exposure to $250^\circ C$ for 500 hours, followed by $50^\circ C$ increments for 500 hours each. The films survived with only slight changes in a and e after 500 hours at $350^\circ C$.

We emphasize the fact that films were taken out each 50 hours in order for their reflectance to be measured. All of the up-s and down-s in the temperature during heating and cooling down are additional stress over the tested samples and would have their own influence on the durability. Such ups and downs would correspond to the night and daily changing in the temperatures, so this kind of testing is very much alike the real working conditions test.

Selective coatings must undergo lifetime testing under a wide range of conditions before they can be qualified for long-period, trouble-free use in commercial collectors. Maximum operating temperatures are but one aspect of the problem; thermal shock from cloud passage, daily cycling or stagnation above the operating temperature due to loss of coolant can also degrade coatings. If meant for evacuated systems, the selective surface should be stable under vacuum, yet survive short periods of atmospheric exposure at operating temperatures.

In conclusion, after passivation and antireflection with Si_3N_4, black molybdenum can operate at temperatures in excess of degradation temperatures of black chrome (Mattox, 1976; Pettit and Sowell, 1976; Lampert and Washburn, 1979). Very simple photothermal solar converter coatings with respectable spectral selectivity and thermal durability at 500°C in vacuum and 350°C in air (Gesheva, Chain and Seraphin, 1981; Chain, Gesheva and Seraphin, 1981b)) have been deposited.

The same technological process has been applied for tungsten based spectrally selective surfaces. The development of black tungsten followed our work on black molybdenum. Tungsten and Molybdenum have similar chemical properties and are both group VI-B refractory metals. In particular, both materials form carbonyl compounds which demonstrate pyrolytically in the same temperature range (Lander and Germar, 1947). The reflectance of bulk tungsten is even lower in the visible than that of Molybdenum and the IR reflectance at the same time is higher so, this difference makes tungsten promising for spectrally selective coatings. The better optical performance of tungsten is defined also by its greater oxidation resistance. Oxidation through pinholes and cracked in the passivation layer is black molybdenum major failure mechanism. If comparing weight increase by oxidation of chromium, molybdenum and tungsten (Savitskii and Burkanov, 1970) we should note that tungsten oxidizes at a lower rate than molybdenum. Also, the transition from parabolic to linear rate of oxidation appears for tungsten at over 600° C compared to 500°C for the molybdenum. This suggests that exposure of tungsten films to air would be less harmful than for molybdenum. In addition, tungsten has the highest melting point and lowest vapor pressure of all metals and shows excellent corrosion resistance. Similar full studying of technology and structural and optical characterization has been made by L. K.Thomas and E.E.Chain, Thin Solid Films, 104 (1983); and later by Gesheva K.A., Gogova D., G.Stoyanov, J.de Physique III, 2(8), (1453) 1992;

RECENT DEVELOPMENT IN THE AREA OF SPECTRALLY SELECTIVE SURFACES

Most solar absorbers today are normally designed as an absorber-reflector tandem which consists of one surface with highly solar absorbing properties and one with high reflectance in the infrared wavelength range so that the thermal losses can be minimized. The requirement for high thermal conductivity implies that metal plates of cooper or aluminum are most often used in the absorber. These two metals have, because of the same fundamental reason, a high infrared reflectance that also makes them suitable as the infrared reflector in his tandem stack. The drawback is, however, that both these metals are sensitive to corrosion and if the absorbing layer does not function as a protecting layer, a specific protection layer has to be added.

The surface coating is often produced by chemical conversion by applying a paint layer or by depositing a coating using physical vapor deposition technology E.Wackelgard, G.A. Niklasson, and C.G. Granquist, in Solar Energy-the state of the art, edited by J.Gordon (JamesandJames, London, 2001),Ch3.; C.G.Granquist, Materials Science for Solar Energy Conversion Systems (Pergamon Press, Oxford, 1991). Chemical conversion has been most widely used technics (G.E. McDonald, Solar Energy 17,119 (1975); E. Wackelgard, Journal of Solar Energy Materials and Solar Cells, 56 35 (1998) but the process is space and material consuming and produces a lot of chemical waste materials. Absorbing paint coatings (Zorica Crnjak Orel, Solar Energy Materials and Solar Cells 57 (1999) 291-301;M.K. Gunde, Z.C.Orel, M.G.Hitchins, Solar Energy Materials and Solar Cells 80 (2003) 239-245; Z.C.Orel, M.K.Gunde Solar Energy Materials and Solar Cells 61 (2000) 445-450; M.K.Gunde and Z.C.Orel, Applied Optics/vol.39, N.4/1 February 2000.;B.Orel, Z.C.Orel, Journal of Solar Energy Materials 18, 97 (1988) can be produced cheaply when simpler deposition technique such as spraying or coil coating is used. Today, vacuum deposition techniques (E.Wackelgard, and G.Hultmark, Journal of Solar Energy Materials and Solar Cells 54, 165 (1998); W.Graf, F.Brucker, M.Kohl, T.Troscher, V.Wittwer and L.Herlitze, Journal of Non-Crystalline Solids 218, 380 (1997) have become established in commercially absorber production. Production units for vacuum deposition are relatively small and the process consumes small amounts of energy and materials. These factors make the production inexpensive and environmentally friendly.

High solar absorbing properties can be achieved with graded optical index coating. A graded optical index profile can be obtained for a dielectric medium filled with a gradually increasing fraction of metal from the base to the top of the layer. The absorbing layer usually consists of a transition metal such as nickel or chromium in an oxide of for example nickel oxide (E.Wackelgard, and G.Hultmark, Journal of Solar Energy Materials and Solar Cells 54, 165 (1998); chromium oxide (J.C.C. Fan and S.A. Spura, Applied Physics Letters 30 511 (1977) or aluminum oxide (G.A. Niklasson and C.G. Granquist, Journal of Applied Physics 55, 3382 (1984);. An antireflecting coating can be added on top to further improve the absorbing properties. The antireflecting coating decreases front surface reflection between air and the absorbing layer and optical interference effects are suppressed within the absorbing film.

Recently solution-chemically derived nanoparticles embedded in a dielectric matrix of alumina (T. K.Bostrom, E.Wackelgard, Gunnar Westin, SEM&SC (2005) 89, 197-207) solar absorbers have undergone a test by increasing the temperature gradually up to $580^{\circ}C$ for 150 hours. If alumina antireflecting coating is deposited over the absorbing coating, the test has been unsuccessful, and if other materials silica and silica-titania mixtures antireflecting coatings are applied no visible degradation of the sample surface has been observed even after 600 hours of testing. Different other types of selective absorbing coatings have been employed: mechanically manufactured selective solar absorbers based on mechanically grooved aluminum surface, then covered by Al_2O_3. During the grinding of the surface some carbon gets absorbed in result of acid bath applied to mechanically operated grinding alone. Such an absorbing surface has solar absorptance of 0.90 and thermal emittance 0.25.Authors do not show the temperature for which the thermal emittance is calculated (P.Kontinen, P.D.Lund, R.J.Kilpi, SEM&SC 79, 2003, 273-283).

A novel design in composites of various materials for solar selective coatings has been developed (M. Farooq, M.G.Hutchins, SEM&SC 71 (2002) 523-535). Developed are multilayered systems of metal-dielectric graded index solar selective coatings with solar absorptance of 0.96 with less than 0.07 thermal emittance at 300 K for 200 nm thick, 4-layer system of V: Al_2O_3 composites. Here there is metallic gradation through the film thickness for the solar flux to meet gradually optically more dense material and thus to get absorbed more smoothly without losses. Other interesting spectrally selective surfaces have been developed by pyrolytically depositing conducting tin oxide on nickel-pigmented anodized aluminum. Cooper and aluminum substrates (50 cm^2), properly cleaned at room temperature by 10 M NaOH and 20 % diluter HNO_3 respectively, onto which composite films of metal (nickel or vanadium) and insulator (quartz or alumina) were deposited. Solar absorptance of 0.96 has been achieved by simulation and experimentally findings with less than 0.07 thermal emittance for 300 K for 200 nm thick, four layer-system of V:AL_2O_3. The early achievements are the so-called AMA selective surfaces, as Walter F.Bogaerts writes in his review on materials for photothermal solar energy conversion based on Al_2O_3/Mo/Al_2O_3 (R.F.Peterson and J.W.Ramsey, J.Vac.Sci.Technol. 12,1975), which is being stable at more than 800°C and having an absorptance of 0.85 and emittance at 500°C of 0.11 (0.07 at 20°C). Based on previously published papers by Tadanaga (K.Tadanaga, K.Iwashita, T.Minami, N.Tohge in J.of Sol-gel Eci.Technol. 6 (1996) 107-111 for sol-gel techniques for producing silica and by Dawnay et.al. E.J.C. Dawnay, M.A. Fardad, M.Green, F.Horowitz, E.M. Yeatman, R.M. Almeida, H.C. Vasconcelos, M.Guglielmi, A.Martucci, Adv.Mater.Opt. (1995) 55-62 for silica and respectively silica-titania mixtures, the authors Tobias K.Bostrom, Ewa Wackelgard, Gunnar Westin SEM&SC (2005) 89, 197-207) develop a promising novel solution-chemical method to fabricate selective solar absorber coatings. The optimal selective optical coating has achieved absorptance value of 0.91 with thermal emittance at room temperature of 0.03.

The interest continues on black copper by texturing the coatings being prepared by chemical conversion process in an alkaline bath of potassium persulphate ($K_2S_2O_8$). As-deposited, air and vacuum annealed and 18 months external environment exposed black cooper coatings have been analyzed using AES and XPS techniques. The results have shown that black cooper is a multicomponent system consisting of Cu and some of its carbonate, sulfate, or hydrate compounds forms. The system as a whole can sustain temperatures of 250°C in air and under vacuum conditions. Koltun M.M. et al, SEM&SC 33 (1994) 41- 44, produced by electrodeposition technique black nickel with solar absorptance of 0.92 and thermal emittance less than 0.15. J.H.Schon, G.Binder, E.Bucher SEMandSC 33 (1994) 403-416 examined two systems of selective absorber surfaces based on metal/dielectric multilayers of Al_2O_3/Pt and Al_2O_3/$MoSi_2$ produced by rf-sputtering. Absorptance of 0.95 and emittances from 0.08 to 0.2 (calculated for T = 1100 K) for Pt/Al_2O_3 and a = 0.92 and thermal emittance of e = 0.14 for $MoSi_2$/Al_2O_3 system have been achieved. These authors continue studying of high-temperature coatings (T ≥ 500°C) based on alumina/noble metal (Rh, Ir) system with absorptance 0.94 and emittance calculated for 1100 K of 0.16. Oxides of these metals could enhance the absorptance as the simulations show to values of 0.96 for the both systems.

REFERENCES

Agnihotri, O.P., and B.K.Gupta, Solar Selective Surfaces (John Wiley and Sons, New York, 1981)

Aspnes, D.E., Proceedings of the Society of Photo-Optical Instrumentation Engineering, Vol. 276, 188 (1981)

Aspnes, D.E. *Thin Solid Films.* 89, 249 (1982a).

Aspnes, D.E., *Phys. Rev.* B 25, 1358 (1982b).

Bennett, H.E., J.M. Bennett, and E.J. Ashly, *J.Opt.Soc.Am.* 52, 1245 (1962).

Bergman, D.J., *Phys. Rev.* B 23, 3058 (1981).

Blocher, J.M. Jr., *J. Vac. Sci. Technol.* 11, 680 (1974).

Bruggeman, D.A.G., *Ann. Phys. Leipzig.* 24, 636 (1935).

Bunshah, R.F., *Vacuum.* 27, 353 (1977).

Bostrom T. K., E.Wackelgard, Gunnar Westin, SEM$SC (2005) 89, 197-207)

Carver, G.E., *Solar Energy Materials.* 1, 357 (1979).

Carver, G.E., Ph.D. dissertation (university of Arizona, Tucson, AZ, 1981) p.78.

Carver, G.E., and E.E.Chain, *J.de Phys.* 42, C1 – 203 (1981).

Carver G.E. and B.O.Seraphin, *NBS Special Publ.*568 (1979)

Chain, E.E., G.E. Carver and B.O.Seraphin, *Thin Solid Films.* 72, 59 (1980).

Chain, E.E., K.A.Gesheva, and B.O.Seraphin, *Thin Solid Films.* 83, 387 (1981a).

Chain E.E., K.A.Gesheva, and B.O.Seraphin, Proceedings of the Eighth International Conference on Chemical Vapor Deposition, Gouviex, France, J.M. Blocher, Jr., G.E. Vuillard, and G.Wahl, eds.(The Electrochemical Society, Pennington, N.J., 1981b) p.461.

Chain, E.E., K.Seshan, and B.O.Seraphin, *J. Appl. Phys.* 52, 1356 (1981).

Chase, L.L., *Phys.Rev.* B 10, 2226 (1974)

Childs, W.J., J.E. Cline, W.M. Kisner, and J.Wulff, Trans. Of the A.S.M. 43, 105 (1951).

Cullity, B.D., Elements of X-ray Diffraction, Second Edition (Addison-Wesley, Reading, Mass., 1978) p.45.

Cuomo, J.J., J.F. Ziegler, and J.M. Woodall, *Appl. Phys. Lett.* 26, 557 (1975).

Dawnay E.J.C., M.A. Fardad, M.Green, F.Horowitz, E.M. Yeatman, R.M. Almeida, H.C. Vasconcelos, M.Guglielmi, A.Martucci, *Adv. Mater. Opt.* (1995) 55-62

Dissanayake, M.A.K.L., and L.L.Chase, *Phys. Rev.* B 18, 6872 (1978)

Fan C.C. and S.A. Spura, *Applied Physics Letters.* 30 511 (1977)

Farooq M, M.G.Hutchins, SEM and SC 71 (2002) 523-535).

Garnett, J.C.M., *Phil. Trans. Roy.Soc.* London A 203, 385 (1904).

Garnett, J.C.M., *Phil. Trans. Roy.Soc.* London A 205, 237 (1906).

Gesheva K.A., E.E.Chain, and B.O.Seraphin, *Solar Energy Materials.* 3, 415 (1980).

Gesheva, K.A. K.Seshan, and B.O.Seraphin, *Thin Solid Films,* 79, 39 (1981).

Gesheva K.A., Gogova D., G.Stoyanov, *J. de Physique III,* 2(8), (1453) 1992;

Gibson, U.J., H.G. Craighead, and R.A. Buhrman, *Phys. Rev.* B 25, 1149 (1982).

Goldstein, J.I. and H.Yakowitz, *Practical Scanning Electron Microscopy.* (plenum Press, New York, 1976)

Granqvist, C.G., and Hunderi, *Thin Solid Films.* 57, 15 (1978).

Granquist C.G, Materials Science for Solar Energy Conversion Systems (Pergamon Press, Oxford, 1991

Gurev, H.S., G.E.Carver, and B.O.Seraphin, *Solar Energy.* (The Electrochemical Society, 1976) p.36.

Gunde M.K, Orel Z.C., M.G.Hitchins, *Solar Energy Materials and Solar Cells.* 80 (2003) 239-245.

Gunde M.K. and.Orel Z.C, *Applied Optics*/vol.39, N.4/1 February (2000).

Hahn, R.E., and B.O.Seraphin, Physics of Thin Films, Vol. 10, G.Hass and M.H. Francombe, eds. (Academic Press, New York, 1978) p.1.

Hashin, Z., and S.Shtrikman, *J. Appl. Phys.* 33, 3125 (1962).

Hass, G. H.H. Schroeder, and A.F. Turner, *J. Opt. Soc. Am.*, 46, 31 (1951).

Holzle, R.A. ,Techniques of Metals Research, Vol. 1, R.F. Bunshah, ed. (Interscience, New York, 1968) p. 1377.

Ognatiev, A., P.O,Neill, and G.Zajac, *Solar Energy Materials.* 1, 69 (1979).

Jackson, J.D., Classical Electrodynamics, Second Edition (Wiley, New York, 1975) p.146.

Jacobson, M.R., Proceedings of the Society of Photo-Optical Instrumentation Engineers, Vol. 325, 188 (1982)

Jacobson, M.R. , and D. Lamoreaux, Proceedings of the Second Annual Conference on Absorber Surfaces for Solar Receivers, P.J. Call, ed., Boulder, Col., 1979 (D.O.E.-S.E.R.I.) p. 151.

Joshi, A., L.E. Davis, and P.W. Palmberg, Methods of Surface Analysis, A.W. Czandorna, ed. (Elsevier Scientific Publishing Company, New York, 1975).

Killeffer, D.H., and A.Linz, Mlybdenum Compounds, Their Chemistry and Technology (Interscience Publishers, New York, 1952) p.192.

Kirillova, M.M., L.V. Nomerovannaya, and M.M.Noskov, *Soviet Physics JETP 33,* 1210 (1971).

(Kontinen P, P.D.Lund, R.J.Kilpi ,*SEM and SC.* 79(2003) 273-283).

4-M.Koltun, G.Gukhman, and A.Gavrilina, *Journal of Solar Energy Materials and Solar Cells.* 33, 41 (1994);

Lamb, W., D.M. Wood, and N.W. Ascroft, *Phys. Rev.* B 21 , 2248 (1980).

Lampert, C.M., *Solar Energy Materials.* 1, 81 (1979).

Landauer R., Electrical Transport and Optical Properties of Inhomogeneous Media, vol. 40, ed. J.C. Garland and D.B. Tanner (American Institute of Physics, New York, 1978) p.2.

Lander, J.J., and Germer, *Trans. AIME* .175, 648 (1947)

Liebert, C.H., and R.R. Hibbard, *Sol..Energy.* 6, 84 (1962)

Macleod, H.A., *personal communication*

McDonald G.E., Solar Energy 17,119 (1975); M.Koltun, G.Gukhman, and A.Gavrilina, *Journal of Sol. Energy Materials and Solar Cells.* 33, 41 (1994); Mattox, D., J.Vac. Sci. Technol. 13,127(1976).

Meinel, A.B., and M.P. Meinel, *Applied Solar Energy,* (Addison Wesley, Reading, 1976).

Milton, G.W., *Appl. Phys.Letters.* 37, 300 (1980).

Milton, G.W. *Phys. Rev. Lett.* 46, 542 (1981)

Movchan , B.A. and Demchishin, *Fiz. Metal. Metalloved.* 28, 653 (1969).

Niclasson, G.A., and C.G.Granquist, *Solar Energy Materials.* 5, 173 (1981)

Niklasson G.A. and C.G. Granquist, *Journal of Applied Physics.* 55, 3382 (1984);T.S.

(Orel Zorica Crnjak, *Solar Energy Materials and Solar Cells.* 57 (1999) 291-301;

Orel Z.C., M.K.Gunde, *Solar Energy Materials and Solar Cells.* 61 (2000) 445-450;

Orel B,.Orel Z.C, *Applied Optics Journal of Solar Energy Materials.* 18, 97 (1988).

Pettit, R.B., and R.R. Sowell, *J.Vac. Sci. Technol.* 13, 596 (1976).

Phillips, B., and L.L.Y. Chang, *Trans.of the A.I.M.E.* 230, 1203 (1964)

Ping Sheng, *Phys. Rev.Lett.* 45, 60 (1980).

Pivovonsky, M., and M.R. Nagel, Tables of Blackbody Radiation Functions (Macmillan, New York, 1961)

Peterson R.F. and J.W.Ramsey, *J. Vac. Sci. Technol.* 12 (1975) 174,

Powell, C.F., Vapor Deposition, C.F. Powell, J.H.Oxley, and J.M. Blocher, Jr.,eds. (John Wiley and Sons, Inc., New York, 1966).

Rouard, P., and A.Meessen, Progress in Optics, vol. 15, ed.by E.Wolf (North-Holland, Amsterdam, 1977) p.79.

Sathiaraj, and O.P. Agnihotry, *Journal of Solar Energy Materials.* 18, 343 (1989).

Savitzki, E.M., and G.S. Burkanov, Physical Metallurgy of Refractory Metals and Alloys (Consultants Bureau, New York, 1970).

Seraphin, B.O., Proceedings of the Symposium of Material Science Aspects of Solar Energy Conversion, Tucson, AZ, ed. B.O. Seraphin, 1974, p.18.

Seraphin, B.O. Topics in Applied Physics, ed. B.O.Seraphin (Springer Verlag, Berlin, 1979) p.5.

Seraphin, B.O. *Thin Solid Films.* 90 395 (1982).

Seraphin, B.O., and Meinel, Optical Properties of Solids – New Developments, ed. B.O. Seraphin (north-Holland, Amsterdam, 1976) p.927.

Seshan, K., P.D.Hillmannn, K.A.Gesheva, E.E. Chain, and B.O.Seraphin, *Mater. Res. Bulletin.* 16, 1345 (1981).

Seto, D.K., V.Y.Doo, and S.Dash, Proceedings of the Third International Conference on Chemical Vapor Deposition, F.A.Glaski, ed. Salt Lake City, (Electrochemical Society, Pennington, N.J., 1972) p. 659.

J.H.Schon, G.Binder, E. Bucher *SEM and SC.* 33 (1994) 403-416

Shardein, D.J., and R.D. Lioyd, Ger.Offen. 2, 811, 393 (September 27, 1979, assigned to Reynolds Metals Co.).

Smith, G.B., and A.Ignatiev, *Solar Energy Materials.* 2, 461 (1980).

Stroud, D. and F.P. Pan, *Phys.Rev.* B 17, 1602 (1978).

Tabor, H. Bull. Res. Council Israel 5A, 119 (1956).

Tabor, H. Proceedings of the International Solar Energy Society Congress, Vol. 2, F. deWinter and M.Cox, eds., New Delhi, 1978 (Pergamon Press, New York), p.829.

Tobias K.Bostrom, Ewa Wackelgard, Gunnar Westin, *SEM and SC.* (2005) 89, 197-207) .

Thomas, A. (Mechericunnel), and J.C. Richmond, Proceedings of the Seminar on Testing Solar Energy Materials and Systems, Gaitthersburg, Md., 1978 (Inst. Of Enviromental Sciences) p.83.

L. K.Thomas and E.E.Chain, *Thin Solid Films,* 104 (1983);

Thornton, J.A., *J. Vac. Sci. Technol.* 11, 666 (1974).

Trotter, D.M., Jr., and A.J. Sievers, *Appl. Phys. Lett.* 35, 374 (1979)

Wackelgard E., G.A. Niklasson, and C.G. Granquist, in Solar Energy-the state of the art, edited by J.Gordon (JamesandJames, London, 2001),Ch3; Wackelgard E., and G.Hultmark, Journal of Solar Energy Materials and Solar Cells 54, 165 (1998); W.Graf, F.Brucker, M.Kohl, T.Troscher, V.Wittwer and L.Herlitze, *Journal of Non-Christalline Solids.* 218, 380 (1997).

Wackelgard E. and G.Hultmark, *Journal of Solar Energy Materials and Solar Cells*. 54, 165 (1998).

Wackelgard E., *Journal of Solar Energy Materials and Solar Cells,* 56 35 (1998).

Weaver, J.H., C.G. Olson, and D.W. Lynch, *Phys. Rev.* B 12, 1293 (1975).

Wood, D.M., and N.W. Ashcroft, *Phil. Mag.* 35, 69 (1977).

Wyant, J.C., *Laser Focus.* 18, 65 (1982).

Yih, S.W.H., and C.T. Wang, Tungsten Sources, Metallurgy, Properties and Applications (Plenum Press, New York, 1979) p. 313.

Chapter 2

ENERGY CONTROL COATINGS ELECTROCHROMISM, MATERIALS AND APPLICATIONS

K. A. Gesheva
Central Laboratory of Solar Energy and New Energy Sources,
Bulgarian Academy of Sciences, Sofia, Bulgaria

2.1. ELECTROCHROMISM

Electrochromic Materials, Basis and Applications

Transition-metal oxides are one of the most interesting classes of inorganic solids, exhibiting wide variety of structures, properties, and phenomena. Their unusual properties are typically attributed to the unique nature of the outer d electrons [1,2] and therefore the d-band defines the interesting electrophysical and optical properties of transition metal oxides. The chromogenic behaviour presents a specific characteristic of the transition metal oxide films. The films can modulate reversibly their optical characteristics under influence of external factors. The induced change in the optical absorption retains even when the excitation source is removed. Electrochromism has been discovered in several different metal oxides and all the pertinent metals are localized in a well-defined area of Periodic table [3]. All these metals belong to the transition series. This experimental fact suggests a closed relation between the electrochromic properties and the electronic structure. Electrochromic materials are exclusively interesting due to the variety of potential applications, including elements for information displays, antiglare rearview mirrors, sunroofs and smart windows.

The corresponding optical states: coloring (high absorptive state) and bleaching (transparent state), as been mentioned above, can be induced by various types of external factors [4-6]. The insertion and extraction of ions (causing the color change in the materials) are often called intercalation and deintercalation, respectively. Due to well established research done on electrochromic effect and materials for optical applications, the terms "colored" and "bleached" are commonly used in literature.

A) Electrochromism

Electrochromism is a phenomenon connected with a color modulation caused in the material by an external applied electric field or voltage [4-8]. The concept of this electrooptical effect is similar to the well studied electroluminescence process with an exception that for the electrochromic process, a creation of metastable absorption center has been assumed, there is not emission.

Nowadays, the investigations and studies reveal the principal essentiality of electrochromism: from one hand, the basic definitions have been determined as well as the groups of materials, which exhibit EC properties and from other hand established is the place of electrochromism as a subject of solid state ionics. More details for electrochromism will be described later in this chapter.

B) Photochromism

When the absorptive states are changed reversibly and this has been induced by light radiation, the materials show photo chromic properties. For example, after radiating thin films of MoO_3 and WO_3 [9, 10] with UV light this leads to color change. The electron-hole couples appear into these materials. The photo generated holes react with the surface adsorptive substances and cause a negative charging of the thin films. The positive ions on the surface (in the most cases, they are protons) are injected into the film by Coulon interaction and a new substance called bronze is created - H_xMoO_3. The new material exhibits broad absorption bands in the visible spectral region. The absorption is due to charge transfers from Mo^{5+} states towards the oxide ligands that create Mo^{6+} levels. Generally, the photochromic materials in thin film form are with enlarged surface area (porosity), capable of adsorbing organic molecules [11]. Several organic compounds can be applied for this purpose: aldehydes, alcohols, organic acids etc.

Photochromic materials and devices possess the advantages due to the fact that they have no necessity of external electrical field and the coloring is uniform independent on the film area differing from the electrochromic devices, where the uniform coloring of large area is still difficult to achieve [12]. So, the photochromic materials are, as the name implies, darken under the direct action of sunlight [13, 14]. They can not be considered as versatile as electrochromics because they cannot be manually controlled and because an optimum energy performance requires the consideration of temperature conditions as well as solar radiation. For example, a photochromic window may darken on a cold sunny day when more solar heat gain is desireable. The photochromic materials finds applications in automatically darkening sunglasses.

C) Gasochromism

The gasochromic sensitive materials are those that change their optical transparency in the presence of a proper gas. The gasochromic systems are consisted of a chromogenic layer and a very thin catalytic film. The hydrogen gas is dissociated on the catalysts as atomic hydrogen H, which enters the structure of the chromogenic film, resulting in film coloring [15-17]. Gasochromic windows possess simple configuration of thin films and high solar transmittance.

The gasochromic switching device is claimed to be inexpensive because only a single WO_3 sputtered layer [15] covered with a few nanometer thick metal layer (Pd or Pt) is sufficient to obtain a gasochromic effect (see figure 1). The bleached state is then achieved by flushing the device with air. The long term cycling stability of a gasochromic device is difficult to attain unless there is strict control over the H_2 gas humidity, the presence of gases which could poison the catalyst (e.g., CO, NO_2) and the concentration of H_2. One of the advantages of gasochromic devices is that it can switch within seconds.

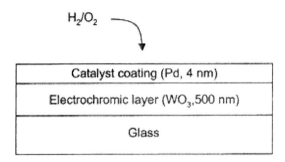

Figure 1. Schematic diagram of gasochromic device.

However, despite the recent detailed studies in literature, the physical mechanism of the gasochromic color change is still not fully understood. The color change in the electrochromic films is believed to be directly related to the double injection/extraction of electrons and ions in the films [16-18]. This double injection model has also gained a wide acceptance for the explanation of the gasochromic mechanism. Recently, Georg et al. [15] proposed a new model to interpret the different bleaching speeds in Ar and O_2 atmospheres. According to Georg's model, H_2 dissociated on the platinum, is transferred into a pore or a grain boundary of WO_3 and subsequently creates water molecules and an oxygen vacancy. The oxygen (O) vacancy diffuses into the interior of the grain and the water desorbs slowly from the film. The Georg's model is different to the concept assumed in the double injection model. Although both of the two models can theoretically explain the coloration of the gasochromic devices by exposure to diluted hydrogen gas, there is no direct evidences that support either of these two models. Gasochromic thin film materials can be applied as hydrogen sensors [19].

D) Termochromism

The phenomenon of some transition metal oxides to reversible transform simultaneously from a metallic to a semiconductor state, induced by a temperature change is called thermochromism. Varying the temperature, the thermochromic materials modulate their optical behavior, as well [20, 21]. The material known to exhibit the best thermochromic properties is the vanadium dioxide. Its critical temperature of the phase transition is $68^{\circ}C$, a value that is in the convectional range of the technological applications [22, 23]. The high temperature metallic phase with tetragonal structure shows good IR reflectivity, meanwhile the low temperature semiconductor monoclinic modification exhibits a high transparency of IR radiation. The phase change is manifested with change of the electrical conductivity, increasing several times with heating.

Termochromic properties are reported for the thin films of MoO_3 and WO_3 [24], suggesting different mechanisms of thermochromic process. In the case of tungsten oxide, the thermochromic coloring results from the charge optical transitions among tungsten ions, varying in valences. But for MoO$_3$ films, another mechanism is proposed, a formation of color centers in the film structure as a function of the annealing temperatures.

Thermochromic effect has also been observed in pigments. Color changes of thermochromic pigments are induced by a temperature variation. The thermochromic pigments are composed of microcapsules, possessing reversible optical modulation. After the temperature approaches a specific temperature, then the color changes and it starts fading back to the original color as the pigment is cooled down.

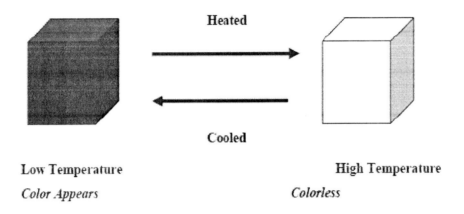

Figure 2. An example of thermochromic pigments.

Reversible and irreversible temperature indicators have been developed. These indicators undergo a color change when a certain predetermined temperature (or temperature range), usually above the room temperature, is reached. They can be applied for detecting a forgery and as security inks. These indicators differ from colorless-to-red, colorless-to-green and blue-to-red upon heating and they can be easily incorporated into the commercially available ink vehicles, e.g., flexo and gravure.

Thermochromic materials for window glazing applications have been investigated [25]. Thermochromic materials have been studied by a number of research groups. The coating techniques employed were PVD or sol - gel based methods. The technology advances were not transferable to commercial applications. The primary reasons are summarized as follows:

- the transition temperatures are too high than the room or the typical ambient temperatures. The thermochromic materials, which phase-change in the visible possess very unattractive or too dark colors.
- there is still a lack of in-depth understanding of how thermochromics could work as energy efficient/comfort enhancing glazings, about the material properties, film structure, and stoichiometry control.
- the thermochromic process appears to be very complex and /or difficult to control the film manufacture.
- there exist significant film durability limitations (e.g. "soft" and moisture sensitive).

2.2. ELECTROCHROMIC CHARACTERISTICS AND MECHANISMS

Electrochromism can be assigned to such phenomena, for which the color modulation is effected by an applied external voltage [26, 27]. The main characteristics describing EC effect are the following:

Color efficiency (CE) - this quality can be regarded as the electrode area which may be colored to unit absorbance by unit charge [4]. CE value has a positive sign for cathodic induced coloration and a negative one for anodic electrochroms. Experimentally, color efficiency qualifies the modulation of the optical properties. This quantity is strongly dependent on the wavelength and the changes in the optical density (ΔOD). It can be determined by the equation:

$$CE = \Delta OD/Q, \tag{1}$$

Where ΔOD represents the change in single-pass, transmitted optical density at the wavelength of interest λ, because of transfer of charge Q.

Contrast Ratio (CR) - a quantitative measure of the intensity of color change. That is the ratio of the intensity of light diffusely reflected through the colored state towards the intensity of light diffusely reflected from the bleached state. CR should be referred to a specific wavelength or related to an integral value for the white light.

Write - Erase Efficiency - the percentage of the originally formed coloration that may be subsequently electrobleached, it can be expressed as a ratio of the absorbance changes.

Response Time - the time required for an electrochromic film to color from its bleached state (or vice versa) is termed response time or switching time. The response time depends on the specific materials, the sizes of the EC devices and the temperature. It was found out that the larger windows switched more slowly (14 to 26 min) than the smaller windows (8 to 18 min). Another interesting feature is that all types of EC smart windows switched faster from colored to bleached state than bleached to colored. It is worth mentioning that the response time requirements differ depending on the applications.

Cycle Life - this is a measure for the stability of the electrochromic film or the device, being the number of cycles possible before occurring failure or misfunction.

The Insertion Coefficient - Electrochromic effect occurs when charge as electrons and positive ions (M^+) are inserted into the solid film (MeO_3). The reaction product is M_xMeO_3, where x is the number of ions entering the film structure and is called insertion coefficient. The value of x depends on the charge passed and hence on the extent of electrode reaction.

Intensive investigations are carried out to determine the nature of electrochromic effect. The working mechanisms have not been understood in full extent, several models have been proposed for the essence and the reasons of electrochromism:

Formation of color centers – Definition of the color center is as certain defects, localized into material lattice, absorbing light. Their formation is manifested by changing the absorption of the material (its coloration). Color centers are formed at negative ion vacancies. Although **F** color centers are found in alkali halide crystals with lattice defects, that absorb visible light, they have been also suggested to be found in amorphous transition metal oxides [28].

Small polarons model - Electrochromic effect occurs after intercalation / deintercalation of small ions and electrons into film structure. It is suggested that the inserted electrons are localized at the metal sites. Then small polarons are created by polarization of their surroundings by the inserted electrons and the localization of the wavefunction takes place predominantly to one lattice site. The polaron is defined as a quasiparticle that consists of an electron or a hole and a force field of phonons. A single polaron is assumed as large when the size of distortion is much larger than the lattice parameter. On the other hand, a polaron is small, when the excess charge, hole or electron, only distorts the nearest neighbors. A small overlap between wavefunctions corresponding to adjacent sites is conductive to the polaron formation [29]. This concept offers an opportunity for a quantitative theory of the optical absorption. The polaron absorption has been proposed as predominant for cathodic EC materials, meanwhile it is negligible for anodic electrochromics.

Intervalence charge transfer model – The complete mechanism of the optical absorption in the disordered transition metal oxides is still controversial. Coloration results from simultaneous insertion of cations and electrons into film. This reversible reaction can be written as follows:

$$MeO_3 + xe^- + xM^+ \leftrightarrow M_xMeO_3, \tag{2}$$

Where M^+ represents protons or alkali ions (H, Li or K) and Me is metal Mo or W.

The charge balancing electrons accompanying the ions can be captured by metallic sites and the valence has been reduced from M^{6+} to M^{5+} and as a result, the optical transitions provoke the coloration of the materials [30] in the presence of light, then the valence electron absorbs a photon and jumps from one site to the neighboring one. The charge transfer is the basic model that explains the optical absorption in disordered WO_3 films [1].

Several different models can be as well proposed for mechanisms of the electrochromic effect:

- charge transfer from valence band to splitted M^{5+} state;
- band-to-band transitions from a filled conductance band to higher energy band;

Electrochromism has been discovered in various metal oxides and all the metal elements, which oxides exhibit EC properties belong to the transition metal series. Cathodic coloration is observed for oxides of Ti, Nb, Mo, Ta and W and anodic coloration reveals in Cr, Mn, Fe, Co, Ni, Rh and Ir oxides.

Electrochromism can be described as a characteristic of a multilayer system – a device, although the device optical function is dominated by one functional active layer [4]. Many different EC device constructions are possible, but all they can be described only as variants of one basic design, presented in figure 3 [1]. The most general description is that electrochromic cell is consisted of two glass substrates covered with transparent and electrically conducting film. Between them is situated an active working electrode (EC film), followed by ion conductor (or electrolyte) and a counter electrode.

Transparent electrodes are the part of EC devices that offer comparatively small problems. The conducting films at the glass interface can be any high transparent IR reflectors such as the doped oxide semiconductors: In_2O_3:Sn; SnO_2:F; SnO_2:Sb, ZnO:Al or Cd_2SnO_4. There are several standard techniques for producing these coatings, which are commercially available [31].

Novel conducting substrates are found for applications as transparent conductors in EC devices, they represent polymeric flexible substrates. Flexible coils are polymer such as poly (ethylene terephthalate) or poly (ethylene nephthalate) known commercially as PET and PEN, respectively [32].

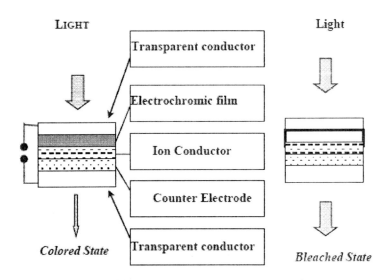

Figure 3. Basic design of an electrochromic device.

Electrochromic(EC) films are deposited on the transparent conductor and must possess a mixed conductivity for ions and electrons. When the intercalation of small ions from the neighboring ion conductor begins, an opposite balancing electron flux will be created, which is assured from these transparent conductors [1, 4]. Electrochromic film acts as a working electrode, after applying the voltage between the two transparent conductors, a spontaneous intercalation of small ions begins from the ion conductor and electrons from the conductors. The electron injection invokes modulation of the optical properties of the EC film, meanwhile the alkali ions balance the charge difference. The electrochromic process is reversible effect, meaning that the change of the polarity leads to return of the initial state of the EC film.

Ion conductors may be liquid, solid or polymeric types. The solid electrolytes are also divided in two groups: a bulk-like and in a form of thin film. The bulk-like electrolytes use proton type conductors such as phosphorous-tungsten acid, zirconium phosphate [4] and $Sn(HPO_4).H_2O$ [33]. Antimonium oxides are also incorporated in EC cell prototypes, many investigations were carried out for $HSbO_3.2H_2O$ or $Sb_2O_5.pH_2O$ [34]. Ion conductors as thin films are very perspective. Another proposed possibility for easily designed EC cell as ion conductors can participate dielectric films with coordinated water. For example, evaporated $MgF_2.H_2O$ [35] films are applied in the electrochromic constructions, known as Deb's devices. But they possess one disadvantage as the most sensible and critical part occurs to be this water containing dielectric thin film, which must be highly porous in order to begin spontaneous adsorption of water when exposed to humidity. Detailed studies are reported for the porous films of SiO_x, LiF, Cr_2O_3, Ta_2O_5 [36].

The liquid electrolytes have the advantages for a faster switching compare with the solid state ones due to the increased ion mobility. On the other hand, these electrolytes are difficult for storage and carrying. Conventional electrolytes consist of sulphur acid solution or lithium perchlorate soluble in propylene carbonate or ethylene glycol [4].

Polymeric electrolytes attain great interest and are widely studied [37]. For example, polyetheric electrolytes: polyethylene oxide, polypropylene oxides or alkali salts.

Generally, the ion conductor role is to allow as faster transfer as possible of the alkali ions or protons between the EC and counter electrodes without permitting electrons to pass through it. The ion conductor must fulfill the following requirements:

- high ionic conductivity;
- high electron resistance;
- transparency in the range of $0.35\mu m \leq \lambda \leq 1.5~\mu m$ [28], if the EC cells will be used in architectural windows.

Counter electrode is a very important part of the EC device. During the past years, most scientific studies are focused on the discovery and investigations of electrochromic properties in different materials in the form of thin films. But the successful work of the EC device requires finding of suitable counter electrodes [4, 39, 40].

The following criteria must satisfy an efficient counter electrode:

- a necessity of reversible storage of a proper charge under the form of injected ions
- to serve as an ion storage;
- to be transparent in the spectral range of the device application;
- under the influence of ion injection to be optically passive or to exhibit electrochromism in a sense opposite to that of working EC electrode;
- To be long-term stable upon cycling.

There are basically two types of materials investigated for counter electrodes in EC devices: anodic electrochromic and transparent passive materials. Anodic electrochromic materials detailed studied are $IrO_2.xH_2O$ [41, 42] and $NiOOH$ [43, 44]. Thin films of nickel oxide possess anodic electrochromic properties [4], with visual appearance changing between transparent and brown upon charge insertion/extraction. In electrochromic devices incorporating W oxide, the brown color of Ni oxide is complementary to the blue color of W oxide, both together yielding a neutral (grey) color in transmittance of the colored (darkened) state of the electrochromic device. Other anodic EC materials, which have been less investigated are Prussian Blue [45, 46], Cr_2O_3 [47] and $LiCoO_2$ [48]. Optically passive counter electrodes require ion storage materials with a large charge capacity and a high transparency. So far, these favorable properties have never been approached by a single stoichiometric compound. For example, cerium oxide is highly transparent, but possesses very low charge capacity. The mixed oxides based on cerium oxide reveal excellent transparency in the visible spectrum and during the intercalation/deintercalation cycles these oxide films are not influenced by the alkali ions and do not change their optical transmittance. Recently, the thin films of ZrO_2 - CeO_2 [49] and TiO_2 - CeO_2 (especially attractive due to their fast response time and high transparency) [50-53] are investigated as counter electrodes in EC cells.

In an electrochromic construction, when the voltage is applied between the two transparent conductors, a spontaneous intercalation of ions starts from the ion conductor and electrons from the transparent conductor are injected in the electrochromic film structure. The electron injection provokes the color, absorption changes, meanwhile the alkali ions M^+ balance the charge difference. In principle, the device coloring is exhibited by monochromatic absorption:

$$A(\lambda) = (A_{EC}^{deintercalated}(\lambda) + A_{CoE}^{intercalated}(\lambda)) + Q_{intercalated}(\eta_{EC}(\lambda) - \eta_{CoE}(\lambda)), \qquad (3)$$

Where with "EC" is marked the electrochromic film, with "CoE" – counter electrode, A – optical absorption, $Q_{intercalated}$ – is the injected charge, η_{EC}, η_{CE} are the color efficiencies of the electrochromic film and the counter electrode, respectively. As can be seen, the device color efficiency depends on the corresponding η_{EC}, η_{CE} of the two films. All the parameters are frequency dependent and the coloring is also determined by the quantity of the charge inserted. From the above relation, it is considered that the counter electrode must possess additional electrochromic response towards the working electrode, meaning color efficiency with opposite sign, so the maximum optical absorption can be approached.

The basic requirements for the properly working EC devices are:

- Proper choice of the EC materials and conductors, the design, including the individual properties and stability, the interface reaction and the operation factors;
- Stability and a working ability in a certain temperature range, low exposition effects towards UV radiation, moisture and other climate conditions;
- Possibility for showing uniform optical properties along the whole surface;
- Need of using of low voltage values in a narrow range;
- Considerable switching time according to the practical applications;
- Long-term stability and economic efficiency;

There are four main applications of EC devices - information displays, mirror with variable specular reflectance, EC smart windows and variable emittance surfaces.

2.3. ELECTROCHROMIC MATERIALS

Electrochromic properties (reversible optical modulations) have been discovered in organic and inorganic compounds [4]. Among the electrochromic materials family, organic compounds can be found such as viologens, phenothiazines, orthotolodine and anthraquinone. On the other hand, the inorganic materials include transition metal oxides, heteropolyacids, indium and tin nitrides, intercalated graphite, Prussian Blue and alternative hexacyanometallates.

A) Organic Compounds, Possessing Electrochromism

Viologens – a halides family based on the specific quaternary bases derived from 4, 4' dipyridinium $(C_2H_4N)_2$ with a general formula V^{2+}, $2X^-$ [4, 54]. These materials usually are colorless. A reduction can proceed on the cathode in one electron step to form V^+ radical, resulting in a deep blue coloring. The second step is the reaction between the radical V^{2+} and X^- that could simultaneously take place if water is used as a solvent. The reaction product is an insoluble purple solid compound VX, which precipitates as a thin film on the cathode surface. A popular EC viologen is the diheptylviologen-dibromide dissolved in water; which presents a maximum light absorption located at 545 nm after applying a voltage 0.66 V (defined voltage vs. SCE (saturated calomel electrode)). Other advantages of the viologen are that its display mode is passive and it can be used either in transmission or in reflection mode. Another good property - the good contrast ratio independent on the viewing angle.

Other modified EC viologen are benzyl-heptyl-viologen and benzyl viologen mixed with polymerized viologen dibromide.

Phenothiazines-the phenothiazines are heterocyclic as the best studied is methylene blue. This material in water seems colorless, when it is in a molecular state it transforms in blue colored in a cationic state. When it has been put in a protonic solution, the methylene blue presents three – color step process from the uncolored state to a red coloring and finally a blue one.

Polyaniline electrochromes-the anodic type polymerization of the colorless aniline in a solution leads to formation of a polyaniline. The electrodes, covered with polyaniline films demonstrate polychromogenic properties and show color changes depending on the applied voltage – the color varies from yellow to green, blue and finally to black [55, 56].

Electrochromic properties are reported for many polymeric organic materials such as polypyrroles, polythiophene and others. There has been much interest in the polymers derived from electrochemical oxidation of thiophene-based monomers that comprise more than one thiophene heterocyclic unit.

Carbozoles – in their neutral form, carbozoles are soluble and colorless, whereas films of radical cation generated oxidatively form a highly colored precipitate on the electrode.

Methoxybiphenyl compounds – the uncharged compounds are colorless, but a film of brilliantly colored radical – cation salt following electro-oxidation.

Ruthenium organic complexes – can be used as electrochromic materials for optical attenuation in the near infrared region. They are unsymmetrical complexes having two different substituents are disclosed, where one substituent is more electron-donating than the other. Complexes which are dimers or trimers (symmetrical or unsymmetrical) are disclosed, as are polymeric complexes. Cross linked polymeric complex films are also disclosed. [57]

Ionically self-assembled monolayers (ISAMs), fabricated by alternate adsorption of cationic and anionic components, yield exceptionally homogeneous thin films with sub-nanometer control of the thickness and relative special location of the component materials. Using organic electrochromic materials such as polyaniline, it has been reported researches of the electrochromic responses in ISAM films. Reversible changes in the absorption spectra have been observed after voltage applying in the order of 1.0 V. Measurements are made using both liquid electrolytes and in all-solid state devices incorporating solid polyelectrolyte such as poly(2-acylamido 2-methyl propane sulfonic acid) (PAMPS).

B) Inorganic Electrochromic Materials

Relatively few investigations have been performed for studying the electrochromic behavior in the inorganic materials such as metal phtalocyanines, Iron(III)-hexacyanoferrate(II) (Prussian Blue) and intercalated graphite. The ferrites, for example the fully oxidized Prussian Blue, are brownish colored in solid state, but in solution they transform in brownish-yellowish and golden yellowish as thin films are deposited. The reduction is as a result of the white Prussian blue, which in the coating form is colorless. Prussian blue presents anodic coloration and is often used as the counter electrode in electrochromic device [4].

Transition Metal Oxides

The transition metals are located in the three rows among the alkali metals (Ca, Sr, Ba) and the noble metals (Cu, Ag, Au) in the Periodic table. The transition metal oxides are wide band semiconductors, displaying a variety of electrophysical properties. The metal ion is very small in size compared to the oxygen ion and therefore it is the oxygen ion that determines the unit cell size. Due to the size difference, the oxygen ions are associated with each other while the metal ions are not. Thus, the oxygen valence electrons, the 2p orbitals, overlapped forming relatively broad, filled energy bands. Since the metal ions are not in contact, their valence electrons, the d-orbitals, form narrow energy bands or localized states. This leads to a high effective electron mass and a low mobility. The transition metal oxides are also known to possess a high hardness and a high melting point. Five transition metals form oxides with cathodic electrochromism, namely Ti, Nb, Mo, Ta and W. Seven others, such as Cr, Mn, Fe, Co, Ni, Rh and Ir have oxides with anodic coloration. The exception is vanadium oxide, due to the fact that in its pentoxide form (V^{5+}) it can reveal anodic or cathodic coloring depending on the wavelengths, meanwhile the vanadium dioxide experiences only anodic electrochromic coloration. One part of the transition metal oxides possesses a full transparency in the visible range, those materials are namely oxides of Ti, Ni, Nb, Mo, Ta, W and Ir. The other metal oxides reveal partial absorption in the VIS part of the spectrum.

The significant feature of the electrochromic metal oxides is their crystallographic structure [58], which is consisted of a general building item, which is MeO_6 octahedron with a metal atom Me located in the center, surrounding by six oxygen atoms. These structural items can be bonded together by shared edges and corners so they can generate a variety of crystal configurations. There exist three main groups of crystal structures, characteristic for the EC transition metal oxides: defect perovskite - like, rutile – types, layered and block structures:

Perovskite structure is often referred to a rhenium oxide configuration. It can be drawn as an infinite array of the corner-sharing octahedrons with a metal ion surrounded by six oxygen ions. Between these octahedrons, tunnels are formed and extended and they can be served as conduits and intercalation sites for the intercalated small alkali ions. The most studied EC material, tungsten oxide possesses a perovskite structure. The crystal structure of WO_3 shows different symmetries. It must be noted that tungsten oxide tends to form substoichiometric (Magneli) phases with large pentagonal or hexagonal cross-sections. Monoclinic MoO_3 also possesses the perovskite structure, and it has been viewed as a metastable analog of WO_3. One interesting fact must be mentioned that although rhenium oxide has a perovskite structure just like WO_3, it does not exhibit electrochromic effect.

Rutile structures or rutile like configurations are presented for the materials showing cathodic or anodic coloration. The rutile ideal structure is built from the octahedral MeO_6 units forming infinite edge-sharing chains, which can be arranged in a way to form an equal number of vacant tunnels. TiO_2, MnO_2 and VO_2 are all rutile types structured and they show some electrochromic properties but not as strong as the exhibited by WO_3. On the other hand, the anatase TiO_2 has attracted much attention in the context of EC properties. Mean while, the ruthenium oxide, RuO_2, has also a rutile structure, allowing ions to move through by insertion and extraction, but it was found out that RuO_2 remains in absorbing state whatever the ionic content is. Other rutile materials are IrO_2 and RhO_2, which clearly possess anodic electrochromism as IrO_2 has been one of the most studied candidates for counter electrode in EC device.

Vanadium pentoxide, hydrated nickel oxide and niobium oxide have *layered or block structures,* all they have a complex chemistry. Very interesting layered arrangement exhibit orthorhombic MoO_3, which will be discussed in details later.

Prussian Blue

Transition metal hexacyanometallates form an important class of insoluble mixed valence compounds. Iron hexacyanoferrate often referred as Prussian Blue (PB) is very well known as the dyeing trade as a pigment for paints, lacquers, printing inks etc. PB has a cubic unit cell with a lattice parameter of 1.02 nm. Fe^{2+} and Fe^{3+} are located alternately on a face centered lattice in such a way that Fe^+ ions are surrounded octahedrally by nitrogen atoms and Fe^{2+} are surrounded octahedrally by carbon atoms, the structure is disordered with one fourth of ferrocyanide sites unoccupied.

The color efficiency of PB was evaluated by Itaya [45] as 81.5 cm^2/C at wavelength 700 nm and 68 cm^2/C - at 633 nm. EC devices can utilize Prussian blue with a K^+ conducting electrolyte. It has been developed a complementary device with an anodically coloring PB film (as CoE) and cathodically coloring tungsten oxide film (as WE). Other EC cells can be constructed by using PB pigments in the polymer matrices.

2.4. APPLICATIONS OF ELECTROCHROMIC DEVICES

The applications of electrochromics are numerous, they have been considered for architectural glazings, ophthalmic products, instrumentation devices and displays. Each application usually insists on the specific electrochromic and material requirements.

EC Ophthalmic Eyewear

Electrochromic devices offer the ability to adjust transmittance of eyeglasses by the user. These glasses will be controllable in the most desirable and comfortable level depending on the incident UV light. The main requirements towards them are the fast switching times and the low leakage current. They must also possess minimum reflectance and to be comparable to commercially available AR-coated lenses in the mechanical durability as well as in the optical performance. The lifetime of these products can be determined at about 5 years as the eyeglasses are very high and fashion oriented items. It may be estimated that if the glasses will be switched 5 times day, then the devices must undergo around 9200 cycles during the five-year period.

Emissivity Modulation Devices

Electrochromic devices can also be used for surface emissivity control (emissivity modulation devices –EMD) [4]. Surface emissivity characteristics are represented by the blackbody spectra at a defined temperature. EMDs are especially of interest for the thermal control in satellites. The thermal state of satellite in a space environment can be monitored through varying the rate of energy absorption (this includes radiation from the sun, earth shine etc), the rate of heat generation (by internal electrical devices) and the rate of energy dissipation by thermal emission from the satellite surface such as IR radiation. In order to compensate the temperature variations, active thermal control systems are used – heat pipes, thermal louvers, heaters. All spacecraft and the instruments they support require an effective thermal control mechanism in order to operate as designed and achieved their expected lifetimes. Traditionally, the thermal design is part of the spacecraft layout determined by all subsystems and instruments. Heat load levels and their locations on the spacecraft, equipment temperature tolerances, available power for heaters, view to space, and other such factors are critical to the design process. Smaller spacecraft with much shorter design cycles and fewer resources require a new, more active approach. All solid state inorganic variable emittance coatings are being considered as a promising technology for thermal control in the space [59].

The heat balance of the satellite or the surface of another component in a low earth orbit or a middle earth orbit will depend upon the total energy absorbed from the environment (solar radiative flux, radiation re-emitted by the earth) the heat generated by the electrical and optical components within the satellite and the radiative heat emitted by the satellite or spacecraft surface. A variable emittance coating become an important part in thermal control systems space – based devices through alteration of the effective optical absorption of the surfaces of the structure at the solar radiative wavelengths or through a modulation of the long-wavelength emittance of the surface, thereby altering of the radiant heat transferred by that surface. EC structures were experimented for this purpose in the spectral range from middle to far IR wavelengths (1- 30 µm) and the modulation reached is 0.60 to 0.68, which is not sufficient for most space applications. The emittance modulation in electrochromic devices is achieved using the crystalline electrochromic materials whose reflectance can be tuned over a broad wavelength (2 to 40 microns) in the infrared spectrum [60].

The modulating device will be attached to the satellite surface and provide a thermal window that can be adapted to the changing conditions in the orbit. EC emittance modulating devices in the end of 1990s become very interesting due to the growing need of smaller systems for thermal control.

Currently satellites are built with a venetian blind type radiator to control their internal temperatures. These radiators function well, but to do so they require the opening and closing of the "blinds". Since electrochromic devices can change their optical properties by means of an applied voltage, the emission of heat could be accomplished without bulky moving parts. The devices for IR modulation have similar configuration as the previously described five layered system. EC devices would allow emittance modulation of surfaces and components without using the bulky blinds with moving parts. Satellite parts could even be coated directly, insuring a good thermal conduction path.

Dynamic Antiglare Automotive Mirrors

The only widespread commercially available devices at present are the car rear-view mirrors (Nikon's costly electrochromically adjustable dark glasses on sale around 1990 were soon withdrawn). The all-solid mirror described in the literature [61] has not yet achieved distribution, but of the liquid-state devices, more than 25 million Gentex mirrors [62] (now 27 million) have been sold since the inception, from the work started in 1974. The reflecting surface adjoins one of the two facing conductive-glass electrodes. These, spaced in parallel, make up the cell, which contains the all-liquid system; the two dissolved electrochromes colorized on oxidation and reduction respectively, one at the each electrode, so that both electrodes are electrochromic. The ionic species is ohmically driven to the cathode, in an ingenious accelerating ploy. The anodically coloring species, which may be of the Wurster's Blue type [63] are neutral to the charge and thus diffuse to the electrode in a chemical-potential gradient created by the driven process at the cathode. The resulting intensely dark blue-green color permits only the outline of otherwise dazzling headlights to appear. Maintenance of the color intensity hence requires a small current, negligible in a car, to continuously generate the colored species. This lack of memory effect, the fading of color on disconnection of current is an advantage: the fail-safe mode required by US law is the clear mirror. The control system ensures that it is only at night that the darkening reaction is effected. For external mirrors a single glass plate is employed, the second electrode being an inert metal, acting also as a reflector. Glare from headlights at nighttime can be greatly reduced.

Many car manufacturers have already begun using them in a large-scale. For example, *Donnelly Corporation* has begun manufacturing interior electrochromic (automatic dimming) mirrors for the Audi A6 luxury sedan. For that company, this marks the beginning of electrochromic mirror production of the series of Audi's upper-class vehicles. The electrochromic mirror production for the additional A-class models began later 1997 year. Donnelly estimates that it will manufacture 60,000 electrochromic mirrors each year for the A-class vehicles. During the night-time driving, the electrochromic mirrors dim automatically when the glare from the headlights of following vehicles is detected. Reducing the glare, the mirrors are improving the forward vision and in that way decrease the rate at which the driver fatigues and increase the level of the comfort and the convenience.

Gentex uses solution-phase electrochromic technology. They have sandwiched an electrochromic gel between two pieces of glass, each of which has been treated with a transparent, electrically conductive coating, and one with a reflector. Then a forward-facing sensor recognizes the low ambient light levels and signals the mirror to begin looking for glare. After that, a rearward-facing sensor then detects glare from the vehicles traveling behind the driver, sending voltage to the mirror's EC gel proportional to the amount of glare detected. The mirror darkens according to the glare and then clarifies when the glare is no longer detected. The interior mirror's sensors and electronics control the dimming of both interior and exterior mirrors.

Displays

In the near future, it is likely that electrochromic devices will be used in displays, possibly replacing LCDs, although the technology isn't quite there yet. By contrast, electrochromic displays appear brighter in sunlight and can be read at almost any angle [64].

Electrochromic displays are produced by coating a transparent electrode (e.g. tin oxide) with a mixture of the composite material dispersed in a solution containing a polymer complex such as a polyionic complex of a macromolecular viologen with macromolecular sulfonic acid. An aqueous solution of sodium sulfate in contact with the composite provides an electrochromic display element. Color changes induced by the voltage are visible through the transparent electrode. The element uses an ionically conducting electrochromic layer and requires the use of a transparent electrode to view the color change.

The technology involves using composite layers in laminates for the electrochromic displays where an ionically conductive layer is in a contact with the electrochromic material. Electrochromic materials can be provided in the laminates as layers between the ionically conductive layer and the composite layer of electrically conductive particles dispersed in a polymer matrix. Alternatively, the electrochromic material can be incorporated in the conductive particles in the polymer matrix, e.g. as titanium dioxide coated with antimony tin oxide coated with polyaniline dispersed in a light transmitting polymer matrix. The materials of this invention allow for the high-speed fabrication of flexible displays, e.g. by printing methods.

Electrochromic systems have been used successfully in the past in the mirrors and windows for the anti-glare and anti-reflective applications. However, other potential applications, including large area privacy/security glazing and high contrast angle-independent dynamic displays have yet to reach the market. This is due to the limitations of the existing electrochromic materials. To date, the commercial electrochromic technologies have relied on either of two, broadly defined, types of architectures. Firstly, there are solution based systems, which rely on the organic electrochromic species dissolved in the electrolyte compartment of an electrochemical cell comprising at least one transparent electrode. The most commonly used electrochromophores, salts of 4,4'-bipyridines, also called viologens, are synthetically tunable which allows for different colors, and have intrinsically high extinction coefficients, yielding excellent coloration intensities. The switching speed depends on the diffusion of the redox active species in the electrolyte to the electrodes and is typically in the order of seconds. Because the redox active species are dissolved in an electrolyte the mobile molecules diffuses to the both electrodes once an appropriate electrical potential has been applied to the circuit. If the potential has been removed, the charged species mix, transfer their charges, and the color dissipates from the system. Therefore, there is no open circuit memory in these devices and the power must be applied continuously to maintain coloration.

The other approach relies on the intercalation, or insertion and bonding of ions into the crystal lattice of materials. This technology is based on electrochromic cells in which at least one of the electrodes is a thin but compact layer of a metal oxide, typically tungsten oxide WO_3. Coloration is achieved when ions, typically H^+ or Li^+, and electrons are injected electrochemically into a WO_3 layer. A strength of this intercalation electrochromic technology compared to the solution based technologies is that it offers a memory capability in the devices on the condition that the complementary redox species are bound to the counter electrode surface or a charge storage layer is available to support the charging of the device without dissipation of its charge upon removal of the applied potential. The intercalation process is slow due to the diffusion of the ions in the electrolyte and into the thin film. In addition, the metal oxides used do not form very highly colored complexes so the contrast ratio is intrinsically lower than for the solution based organic systems. Due to the specific weaknesses of each of these earlier systems, neither technology is suitable for the development of paper quality displays.

Electrochromic displays are reflective displays, using inks which can change color induced by an applied voltage. The displays today cost around 5 US cents per square centimeter of the active area. As an indication of how electrochromic displays can be used imaginatively in low cost devices, Spanish researchers have recently announced artificial fingernails that have multilayered electrochromic surfaces that are electrically switchable to different colors. It is envisaged that they will be sold with a camera that enables the lady to match her nails to her dress at the flick of a button.

The Electrochromic Nail

The electrochromic nails [65] use a clever combination of polymer films to change colour at the flick of a switch. The materials scientists at the Cidetec research centre, which developed the nail, say it can even mimic the colours and patterns of the surrounding surfaces. Passing electricity through these materials can subtly change them, making them absorb different wavelengths of light and then they appear to be different colors. Conducting polymers have been developed to display a variety of colors, including blue, green, orange and purple. The electrochromic nail is built up of these polymers as layers. A separate device uses a camera and a computer chip to convert images of any selected surface into instructions to change color, which are relayed to the nail through attachable electrodes. The scientists say the color can be changed as often as you like.

Sunroofs

Sunroofs are most likely to be the first application of EC glazing in automobiles. They have the advantage in that because of their location in a vehicle they do not have to comply with the high transmission requirements and they are relatively small, in typical sizes less than 0.7 square meters.

Electrochromic Paper

There is a potential application in "electrochromic writing paper" [66, 67], which, as the name implies, incorporates electrochromic materials into paper. The key to the functionality of this paper is the use of a stylus electrode as a writing utensil.

Organic electrochemical transistors and organic electrochromic display cells have been combined to smart pixels arranged in a cross-point matrix. The obtained active matrix display is printed on cellulose-based paper substrate utilizing standard printing techniques.

The goal today is to achieve low-power consuming displays on flexible substrates. Ordinary cellulose-based paper serves as the carrying substrate and the resulting display is made from all-organic materials. Standard printing techniques are used in the manufacturing process, both additive and subtractive printing techniques can be used, which results in an inexpensive process without the requirements of clean room facilities such as evaporation chambers and plasma etching equipment. The vision for the future is to extend the use of paper as a part of electronic circuits, manufactured in reel-to-reel process flows.

The final display is built up from single smart pixels in a cross-point matrix, where every smart pixel is realized through a combination of an electrochemical transistor and an electrochromic display cell. Voltages below 2V, which promises for a low-power consuming display, drive both the electrochemical transistor and the electrochromic display cell. Another thing that reduces the power consumption even further is the fact that both the transistor and the display cell possess bi-stable properties, which might be resembled to a kind of memory effect in the conducting polymer. Upon disconnection of the applied voltage, the polymer remains in this state for a while. The current modulation behavior is used in the electrochemical transistor channel as a 'digital' switch. Both switching speed and bi-stability in transistors and display electrochromic cells are dependent on the ion conductivity of the electrolytes utilized. With a low ionic conductivity in the electrolyte, the switching time increases and the bi-stability is better, therefore, a trade-off between these two characteristics is needed. The area of the active polymer region also influences the switching time, especially in the matter of lateral devices. Electrochemical switching of electrochromic materials results in different optical absorption spectra. Oxidation of PEDOT:PSS (Poly(3,4-ethylenedioxythiophene) poly(styrenesulfonate), which is a polymer mixture of two ionomers) results in an absorption spectra in the near infra red region due to more free charge carriers, i.e. bi-polarons, and the initial sky-blue colored polymer turns to a more transparent state. Upon reduction, on the other hand, PEDOT:PSS absorbs strongly in the red/orange wavelength region and appears dark blue to the human eye. The color shift in PEDOT: PSS can be utilized in electrochromic display cells.

An electrochromic display element consists of at least two conductors, an electrochromic material, and an electrolyte combined on a carrying substrate. The schematic drawing shows an electrochromic paper display made entirely of organic polymers (the electroactive materials) and an organic electrolyte on ordinary double-coated fine paper. The display element is updated by applying a voltage of typically 0.6 to 0.9 V. Switching the color takes about 1 second while the associated memory time is on the order of 10 minutes to several hours. Electrochromic displays are manufactured as follows: First, a thin film of the conducting polymer is coated onto a fine paper. The coated polymer is then patterned, using offset- or screen-printing, and a high contrast electrochromic polymer colorant is deposited on top. A polymer electrolyte layer is printed on the conducting polymer patterns, forming the display elements. As the final step a protective coating is added to seal the display. The resulting paper display is fully flexible and the printed devices are less than 100 micrometer thick. Today, the electrochromic displays can be manufactured in separate printing tools in a handcraft fashion.

At the present time demonstrators have been created proving the printability of electrochromic paper displays. Meanwhile, several applications and potential customers have been identified. However, products and manufacturing design tools do not exist yet.

Electrochromic Nanocrystal Quantum Dots

Electrochromic materials, in which the optical properties change in response to application of an electric current, are widely studied for their use in applications such as displays and solar control windows, with current materials often based on nanocrystalline metal oxides. After injection of an electron into the nanocrystalline matrix, chromatic changes result from optical transitions induced from the newly occupied surface states and/or accumulation of electrons in the conduction band [68]. Despite the nanometer-sized scale providing a large surface area to enhance the efficiency of electron injection, no manifestation of the quantum confinement effect has been observed in these materials, to the best of our knowledge.

II-VI and III-V semiconductor quantum dots such as CdSe, ZnSe, and InP are known for the size-tunable atomic-like properties arising from the quantum confinement in the nanometer scale. Colloidal nanocrystals of these semiconductors display striking absorption features and a strong fluorescence from the ultraviolet (UV) to the infrared (IR) spectral regions arising from their "artificial atom" character. Changing the "oxidation number" of the nanocrystal colloids by placing electrons into the quantum-confined states (rather than in surface states) has only recently been achieved by using an electron transfer from a reducing species in the solution. Furthermore, the electron injection in the quantum confined states has been shown to lead to marked changes in the absorption and fluorescence properties. It has been shown that electrons can be reversibly injected in the quantum-confined states of the nanocrystals by using an electrochemical potential control. Although electrochemical reduction of ZnO nanocrystals has been reported, electrons were most likely placed in the surface trap states [69].

Smart Windows

The future needs of energy efficient buildings put apart requirements and higher standards for the used windows: they must reduce the heat loss, to prevent overheating and to assure comfort daytime lighting. Glazing is increasingly required to meet multiple, and often conflicting, performance objectives, as: (i) to minimize solar gains to reduce cooling energy consumption and peak electric demand, (ii) to transmit daylight by decreasing lighting energy consumption and electric demand, (iii) to improve occupant visual comfort by minimizing the glare and providing access to clear views of the exterior, (iv) to maintain thermal comfort by minimizing heat losses, and (v) reduce heating energy consumption by utilizing solar gains for passive solar heating. The ideal glazing must therefore be able to respond to a wide range of the conditions in order to achieve the various energy and comfort objectives. Conventional glazing, such as clear, reflective, and tinted, offer limited control of the above performance criteria since their properties remain static over the broad range of environmental conditions. Therefore, often, one performance criteria is met at the expense of another.

These criteria can be satisfied by the switching smart windows as described in [70-72]. The latter possess the ability to modulate their optical properties (transmittance and reflectance). Smart windows may be cooperated in cars for cooling as EC roof, which can be deeply colored when the car is parked (protecting the internal parts from heating) and to bleach at driving. The most suitable device configuration as smart windows is the described one on figure 3.

Different types of smart windows exist based on the external factor that induces the optical modulation:

Photochromic windows – the transmittance is influenced by the radiation dose.
Thermochromic windows – transparency depends on the temperature difference.

A thin film of a thermochromic material on an exterior window could modify its reflectance properties, dependant on outside ambient temperature. The solar radiation that is not able to pass through the window when it is in its darkened state must be either reflected or absorbed. Ideally, in a cooling dominated application, the window would pass all or part of the visible radiation incident on the window and reflect the majority the Sun's near infrared radiation. Incident solar radiation, that is not transmitted, is absorbed. Absorption will cause significant heating of the window if left to stagnate in a colored state under conditions of high irradiance. Temperature rises in the window will give rise to a radiant heat source adjacent to the room, potentially leading to thermal discomfort, and will impose additional demands on the temperature stability of the materials used in the smart window. Effective thermochromic window coatings would respond to this heating by increasing their reflectance and "compensating" for the increased heating by reflecting more heat away. Such intelligent coatings could be used in applications including windscreens of automobiles, sunscreens and greenhouses. The development of such coatings will lead to large savings in energy costs (e.g. power to air conditioning units), improved building environments, and environmental benefits (eg. reduced CO_2 emissions).

Electrochromic windows – the optical change is caused by external applied voltage. The most proper construction of EC device applicable as smart windows is the above described five-layered structure. Static state-of-the-art glazings for architectural window applications are reaching their physical limits when it comes to improving the energy efficiency of the building. Using thin film technology it is possible to achieve close to zero thermal emittance leading to "low-e" windows with low U-values. Furthermore, "solar control" windows having twice as high light transmittance (T_{vis}) as solar transmittance (T_{sol}) leading to solar heat rejection with maintained high visual transmittance is manufactured on a regular basis.

A subsequent possible step towards improved glazing energy efficiency may be through the use of switchable (variable transmittance) windows, or smart windows. These windows are desirable to regulate the solar radiation in the range of 300 – 3000 nm. So, it must be pointed out that additional to the color change which can be observed with the eye (that is in the visible spectrum 400 – 700 nm), the electrochromic windows may also have regulated transmission properties in the other parts of the electromagnetic spectrum in the UV range (for wavelength below 400 nm) and near infrared radiation NIR (700 – 3000 nm). Large-area electrochromic windows have very recently become available in very limited quantities. Samples of these windows were installed in two side-by-side private office test rooms, enabling researchers to conduct full-scale monitored tests [73, 74].

The basic properties and requirement that need to be fulfill for large area applications of smart windows [75] are:

- A continuous range in solar and optical transmittance, reflectance, and absorptance in bleached and colored states
- A contrast ratio at least 5:1 (maximum transmittance: minimum transmittance).
- A switching speed (switch from colored to bleached state and vice versa) to be a few minutes.
- Thermal and UV stability. The EC windows must survive temperatures between -80°C and 120°C.
- Switching with small applied voltages in the range of 1 - 5 V.
- Open circuit memory to be several hours (this means the EC device to keep its transmittance state (colored or bleached) without corrective potential impulses).
- Neutrality of the color is particularly desirable for EC architectural windows, where the color of the interior furnishing could be affected by non-neutral transmitted color of the electrochromic windows. Also, the neutrality comforts eyes sighting.
- Large area with excellent optical clarity and uniformity.
- To possess sustained performance for 20 – 30 years (meaning good cyclic lifetime more than 10^7 cycles).
- Mechanical durability [76]
- Acceptable cost.

To improve overall performance and user satisfaction, from the EC windows is expected to be controllable by both the user and an automated system. An automated system cannot fully accommodate occupant preferences for illuminating a given task type, position, and field of view as well as addressing conditions and concerns including glare, direct sun, privacy, view, brightness, color rendition and a multitude of others. Energy-efficiency algorithms are best implemented with permission of the occupant and may require that the occupant relinquish some autonomy regarding the system during critical peak energy use periods. For public (lobbies, glazed hallways, cafeterias, etc.) or open-plan shared spaces, some lack of autonomy regarding the window system may be acceptable to occupants. Multi-pane windows will allow more flexible control than a single-opening window.

Failure analysis has formed a key element in the EC design of the device. Two basic reasons can exist for device failure:

- Overvoltage (thermodynamic limit) - the applied voltage should not be higher than the potential required for the destructive side reactions.
- Excessive charging current (kinetic limit) – high current causes an irreversible Li trapping in the film surface.

Some EC properties of the smart windows as redox potentials, resistance, diffusion coefficients are temperature dependent and required adjustments in switching parameters for changing environment conditions. [77].

EC windows have some particularities which must be studed by large area tests. These tests are essential for understanding how electrochromic windows will perform in actual buildings. Unlike many energy-efficient glazing technologies, electrochromic glazings have size-dependent characteristics. For example, the switching speed of a 1x2 device will be faster than that of a 2x4 device. The optical range and contrast ratio are expected to remain the same with size; however, this has yet to be verified for the full array of devices developed in laboratories. An electrochromic device will switch significantly more slowly when cold than when hot. An automated, integrated window-lighting control system is needed to realize lighting and cooling energy savings. Few electrochromic prototypes have been produced in the large sizes required by the building industry, so few performance studies of them have been conducted.

Some advantages of EC windows are:

- A high value heat protection in summer time: a prevention of an overheating of rooms during strong sunshine levels;
- The degree of tinting can be changed gradually in response to the intensity of the solar radiation;
- The use of different colors gives new possibilities for architecture - ("fancy windows");
- The integration in facility management systems is possible (smart building);
- The electrical energy for switching can be supplied by PV (photovoltaic) modules;
- In vehicle windshields and in rear view mirrors;

Producers of EC Windows

Several different companies produce prototypes of electrochromic windows and there is much work going on for fully commercializing of EC technology.

The company, Schott Donnelly LLC Smart Glass Solutions, USA (www.schottdonnelly.com) provides electrochromic glass for automotive and architectural applications. Electrochromic glass darkens automatically or through controlled dimming to reduce unwanted solar heat and glare.

Gentex Corporation, USA (http://www.gentex.com/) is best known as the pioneer of the electrochromic, automatic-dimming mirror industry. They made nighttime driving safer by providing the worldwide automotive industry with mirrors that detect and eliminate dangerous rearview mirror glare. The auto-dimming mirrors are offered as standard or optional equipment on over 220 vehicle models around the world.

The company Gesimat GmbH, Germany (http://www.gesimat.de/company.htm) works on the development and future production of smart windows with automatic regulation of light and heat transmittance based on electrochromism.

SAGE Electrochromics (USA) (http://www.sage-ec.com/) is a private company for development and commercialization of electrochromic glass products suitable for the building industry. The technology used to make SageGlass® electronically tintable products is an inorganic, solid-state approach. Over seven years ago, SAGE volunteered to have its prototype window samples evaluated by the U.S. Department of Energy. Since then, their glass has undergone testing for its durability, functionality, and performance by other government agencies as well as industry leading glass fabricators and window manufacturers.

Another available very soon application of EC windows will be on the board of new Boeing 787. Dreamliner will be able to darken or lighten their windows at the touch of a button for more comfortable travel. Global aircraft window leader PPG Aerospace has been awarded a contract by the Boeing Company to supply electronically dimmable systems for the windows of the 787's passenger compartment.

A new achievements in the area of EC windows was found in American Society of Heating, Refrigerating and Air-conditioning Engineers, Inc. (ASHRAE) (www.ashrae.org) published valuable work (2006) [82] of a group of scientists from Lawrence Berkely National Laboratory, Berkely, California - Eleanor S.Lee, Denis L. Bartolomeo, Joseph H.Klems, Mehry Zardanian and Stephen E.Szelkovitz "Monitored Energy Performance of Electrochromic Windows Controlled by Daylight and Visual Comfort", ASHRAE Transactions, part 2, p.122-141. The paper presents 20 months field study related to state-of-the-art measurements of energy performance of large-area – tungsten-oxide absorptive electrochromic windows.

2.5. Main Principles of Cyclic Voltammetry

Cyclic voltammetry is a widely used method for obtaining information concerning the electrochemical reactions [78]. The power and the advantages of this measurement technique results from its ability for a quick study and gathering of significant knowledge for thermodynamics of the redox processes and the kinetics of electro transfer reactions or for double chemical reactions or adsorbing processes. Often, the cyclic voltammetry is the first measurement, which is carrying on in one electroanalytical investigation. It is consisted of linear scanning of the potential of stationary working electrode, using the triangle potential wave form. During potential change, a potenciostat measures the current, which is caused as a result of applied voltage. The dependence obtained of current vs. applied voltage is called voltammogram. The obtained cyclic voltammogram represents a complicated time dependent function of a high number of physical and chemical parameters [78].

At the voltammometric measurements, the experimental dependence has been characterized by peaks, where the current peak is determined with the equation defined from Randles and Sevcik:

$$i_p = 0.4463 \, A \, n \, F \, (n \, F/R \, T)^{1/2} \, D_i^{1/2} \, v^{1/2} \, c_i \tag{4}$$

where A is the electrode area, c_i is ion concentration in electrolyte, D_i is diffusion coefficient of ions. It must be noted that this equation validates only for the reversible systems.

The rate of an electron transfer is a function of the gradient of the applied voltage towards the electrode and follows the expression of Butler - Volmer:

$$i = n \, F \, A \, k_f \, c_R \, (\exp(-\alpha_f n \, \Phi \, \eta)) - n \, F \, A \, k_b \, c_O \, (\exp(\alpha_b \, n \, \Phi \, \eta)) \tag{5}$$

where $\Phi = F/RT$, c_O is the concentration of the oxidized form of electroactive materials – the initial one; c_R is the respective concentration of the reduced form and α is a fraction defined as a transfer coefficient and indexes show the reaction direction – straight or backwards. This is a kind of symmetry measure of the energy barrier of electron transfer. Also, η is an overpotential determined from $(E - E_{oc})$, where E is the applied voltage and E_{oc} is the potential at zero current; k_f and k_b are the rate constants of the electron transfers at the right and reverse reaction.

Assuming that for a certain system we have $\alpha_b = (1 - \alpha_f)$ and $c_O = c_R = c$, which means that the two speed rate constants are equal and can be presented as κ_S. When the speeds are equal in the both directions, there will not be electrical current, $\eta = 0$ and in that case a standard change current is defined: $i_o = (n\ F\ A\ k_s\ c)$ and this is a measure for the speed of electron transfer to electrode.

In the electrochromic cells designs, the substrate used is usually a glass covered with a transparent conductive layer, possessing a proper thickness and conductance. These characteristics influence on the response time of the electrochromic cell.

The current flux j of electrons through the conductor and the corresponding electrical current i are determined by these equations:

$$i = n A v e \qquad (6)$$

$$j = n v \qquad (7)$$

n is the number of charge carriers, A is the cross section of the conductor, v is electronic velocity and e is the electronic charge.

Before the electron transfer reaction can occur, there must be a mass transfer, processing by three main mechanisms: migration, convection and diffusion. The mass transfer is formally defined as the flux j_i of electroactive species i to an electrode as defined in the Nernst – Planck equation:

$$j_i = \mu_i\ c_i\ (\partial\Phi/\partial x) + c_i\ v_i - D_i\ (\partial c_i/\partial x) \qquad (8)$$
$$\text{migration} \qquad \text{convection} \qquad \text{diffusion}$$

μ_i is the ionic mobility of species i, Φ – the strength of the electric field, v_i is the vectrorial velocity of solution and D_i and c_i present the diffusion coefficient and the concentration of species, respectively.

The *convection* will not be concerned in the case of electrochromic devices as in the solid electrolytes since it is negligible small. *Migration* is the movement of ions through a solution or a solid body as a response of the applied electric field. For liquid electrolytes, containing an excess of unreactive ionic salts, the migration may be neglected, due to the fact that the transport number of the electroactive material becomes negligibly small. But the migration is an important part of the mass transport in ionic movements in solid-state electrolytes where the transport number appreciably increases [4, 79]. Diffusion or more correctly the diffusion coefficient has been of a particular interest for every kinetic investigation. *Diffusion* is obeyed to the Fick law:

$$j_i = - D_i\ (\partial c_i / \partial x) \qquad (9)$$

$(\partial c_i/\partial x)$ is the concentration gradient, the change of the concentration per unit area. In electrochemical reactions, this gradient is not zero and it is increased due to the fact that we have a consumption of electroactive substances around the electrodes. In that way, diffusion causes additional concentration gradient.

The second law of Fick describes the time dependence of the migration:

$$(\partial c_i/\partial t) = D_i (\partial^2 c_i/\partial x^2) \tag{10}$$

A useful approximation solution to Fick's second law is:

$$l \approx \sqrt{D_i t} \tag{11}$$

where t is the time required for species to move a distance l and D_i is the diffusion coefficient of species.

Another characteristic for the ionic movement is the ionic mobility μ (velocity v divided by the driving field), which is related to the diffusion coefficient by the Nernst – Einstein equation:

$$D/\mu = k_B T/z\, e, \tag{12}$$

k_B is the constant of Boltzmann, T is the absolute temperature in Kelvin, and z is the charge number of the electrode reaction (which is the number of moles of electrons involved in the reaction).

When the applied voltage of an electrode is stepped from a value giving zero current to a maximum, then all the electroactive material at the electrode/solution interface will undergo electrochemical change. The electroactive material will move towards the electrode, diffusing from long distances. The Cottrell equation describes the current/time response as

$$i = n F A \sqrt{D/\pi t} \tag{13}$$

where A is the electrode area, F is the Faraday constant.

Cyclic voltammometry is widely used for investigation of the electrochromic properties of thin films [80].

At the start of the voltammetric experiment, the bulk solution contains only the reduced form of the redox couple (R) so that at potentials lower than the redox potential, i.e. the initial potential, there is no net conversion of R into O, the oxidized form (point A). As the redox potential is approached, there is a net anodic current which increases exponentially with potential. As R is converted into O, concentration gradients are set up for both R and O, and diffusion occurs down these concentration gradients. At the anodic peak (point B), the redox potential is sufficiently positive that any R that reaches the electrode surface is instantaneously oxidised to O. Therefore, the current now depends upon the rate of mass transfer to the electrode surface and so the time dependence is qt resulting in an asymmetric peak shape.

A typical cyclic voltammogram is presented on figure 4. Upon reversal of the scan (point C), the current continues to decay with a qt until the potential nears the redox potential. At this point, a net reduction of O to R occurs which causes a cathodic current which eventually produces a peak shaped response (point D).

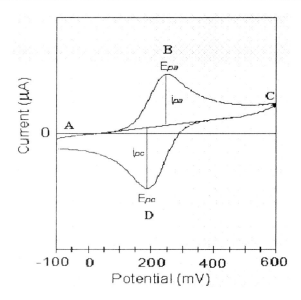

Figure 4. Cyclic voltamogramm - a general case of a reversible system.

If a redox system remains in equilibrium throughout the potential scan, the electrochemical reaction is said to be reversible. In other words, equilibrium requires that the surface concentrations of O and R are maintained at the values required by the Nernst Equation. Under these conditions, the following parameters characterize the cyclic voltammogram of the redox process:

- the peak potential separation (E_{pa} - E_{pc}) is equal to 57/n mV for all scan rates where n is the number of electron equivalents transferred during the redox process.
- the peak width is equal to 28.5/n mV for all scan rates.
- the peak current ratio (i_{pa}/i_{pc}) is equal to 1 for all scan rates.
- the peak current function increases linearly as a function of the square root of v.

The situation is very different when the redox reaction is not reversible. In fact, it is these "non-ideal" situations which are usually of greatest chemical interest.

2.6. EXPERIMENTAL CONDITIONS FOR ELECTROCHROMIC INVESTIGATIONS OF CVD THIN FILMS

Cyclic voltammetric experiments for investigation of CVD obtained metal oxide films were performed in a standard three-electrode arrangement. The cell was adopted with Pt as a counter electrode and a saturated calomel electrode (SCE) as a reference electrode. The sweeping potentials were provided by a Bank - elektronik Potentiostat under a computer control. The electrodes were immersed in various electrolytes such as 0.3mol/l $LiClO_4$ and 1mol/l $LiClO_4$ dissolved in propylene carbonate (PC). In these cases, Li intercalation and deintercalation causes electrochromic effect. The proton mechanism of EC process has been studied by acid electrolytes using 1M H_2SO_4 + 50% glycerin or 1 M H_3PO_4 which were the experimental solutions for the voltammometric measurements.

Electrochromic effect in the case of transition metal oxide is proposed to be as follows:

$$MeO_3 + xe^- + xM^+ \leftrightarrow M_xMeO_3 \tag{14}$$

where M^+ can be H^+, Li^+, Na^+ etc. – small ions, especially of alkali elements and Me is a transition metal.

In order to estimate the color change, the cyclic voltammetry equipment was upgraded with an optical system, containing a chopped light source and a lock-in amplifier (figure 5). Light of given wavelength was obtained by means of a halogen lamp and a grating monochromator. A quartz cuvette was used for the electrochemical cell, in which the three electrodes were immersed.

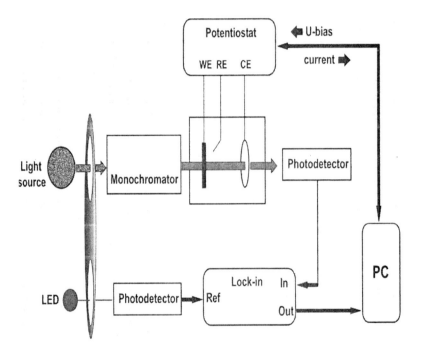

Figure 5. Experimental set-up for electrochromic effect measurements.

The photodetector signal has been measured by the lock-in voltmeter simultaneously with the cell-current as a function of the applied voltage. An IR-LED and a second photodetector have been used to provide a reference signal for the lock-in. Thus, the current, the passed charge and the light transmittance were measured as a function of applied voltage at different wavelengths.

Current density vs. voltage voltammograms were registered between -1 V and +1.5 V at different scanning rates ranging from 5 to 50mV/s, automatically, the values of current (I) and charge (Q) are measured in dependence on applied voltage.

The diffusion coefficient D of the diffusing species is of particular interest to any study, being representative of its spontaneous motion. The diffusion coefficient can be determined from the formula given in Ref. [81]:

$$i_p = 2.72 \times 10^5 n^{3/2} D^{1/2} C_o v^{1/2} \tag{15}$$

where D is measured in [cm^2/s], Li concentration C_o - [mol/cm^3], sweep rate v [V/s], number of electrons n is suggested to a unit, current density peak i_p - [A/cm^2].

Color Efficiency

Color efficiency (CE) experimentally qualifies the modulation of the optical properties of electrochromic materials.

In a transmission mode, the optical absorption is connected with the injected charge for an area unit Q in a equation similar to Beer – Lambert law:

$$A = log(I_o/I) = CEQ \tag{16}$$

In (17) *CE* is a symbol for the color efficiency. This characteristic is related to the optical absorption change ΔA, the linear absorption coefficient α, film thickness d and charge for an area unit Q:

$$CE = (\alpha d)/Q = \Delta A/Q \tag{17}$$

It must be pointed out that these equations are valid only when the processes are absorptive and the Beer – Lambert law is obeyed.

Color efficiency can be also described as an electrode area that colors for an absorption unit for a charge unit. For cathodic type of electrochromism, the sign of the color efficiency is set to be positive (+) and for anodic electrochromic materials the value of *CE* is taken as negative with sign (-). Color efficiency strongly depends on the wavelength and the changes in the optical density *(ΔOD)*. The optical density represents the absorption, related with the extinction and film thickness:

$$OD = (4\pi/\lambda) kd \tag{18}$$

The relation between the optical density and CE is defined as follows [3-5]:

CE (most often used expression of color efficiency) $= \Delta(OD)/\Delta Q$

Experimentally, $\Delta(OD)$ is given by the expression:

$$\Delta(OD) = -\log_{10}(T_{colored}/T_{bleached}) \tag{19}$$

It must be pointed out that color efficiency is spectrally dependent, and comparison is possible only for CE values obtained for one and the same wavelength.

The next part of the book will present the research done on transition metal oxides based on W and Mo, as well as Cr, the first two in form of pure component oxide, and in mixed oxide form. Separate chapters will describe the technology and investigation of WO_3, MoO_3, mixed Mo/W oxides and CrO_3, all produced by the APCVD method.

REFERENCES

[1] C. N. R. Rao and B. Raveau. Transition Metal Oxides: Structures, Properties and Synthesis of Ceramic Oxides. New York Wiley-VCH, 1998
[2] C.G. Granqvist, *Mater. Sci. Eng.* A 168 (1993) 209.
[3] C.G. Granqvist, "Handbook of Inorganic Electrochromic Materials", Elsevier Science, 1995.
[4] P.M.S. Monk, R.J. Mortimer, P.R. Rosseinsky, "Electrochromism: Fundamentals and Applications", VCH, 1995.
[5] S.K. Deb, *Applied Opt. Suppl.* 3 (1969) 192.
[6] S.K. Deb, Solar Energy Mater. *Solar Cells.* 25 (1992) 327.
[7] A. Avendano, L. Berggren, G.A. Niklasson, C.G. Granqvist, A. Azens, *Thin Solid Films.* 496 (2006) 30.
[8] A. Donnadieu, Materials Science and Engineering B: Solid-State Materials for Advanced Technology. B3 (1-2) (1989) 185.
[9] C. Bechinger, S. Hermingaus, W. Petersen, P. Leidener, *SPIE.* 2255 (1994) 467.
[10] M. A. Quevedo-Lopez, R.Ramirez-Bon, R.A. Orozco-Teran, O. Mendoza-Gonzales, *Thin Solid Films.* 343-344 (1999) 202
[11] A. Gavriyuk, Electroch. *Acta.* 44 (1999) 3027.
[12] G. Teowel, T. Gudgel, K. McCarthny, A. Agrawal, P.A. Allmand, J. Cronin, *Electroch. Acta.* 44 (1999) 3017.
[13] A. Klukowska, W. Posset, G. Schottner, M.L. Wis, C. Salemi-Delvaux, V. Malatesta, *Materials Science.* 20 (2002) 95.
[14] M. Sun, N. Xu, Y.W. Cao, J.N. Jao, E. G. Wang, *J. Mater. Sci.* 15 (4) (2000) 927.
[15] A. Georg, W. Graf, R. Neumann, Y. Wittwer, *Solid State Ionics.* 127 (2000) 319.
[16] S.-H. Lee, H.M. Cheong, P. Lin, D. Smith, C.E. Tracy, A. Mascarenhas, J.R. Pitts, S.K. Deb, *Electroch. Acta.* 46 (2001) 1995.
[17] D.P. Haberman, *Proc. SPIE.* 3535 (2000) 185.
[18] B.W. Faughman, R.S. Grandall, P.M. Heyman, *RCA Reviews.* 36 (1975) 177.
[19] H. Cheong, H. Ch. Jo, K.M. Kim, S.-H. Lee, *J. Korean Physical Society.* 46 (2005) S121.
[20] P. Jin, S. Tanamera, *Thin Solid films.* 281-282 (1996) 239.
[21] J.M. Leger, A.L. Holt, S.A. Carter, *Appl. Phys. Lett.* 88 (2006) 111901.
[22] K.A. Khan, G.A. Niklasson, C.G. Granqvist, *J. Appl. Phys.* 64 (1988) 3327.
[23] F. Beteille, J. Livage, *J. Sol-Gel Sci. Technology.* 13 (1-6) (1998) 915.
[24] A.Siokou, G. Leftheriotis, S. Papaefthimiou, P. Jianoulis, *Surface Science.* 482-484 (2001) 294.
[25] G. Xu, P. Jin, M. Tazawa, K. Yoshimura, Solar Energy Mater. *Solar Cells.* 83 (2004) 29.
[26] C.G. Granqvist, Solar Energy Mater. *Solar Cells.* 68 (2001) 401.
[27] N. Ozer, J.P. Cronin, Key Engineering Materials. 264-268 (2004) 337.
[28] S.K. Deb, Solar Energy Mater. *Solar Cells.* 68 (2001) 248.
[29] G.C. Granqvist, *Appl. Phys.* A 57 (1993) 285.
[30] J.M. Bell, J.P. Mathew, Solar Energy Mater. *Solar Cells.* 68 (2001) 285.
[31] J.S.E.M. Svensson, C.G. Granqvist, *Solar Energy Materials.* 12 (1985) 391.

[32] S.I.Cho, D.H. Choi, S.-H. Kim, S.B.Lee, *Chemistry of Materials.* 17 (18) (2005) 4564.
[33] K. Kuwabara, K. Sugiyama, M. Ohno, *Solid State Ionics.* 44 (1991) 313.
[34] K. Kuwabara, Y. Noda, *Solid State Ionics.* 61 (1993) 303.
[35] C. Bechinger, J.N. Bullock, J.-G. Zhang, C.E. Tracy, D.K. Benson, S.K.Deb, H.M. Branz, *J. Appl. Phys.* 80 (1996) 1226.
[36] E.O. Zayim, F.Z. Tepehan, *Key Engineering Materials.* 264-268 (I) (2004) 435.
[37] A Bessière, C. Duhamel, J-C. Badot, V. Lucas, M-C. Certiat, *Electrochimica Acta.* 49 (12) (2004) 2051.
[38] Bouessay, A. Rougier, J.-M. Tarascon, *J. Electrochemical Society.* 151 (6) (2004) H145.
[39] Ceccato, R., Carturan, G., Decker, F., Artuso, F. , *J. Sol-Gel Science and Technology* 26 (1-3) (2003) 1071.
[40] A. Pawlicka, C.O. Avellaneda, Molecular Crystals and Liquid Crystals Science and Technology, *Section A: Molecular Crystals and Liquid Crystals.* 354 (2000) 463.
[41] P.S. Patil, R.K. Kawar, S.B. Sadale, *Applied Surface Science.* 249 (1-4) (2005) 367.
[42] J. Backholm, A. azens, G.A. Niklasson, Solar Energy Mater. *Solar Cells.* 90 (2006) 414
[43] Bouessay, A. Rougier, P. Poizot, J. Moscovici, A. Michalowicz, J-M. Tarascon, *Electrochimica Acta.* 50 (18) (2005) 3737.
[44] S.R. Jiang, P.X. Jan, B.X. Jan, B. Feng, X. Cai, *Materials Chem. Physics.* 77 (2002) 384.
[45] T.-S.Tung, K.-C. Ho, Solar Energy Mater. *Solar Cells.* 90 (4) (2006) 521.
[46] R.J. Mortimer, J.R. Reynolds, *J. Materials Chemistry.* 15 (22) (2005) 2226.
[47] A. Azens, G. Vaivars, L. Kullman, C.G. Granqvist, *Electroch. Acta.* 44 (1999) 3059.
[48] G. Wei, T.E. Haas, R.B. Ronald, Proceedings - *The Electrochemical Society.* 90 (2) (1989) 80.
[49] M. Veszelei, L. Kullman, M. Stromme-Mattson, A. Azens, C.G. Granqvist, *J. Appl. Phys.* 83 (1998) 1670.
[50] F. Ghodsi, F. Tepehan, C. Tepehan, Solar Energy Mater. *Solar Cells.* 968 (2001) 355.
[51] A. Verma, A.K. Srivastava, A.K.Bakhshi, R. Kishore, S.A. Agnihotry, *Materials Letters.* 59 (27) (2005) 3423.
[52] A. Verma, D.P. Singh, A.K.Bakhshi, S. A.Agnihotry, *J. Non-Crystalline Solids.* 351 (30-32) (2005) 2501.
[53] C. O. Avellaneda, L.O.S. Bulhões, A. Pawlicka, *Thin Solid Films.* 471 (1-2) (2005) 100.
[54] A. B. Bordyuh, G.V. Klimusheva, A.P. Polishchuk, S. Bugaychuk, T.A. Mirnaya, G.G. Jaremchuk, *Proceedings of SPIE.* 6023 (2005) art. no. 60230B.
[55] C.-H. Yang, Y.-K Chih,W.-C. Wu, C.-H. Chen, *Electrochem. Solid-State Letters.* 9 (1) (2006) C5.
[56] A.A. Argun, P.-H. Aubert, B.C. Thompson, I. Schwendeman, C.L. Gaupp, J. Hwang, N.J. Pinto, J.R. Reynolds, *Chemistry of Materials.* 16 (23) (2004) 4401.
[57] Y. Wang, US patent. 6815528, 2004
[58] C.G. Granqvist, Solar Energy Mater. *Solar Cells.* 32 (4) (1994) 369.
[59] P. Chandrasekhar, B.J. Zay, T. McQueeney, A. Scara, D. Ross, G.C. Birur, S. Haapanen, D. Douglas, *Synthetic Metals.* 135-136 (2003) 23.
[60] A. G. Dark, R. Osiander, J. Champion', T. Swanson, D. Douglas, Space Technology and Applications International Forum-2000, edited by M. S. El-Genk, p. 803.

[61] G. Liu, T.J. Richardson, Solar Energy Mater. *Solar Cells.* 86 (1) (2005) 113.
[62] www.gentex.com/automotive.html
[63] D. R. Rosseinsky, R. J. Mortimer, *Advanced Materials.* 13(2001) 783.
[64] Corr, D. International Conference on Digital Printing Technologies, 2005, pp. 47.
[65] www.basqueresearch.com/ and see also, A. Argun,P.-H. Aubert, B. C. Thompson, I. Schwendeman, C. L. Gaupp, J. Hwang, N. J. Pinto, D. B. Tanner, A. G. MacDiarmid, J. R. Reynolds, *Chem. Mater.* 16 (2004) 4401
[66] P. Andersson, D. Nilsson, P.-O. Svensson, M. Chen, A. Malenstrom, Granmar and Cho, *Science.* 6 May 2005: 785-786
[67] C. Wang, M. Shim, Ph. Guyot-Sionnest, *Science.* 291 (2001) 2390.
[68] C. Wang, M. Shim, Ph. Guyot-Sionnest, *Applied Phys. Lett.* 80 (2002) 4.
[69] U.O. Krasovec, A.S. Vuk, B. Orel, *Solar Ener. Mater. Solar Cells.* 73 (2002) 21.
[70] E.S.Lee, D.L.DiBartolomeo, *Solar Energy Mater. Solar Cells.* 71 (2002) 465.
[71] http://eetd.lbl.gov/BT.html
[72] A. Pennisi, F. Simone, G. Barleha, G. Di Marco, M. Lanza, *Electroch. Acta.* 44 (1999) 3237.
[73] Kraft, M. Rottman, K.H. Heckner, Solar Energy Mater. *Solar Cells.* 90 (2006) 469.
[74] A.W. Czanderna, D.K. Benson, G.J. Jorgensen, J. Zhang, C.E. Tracy, S.K. Deb, Solar Energy Mater. *Solar Cells.* 59 (1999) 419.
[75] N.A. O'Brien, J. Gordon, H. Mathew, B.P. Hichwa, *Thin Solid Films.* 345 (1999) 312.
[76] G. Tulloch, I. Skryabin, G. Evans, *J. Bell.* SPIE 3136 (1997) 426.
[77] www.epsilon-web.net/Ec/manuual/ Techniques/CycVolt/cv_analysis.html
[78] http://nanonet.rice.edu/research/CVtutorial2/
[79] Wu, P., Shi, Y., Cai, C., *J. Solid State Electrochemistry.* 10 (5) (2006) 270.
[80] M. Habib, p. Glueck, *Solar Energy Materials.* 18 (1989) 127.
[81] Eleanor S.Lee, Denis L. Bartolomeo, Joseph H.Klems, Mehry Zardanian and Stephen E.Szelkovitz "Monitored Energy Performance of Electrochromic Windows Controlled by Daylight and Visual Comfort", *ASHRAE Transactions*, part 2, p.122-141.

Chapter 2A

TECHNOLOGY DEVELOPMENT AND PROPERTIES OF APCVD WO$_3$ ELECTROCHROMIC THIN FILMS

D. S. Gogova and K. A. Gesheva

Central Laboratory of Solar Energy and New Energy Sources,
Bulgarian Academy of Sciences, Sofia, Bulgaria

The aim of this chapter is to explore the high application potential of the APCVD technique for growth of high quality WO$_3$ thin films with properties suitable for electrochromic window utilization in innovative solar energy systems.

2.A.1. TECHNOLOGY DEVELOPMENT OF LOW-TEMPERATURE APCVD CARBONYL PROCESS FOR DEPOSITION OF WO$_3$ THIN FILMS

The optical properties of WO$_3$ thin films strongly depend on the deposition method employed (PVD, CVD, sol-gel, electrochemical deposition) and on the specific growth parameters, determining the film structure and chemical composition. The electrochromic properties of these films (coloration efficiency, optical modulation, switching time, memory) depend on the degree of thin film crystallinity, porosity, stoihiometry, water content in the film composition, etc. [1,2].

Two approaches are known from the literature for APCVD deposition of WO$_3$ thin films: A two stage approach consisting of:

1) deposition of thin metallic film in result of thermal decomposition of tungsten hexacarbonyl (W(CO)$_6$) vapors in Ar environment according to the following chemical reaction:

$$W(CO)_6 \rightarrow W + 6\,CO \quad T_{depos.} = 400°C; \qquad (1)$$

2) followed by an oxidizing of the metallic W film in oxygen rich atmosphere (oxygen or a mixture of oxygen and argon) in the temperature range 450 – 600°C.

$$2W + 3O_2 \rightarrow 2WO_3 \quad (2)$$

The second approach consists of two stages also:

1) stage of deposition of thin film of tungsten dioxide, according to the oxidizing pyrolitical reaction:

$$W(CO)_6 + 4 O_2 \rightarrow WO_2 + 6 CO_2 \quad T_{depos.} = 400°C \quad (3)$$

2) stage of additional oxidation in oxygen containing atmosphere:

$$2WO_2 + O_2 \rightarrow 2WO_3 \quad Tanneal. \geq 450°C \quad (4)$$

The structure, optical and electrochromic properties of the WO_3 films obtained by the two stage process are thoroughly investigated by Donnadieu et al. [3-5]. However, the two stage approach has some important disadvantages, namely a long-term additional treatment at temperatures about or above 500°C when diffusion of alkali ions from the soda-lime glass substrates is very intensive. Moreover, the material obtained is always polycrystalline, which is a certain limitation from the point of view of electrochromic application, since it is generally believed that the amorphous tungsten oxide films have larger coloration efficiency than the polycrystalline ones [2].

Therefore, we have developed a low-temperature carbonyl technological process permitting in-situ deposition of tungsten oxide films, according to the following reaction:

$$W(CO)_6 + 9/2 O_2 \leftrightarrow WO_3 + 6CO_2\uparrow \quad (5)$$

Thin films of WO_3 were deposited at 200, 300 and 400°C by pyrolysis of tungsten hexacarbonyl vapours in an oxygen stream in a CVD horizontal reactor with cold walls (The APCVD system used is the same as the presented one in chapter 1). The sublimator containing the $W(CO)_6$ powder was immersed in a silicon oil bath, the temperature of which (T_{subl}) is maintained at 90-110°C and controlled with an accuracy of ± 1 K. This temperature provides sufficient vapour pressure. A very detailed analysis of the concentration unsteadiness of CVD with a precursor sublimated from packed bed of solid source was done by Peev and Tsibranska [6].

Silicon wafers and soda-lime glass were used as substrates for deposition of samples intended for different measurements. The substrates were placed on a graphite susceptor, which was heated by a high frequency generator. A thermoregulator controlled the substrate temperature, which is practically the deposition temperature ($T_{depos.}$). The gas lines made of "Galtek" type teflon were heated to a temperature equal to the source chamber temperature or a bit above it, to prevent the vapours from condensation. The oxygen flow was varied from 0.001 to 2.4 l/min and the argon flow through the source was 0.024 - 0.2 l/min. For oxygen source in the experiments performed we used diluted in argon O_2, N_2O, H_2O – vapors or their combinations.

Possible are some other chemical reactions proceeding together with the basic one [7], given above, in result of which the composition of the deposited film differs from the stoihiometric WO_3 composition. In the composition of films, deposited by carbonyl CVD process, very small amount of carbon might be present due to the reaction:

$$2(CO) \rightarrow C + CO_2 \uparrow \qquad (6)$$

Our purpose was to establish the range of technological parameters values, in the frame of which it is possible to deposit highly transparent in the visible range of spectrum films. Furthermore, our technology development was focused on tailoring of the optical and structural properties and in result of the electrochromic properties in order to suit certain applications in solar energy systems.

For investigation of the influence of different APCVD technological parameters on the structure and optical properties of WO_3 films, these parameters were changed in the following range:

- Temperature of deposition, $T_{depos.}$ – from 200 to 400°C;
- Sublimation temperature of $W(CO)_6$, $T_{subl.}$ – from 80 to 110°C;
- Rate of the oxygen flow (O_2, N_2O, H_2O-vapors + Ar) – from 0.008 to 2.4 l/min;
- Flow rate of the argon gas, passing through the sublimator, $Ar_{W(CO)_6}$ - 0.0024 – 0.4 l/min;
- Annealing temperature, $T_{anneal.}$ – 450°C to 500°C

As CO (from $W(CO)_6$) acts as a classic reductor, a partial reduction in the course of deposition is possible and, hence, a deviation in the obtained layer stoichiometry. The light bluish colour of some of our in-situ obtained CVD-WO_3 layers suggests their non-stoichiometry. Therefore, these as-deposited films were additionally annealed in oxygen-containing environment at 400 and 500 °C to obtain stoichiometric films.

The film thickness was measured by Talystep profilometer and as well as determined on the basis of the ellipsometrical measurements. By weighting of the substrate before and after deposition of the film on it, film densities were calculated. The polycrystalline WO_3 coatings have higher densities than the amorphous ones, but in all cases the densities are much lower compared to bulk crystalline monoclinic WO_3 (7.16 g/cm^3) [8]. The porosity values were calculated on the basis of the density of the crystalline bulk material.

At oxygen flow-rates below 0.008 l/min the dominant phase in the film structure is the nontransparent WO_2 phase. More details for the preparation and properties of such black tungsten coatings finding application in photothermal solar energy collectors can be seen in Ref. 9.

At deposition temperatures larger than 400°C the chemical reaction is mainly in the vapor phase and therefore very little material is deposited on the substrate. The temperature of deposition of 200°C was the lowest possible for decomposition of the tungsten hexacarbonyl at atmospheric pressure.

The first investigations of carbonyl amorphous tungsten oxide (a-WO_3) films, obtained by APCVD are these of Olevskii et al.[10], who studied the film structure and these of Maruyama and Arai [11], who studied the electrochromic properties. However, there are no technological details in these publications. Therefore, we have developed a low temperature deposition process and have done a systematic investigation of the influence of the CVD process parameters ($T_{depos.}$ and $T_{subl.}$, ratio between the reactive gases $Ar_{W(CO)_6} : O_2$) on the growth rate, density and optical quality of the films deposited. The approach applied to study the influence of different process parameters on the film properties was the following: varying the value of one parameter and keeping all the other parameters constant. The results obtained are summarized in table 1. All data given in the table correspond to $T_{subl.} = 110°C$ and a film thickness of 200 nm.

On the basis of the experimental results obtained the following conclusions could be made:

The most important parameter governing the CVD process thermodynamics is the temperature. It influences the film growth rate, the structure (degree of crystallinity and porosity) and the stoichiometry of the thin films.

Table 1. Influence of the technological parameters (deposition temperature, argon and oxygen flow rates and reaction gas flow rates ratio $Ar_{W(CO)_6} : O_2$ on the film growth rate ($v_{gr.}$), film density (ρ), and optical quality – visual film transparency

№	$T_{deposition}$ (°C)	$Ar_{W(CO)_6}$ (l/min)	O_2 (l/min)	$Ar_{W(CO)_6} : O_2$	$V_{gr.}$ (nm/min)	ρ (g/cm³)	Optical Quality
1	400	0.024	0.008	3:1	345	6.10	Dark blue
2	400	0.050	0.050	1:1	323	6.05	Blue
3	400	0.100	0.300	1:3	354	5.71	Light blue (bluish)
4	400	0.100	0.400	1:4	367	5.50	Transparent
5	400	0.100	0.600	1:6	391	5.20	Transparent
6	400	0.200	1.200	1:6	398	5.15	Transparent
7	400	0.100	1.200	1:12	402	5.12	Transparent
8	400	0.200	2.400	1:12	404	5.08	Transparent
9	400	0.100	1.400	1:14	408	5.02	Transparent
10	300	0.024	0.008	3:1	50	5.50	Dark blue
11	300	0.050	0.050	1:1	45	5.31	Blue
12	300	0.100	0.300	1:3	32	5.10	Blue
13	300	0.100	0.400	1:4	34	5.02	Bluish
14	300	0.100	0.600	1:6	36	4.90	Transparent
15	300	0.200	1.200	1:6	38	4.85	Transparent
16	300	0.100	1.200	1:12	40	4.55	Transparent
17	300	0.200	2.400	1:12	42	4.50	Transparent
18	300	0.100	1.400	1:14	45	4.48	Transparent
19	200	0.024	0.008	3:1	0.5	5.03	Dark blue
20	200	0.050	0.050	1:1	0.8	4.78	Dark blue
21	200	0.100	0.300	1:3	1.3	4.35	Blue
22	200	0.100	0.400	1:4	1.8	4.25	Bluish
23	200	0.100	0.600	1:6	2.0	4.10	Light bluish, transparent
24	200	0.200	1.200	1:6	2.3	4.05	Transparent
25	200	0.100	1.200	1:12	3.0	4.03	Transparent
26	200	0.200	2.400	1:12	3.1	4.01	Transparent
27	200	0.100	1.400	1:14	3.2	3.59	Transparent

1. The growth rate depends on:

- deposition temperature;
- reactive gases flow rate and total gas flow rate.

An increase of the $T_{depos.}$ with 100°C leads to increase of the growth rate with one order of magnitude. The highest value of the growth rate of the APCVD-WO$_3$ films achieved is 405 nm/min and it is reached at $T_{depos.}$ = 400°C. Larger growth rates are obtained by Plasma-enhanced CVD from WF$_6$ precurssor, only [12]. An increase of 10°C in the $T_{subl.}$ leads to increase of the vapor pressure of the tungsten hexacarbonyl ($P_{W(CO)_6}$) about two times (see for details Ref. 13) in result of that the growth rate increases approximately twice, also. The flow rates of the reactive gases influence on the film growth rate to a smaller extent than the temperature but they affect the optical quality of the films.

2. The stoichiometry of the CVD-WO$_3$ depends on:

- ratio of the flow rates of the reactive gases;
- temperature;

Low values of the oxygen flow rate lead to deficiency of oxygen in the films, which appears as a dark blue color. An ratio $Ar_{W(CO)_6}:O_2 < 1:3$ is necessary in order to deposit films with close to the stoihiometry composition and in result of this with good transparency in the visible range. Keeping constant the $T_{subl.}$ = 110°C and $T_{depos.}$ = 400°C and changing the ratio of the reactive gases $Ar_{W(CO)_6}:O_2$ in the order 1:3, 1:6 and 1:12 leads to an increase of the transmittance value at λ = 500 nm of WO$_3$ films with a thickness of 100 nm, deposited on glass substrates in the following order: 72→79→88%. Lower concentration of the hexacarbonyl vapors and lower $T_{subl.}$ at fixed the other process parameters lead to more complete oxidation and thus, the films are optically more transparent. An increase of the $T_{depos.}$ favors the full oxidation reaction and in result of this more transparent films grow. For example, keeping the $T_{subl.}$ = 110°C and $Ar_{W(CO)_6}:O_2$ = 1:12 constant and increasing the deposition temperature from 200 up to 400°C lead to increase in the transmittance values of WO$_3$ films with a thickness of 100 nm at λ = 500 nm in the following order: 77→82→88%.

3. The density of the CVD-WO$_3$ films is influenced by:

- deposition temperature;
- gas flow rates;
- growth rate.

The deposition temperature strongly influences not only the film growth rate, but the film density, as well. At high $T_{depos.}$ = 400°C the CVD-WO$_3$ films grow with the highest density, 5.02 – 6.10 g/cm^3 and the lowest density is usual for the low temperature films, grown at $T_{depos.}$ = 200°C. The values vary in the range 3.59 – 5.03 g/cm^3 within the growth window used. As can be expected, the gas flow rates have much less influence on the film density than the temperature. At low values of the gas flow rates in the order of 0.008-0.050 l/min, the films deposited are with large density values – about 6.0 g/cm^3. By means of increasing of the gas flow rates the growth rate increases to a certain extend and meantime the density of the films decreases. At high growth rates the material obtained has a low density since the WO$_3$ molecules do not have enough time to migrate over the substrate and they form clusters of molecules, which get deposited on the substrate [14]. Thus, the molecules do not arrange very densely, as it is expected for a CVD process, and empty spaces between them remain, which were detected by positron annihilation spectroscopy [15]. The establishment of the range of process parameters within which the CVD films grow with very low density is an important achievement, since the low density (high porosity) is a valuable property for the electrochromic process enhancing its kinetics since it facilitates the easy movement of ions [16]. At one and same ratio of the flow rates of reactive gases, but at different values a difference in the growth rate and the film density (see table 1) is observed. The growth rate of CVD-WO_3 films deposited in result of reaction processing at higher total gas flow rate is higher, and the film density - smaller. This tendency of increasing of the growth rate and density decreasing is observed up to a value of the total gas flow in the order of 2.7 l/min, after which the growth rate drops. The relative density of the CVD-WO_3 films (the film density related to the density of a volume sample – 7.16 g/cm^3 [8]) varies strongly with deposition process parameters for films grown by the other deposition techniques [17]. A.Donnadieu et al. [4] reported for polycrystalline carbonyl WO_3 films, obtained by annealing of metallic tungsten and tungsten dioxide films, density values in the range 0.65 – 0.8. The APCVD carbonyl WO_3 films are with relative density ranging in the limits 0.5– 0.85.

2.A.2. Investigation of Structure and Optical Properties of APCVD WO_3 Films

In this paragraph the correlation between the APCVD deposition parameters and the structural and optical properties of WO_3 films is studied. For that purpose we have optimized the technological conditions and obtained amorphous and polycrystalline layers of good optical quality (high transmission in the VIS-NIR, excellent adhesion and good homogeneity on a large area).

X-ray diffraction (XRD) measurements were performed to assess the structure of the WO_3 films. X-ray Photoelectron Spectroscopy (XPS) and Raman scattering measurements were employed to examine the film bonding and composition. The optical properties were investigated by spectroscopic ellipsometry (SE) and VIS-NIR spectrophotometry measurements.

For Raman scattering measurements the 488 nm line of an Ar ion laser was utilized and the spectra were recorded using a SPEX 1403 double monochromator and a Hamamatsu R 943 PMT in photon counting mode. For determination of the optical constants of the films a Rudolph Research spectroscopic ellipsometer type 436 with PCSA configuration was employed in the spectral region from 350 to 750 nm and at an angle of incidence of 45°. The accuracy of the angles of the polarizer, analyzer and light incidence is within ±0.001°. X-ray photoelectron spectroscopy (XPS) measurements were carried out in the analyzer of an electron spectrometer type ESCALAB II (VG SCIENTIFIC) using a $Mg_{K\alpha}$ (1253.6 eV) X-ray source. The resolution of the analyzer is 0.1 eV with a pass energy of 20 eV.

2A.2.1. Investigation of Composition of CVD-WO_3 Films

Tungsten oxide films usually grow understoichiometrical. Their oxygen deficiency depends on the method of preparation. In figures 1 and 2 XPS spectra typical of WO_3 films deposited under technological conditions given in table 2, are shown.

Table 2. Technological conditions of CVD-WO$_3$ films deposition, studied by XPS analysis

sample set number	T$_{depos}$ (°C)	Ar$_{W(CO)6}$: O$_2$	T$_{anneal}$ (°C)
s. 1	400	1 : 4	-
s. 2	200	1 : 3	-
s. 3	300	1 : 2	-
s. 4	400	1 : 1	-
s. 5	200	1 : 3	500
s. 6	300	1 : 5	-
s. 7	200	1 : 6	-

In figure 1a the XPS spectra of amorphous WO$_3$ film, deposited at 400°C and Ar$_{W(CO)6}$:O$_2$ = 1:4, are given. Two pronounced peaks appear: the spin-orbit doublet due to W4f$_{7/2}$ electrons with binding energy of 36 eV and W4f$_{5/2}$ at 38.15 eV. It is obvious that these peaks represent the W^{6+} state [16]. The ratio of the areas of W4f$_{7/2}$ and W4f$_{5/2}$ peaks equals 0.75 and the difference of 2.15 eV in their binding energies agrees with the theory of 4f level splitting [16]. The peak of oxygen O$_{1s}$ electrons has been observed at 530.8 eV, see figure 1b. The peak of C$_{1s}$ electrons (not shown here) is centered at 285 eV, which shows that carbon is in nonbonded state, since the binding energy of CO is 290.1 eV, and that of WC is in the range 31.2 to31.5 eV [18].

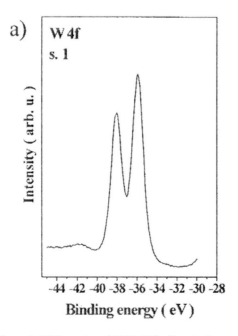
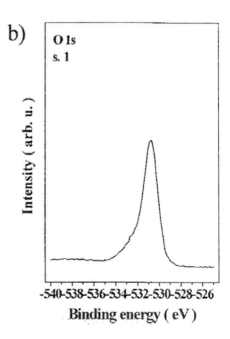

Figure 1. XPS spectra of CVD-WO$_3$ films in the range of binding energies of a) W$_{4f}$ and b) O$_{1s}$ electrons.

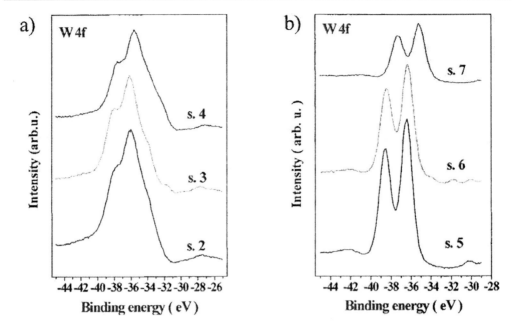

Figure 2. XPS spectra of CVD-WO$_3$ films deposited under different technological conditions, in the range of binding energies of the W$_{4f}$ electrons.

In the XPS spectra of the CVD-WO$_3$ films deposited at low oxygen flow rates (see figure 2a), a broadening of the peak at about 34 eV is observed, which was assigned to the overlapping of the spectra of W^{6+} states with the ones of W of lower valences. In this case the XPS spectrum can be considered as consisting of three doublets. The first one corresponds to the W^{6+} state. The second one due to W4f$_{7/2}$ electrons with a main maximum at a binding energy of approximately 34.7 eV we relate to W^{5+} state, and the third one at 34 eV is related to W^{4+} [16]. In XPS spectra of PVD-WO$_3$ films W^{6+} and W^{4+} states are observed [16], while for PECVD-WO$_3$ films [19] in addition W^{5+} state is also existing.

The XPS spectra show that annealing of the CVD-WO$_3$ in oxygen-containing atmosphere at 450 to 500°C does not change the tungsten to carbon ratio, i.e., it only changes the tungsten to oxygen one, because of the additional incorporation of oxygen atoms in the crystal lattice. The carbon content of the CVD-WO$_3$ films varies in the range of 2 to 8 at.%. The oxygen deficiency of these films depends on the preparation conditions and it is in the range 0.02 to 0.1 for the samples obtained under the technological conditions given in table 2.

Some authors [20] consider that sputtering of the surface with a purpose to clean it from C contaminations (as done in our case) should not be applied in order to avoid possible reduction of the oxide and because this procedure affects the XPS spectra measured.

2.A.2.2. Study of Structure of APCVD-WO$_3$ Films

In figure 3 XRD spectra typical of CVD-WO$_3$ films deposited under equivalent conditions (T$_{depos}$ = 300°C and one and same gas ratio – from 1 : 3 to 1 : 12), but with a different thickness: 200, 600 and 3000 nm, respectively, are shown. It is obvious that the spectrum of the thinnest sample A exhibits a structure which is characteristic of an amorphous phase, the spectrum of the thicker film (spectrum B) corresponds to a mixture of amorphous phase with some separate crystallites. The crystallization of the film starts at a temperature of

300°C. Although „thin" films, deposited at this temperature, are amorphous, they transform in a mixture of amorphous and crystalline phase with increasing of the thickness. At a thickness above 1000 nm the polycrystalline phase becomes predominant. This structure is similar to the one observed with HRTEM by Kaito et al. [21] for vacuum evaporated WO_3 films. It includes randomly distributed microcrystalline domains, which form a homogeneous crystalline film with increasing of the temperature or film thickness. Taylor and Patterson [22] observed the same increase of the crystalline phase of WO_3 films sputtered from W and WO_3 target in Ar/O_2 environment at 200°C.

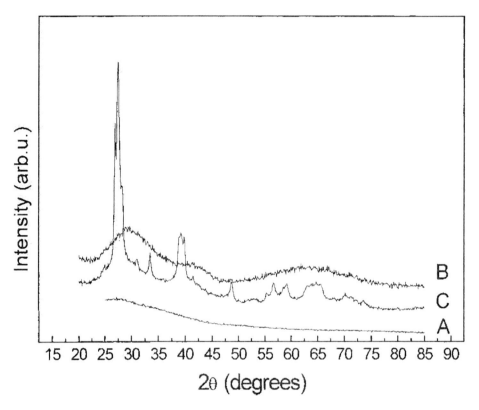

Figure 3. XRD spectra of CVD-WO_3 films deposited at 300°C with different thicknesses: sample A: 200 nm; sample B: 600 nm and sample C: 3000 nm.

In figure 4 the influence of the deposition temperature on the degree of crystallinity of CVD-WO_3 films with a thickness of 2000 nm is presented. The XRD spectrum of the film deposited at 200°C shows an amorphous phase, the one for the film obtained at 300°C exhibits amorphous and crystalline phases, while WO_3 films grown at 400°C have typical polycrystalline structure, demonstrated by strong peaks in the diffraction pattern. A very good agreement has been obtained between the interplanar distances, characteristic of the orthorhombic crystallographic syngony [23], and the experimentally obtained ones for CVD-WO_3 films. The intensity of the XRD peaks increases with increasing of the film thickness. This is most likely related to an increase in the crystallite size with increasing of the film thickness, and was confirmed by SEM observations on film morphology.

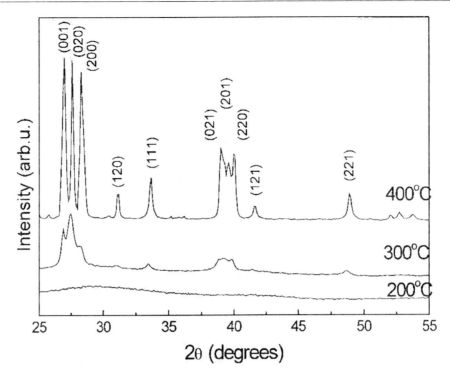

Figure 4. XRD spectra for CVD-WO$_3$ films deposited at different temperatures.

Donnadieu [4] assigned the maxima in the XRD spectra of APCVD-WO$_3$ films, obtained by additional annealing of CVD-deposited metallic tungsten or black tungsten coatings, to the monoclinic WO$_3$ crystallographic modification. According to Taylor and Patterson [22], the maxima in the diffraction spectra of the WO$_3$ films could be assigned either to the monoclinic, orthorhombic, or triclinic syngony. We agree with this conclusion, since the WO$_3$ in all these crystallographic syngonies has one and the same main diffraction peak and additional information is needed for the assignment.

2.A.2.3. Vibrational Properties of CVD-WO$_3$ Films

In figure 5a typical Raman spectra of films obtained at 200, 300 and 400°C, respectively, at one and the same gas ratio (Ar$_{W(CO)6}$: O$_2$ = 1/6) and all with thickness of about 100 nm (samples 1, 2 and 3 in table 3) are shown. In figure 5b Raman spectra for samples obtained under identical conditions as samples 1, 2 and 3, but additionally annealed in oxygen-containing environment at 500°C for 1 h (samples 4, 5 and 6) are presented. Detailed analysis of the Raman spectra involved decomposition into a set of Gaussian and Lorentzian peaks. Table 3 presents the technological conditions of deposition of CVD-WO$_3$ films and fit parameters of the Raman spectra.

The WO$_3$ films deposited at 200, 300 and 400°C and additionally annealed at 500°C for 1 h are crystalline, as can be seen from the narrow peaks at 717 and 809 cm^{-1} (see figure 5b), corresponding to the W—O stretching vibrations [24]. Samples, obtained at 200 and 300°C are amorphous (figure 5a) with a broad band at about 750 cm^{-1} [24,25] and another one at 960 cm^{-1} (observed in the spectrum of the film deposited at 200°C only) characteristic of amorphous WO$_3$ material, which other authors assign to stretching vibrations of W=O terminal bonds [25].

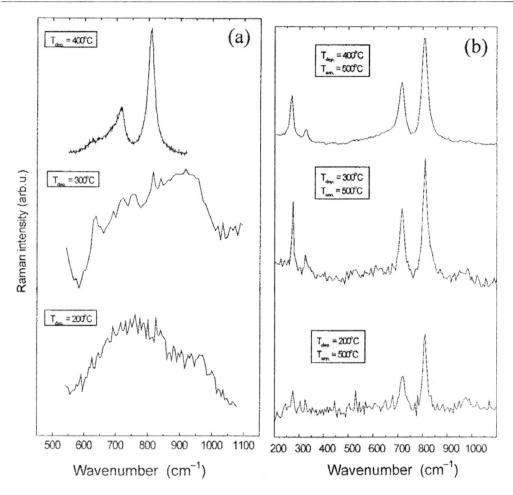

Figure 5. Raman spectra of CVD-WO$_3$ films a) deposited at different temperatures and b) additionally annealed at 500°C.

There is a contradiction in the literature about the interpretation of the 950 cm^{-1} scattering band. Ramans et al. [26, 27] consider that the Raman band at 950 cm^{-1} in a-WO$_3$ film can be assigned to the W=O stretching mode of the terminal oxygen. They propose that the WO$_6$ octahedra with one short W=O bond and an opposite long W—O bond can be accepted as a basic structure unit in a-WO$_3$ films. Therefore, the structure of a-WO$_3$ films consists of clusters built from such deformed octahedra. According to Daniel et al. [28] the terminal oxygen bonds are due to adsorbed water molecules. The W=O bonds are considered as situated outside the surface of the clusters. Granqvist [29] showed for evaporated WO$_3$ films at a low substrate temperature and high water partial pressure the intensity of the peak at 950 cm^{-1} is high, while its width does not depend noticeably on the deposition parameters.

In the Raman spectra of samples grown at 400°C, similarly to the spectra of WO$_3$ obtained by oxidation of thin metallic W films [29] this band is absent. Therefore, it was considered that the CVD-WO$_3$ films do not have double bonds [29]. This conclusion was due to the fact that Raman scattering studies have been performed on a polycrystalline material only. We observe that in amorphous CVD-WO$_3$ this band exists in agreement with the results for sputtered and evaporated amorphous WO$_3$ layers [25,28]. In the Raman spectra of the CVD-WO$_3$ films obtained at 300 to 600°C by decomposition of WF$_6$ in presence of water vapors [30] an absorption band at 950 cm^{-1} is not observed, which shows lack of water molecules in the film structure. The Raman spectrum obtained for our samples and shown in figure 5a is in agreement with this conclusion.

Table 3. Results from Raman study of CVD-WO$_3$ films and the corresponding technological parameters

Sample №	Deposition temperature (°C)	Annealing temperature (°C)	Fit Parameters	Peak Center	Width (cm^{-1})
1	200	no	Amorphous		
2	300	no	Amorphous		
3	400	no	Peak 1	713	W29
			Peak 2	808	W25
			Peak 3	670	W82
			Peak 4	270	W16.7
			Peak 5	326	W19
4	200	500/1 hour	Peak 1	717	W19
			Peak 2	809	W20
			Peak 3	444	W11.8
			Peak 4	326	W11.3
			Peak 5	274	W11
5	300	500/1 hour	Peak 1	716	W20
			Peak 2	809	W20
			Peak 3	326	W10.5
			Peak 4	274	W11
			Peak 5	225	W11.4
6	400	500/1 hour	Peak 1	716	W22
			Peak 2	810	W21
			Peak 3	636	W120-130
			Peak 4	326	W14.6
			Peak 5	269	W14.3
7	300	300/1 hour	Peak 1	711	W36
			Peak 2	809	W32
			Peak 3	664	W98
			Peak 4	270	W19.7
			Peak 5	324	W12.3
8	200	200/1 hour	Peak 1	709	W44
			Peak 2	809	W30
			Peak 3	633	W300
			Peak 4	265	W76
9	400 from W film	500/2 hour	Peak 1	717	W21
			Peak 2	808	W20
			Peak 3	326	W14.2
			Peak 4	275	W14

The two W-O stretching bands of the sample, prepared at 200°C and annealed at the same temperature (sample 8) are similar to the corresponding bands of the sample deposited at 300°C and additionally annealed at 300°C (sample 7), but they are broader than those of the samples 4, 5 and 6 (annealed at a higher temperature). They are broader than the W-O stretching bands of bulk WO_3 material, too. This is an evidence for the smaller crystallites and greater disorder in the sample sets 3, 7 and 8. The lower frequency W-O stretching band position is sensitive to the structural disorder. The position of this band in the spectra of the samples 3, 7 and 8 is shifted to lower frequencies in comparison with those in the Raman spectra of samples 4, 5 and 6, annealed at a higher temperature. In the Raman spectra of the samples 3, 6 and 7 a weak broad band at 630 to 670 cm^{-1} is observed. Its origin is unclear so far. Most probably it is due to WO_3 hexagonal phase [29]. Thermal annealing of WO_3 films in oxygen-containing atmosphere for 1 h at 500°C leads to recrystallization of the films, and independent of the deposition temperature, the dimension of the crystallites is approximately equal for all samples since the halfwidths of the W-O stretching vibration bands in these samples are practically the same.

In the spectra of annealed CVD-WO_3 films, some additional bands were found at 444, 326 and 276 cm^{-1} which according to [27] are related to the $\delta(O-W-O)$ deformation mode. The Raman stretching bands of WO_3 films, obtained by annealing of metallic tungsten films (sample 9), are similar to the corresponding bands of directly deposited polycrystalline ones.

The Raman spectra of amorphous CVD-WO_3 films show strong bonding between the clusters forming the film structure, which is similar to the results of Ramans et al. [26]. However, this conclusion is in contradiction to the model developed by Arnoldussen [31], who considers that the clusters are bonded by bridges formed of water molecules.

2.A.2.4. Optical Properties of CVD-WO_3 Films

The purpose of this paragraph is to gain insight into the optical properties of APCVD tungsten oxide thin films, as well as tailoring of the optical and electro-optical properties to suit application device requirements.

The thickness of the CVD-WO_3 films and their effective refractive index (N_e) were calculated from the experimental ellipsometric data using a single-layer (oxide-substrate) model. In figure 6a the dispersion of the effective refractive index of films grown on Si(111) at three different deposition temperatures (200, 300 and 400°C) and at $Ar_{W(CO)_6} / O_2 = 1/6$ before and after annealing is illustrated. The as-deposited on Si(111) CVD-WO_3 films have values of N_e at 500 nm in the range of 2.2 to 2.5 for amorphous and in the range of 1.9 to 2.0 for polycrystalline films. The values of N_e of films deposited on glass substrates are lower in comparison with the values of N_e for the same films deposited on Si(111) (see figure 6b). It is due to the fact that on crystalline substrate the material deposited by CVD has higher density in certain crystallographic directions compared to that grown on amorphous glass substrates. From figure 6a it can be seen, as well, that a tendency of decreasing of the effective refractive index in the visible region exists with increasing of the deposition temperature. At $T_{depos.}$ of 200°C the refractive index of the CVD-WO_3 films at $\lambda = 500$ nm is equal to 2.50±0.005, at $T_{depos.} = 400°C$ it is $N_e = 2.20$. Obviously, the values of the refractive index of amorphous CVD-WO_3 films deposited at temperatures lower than 400°C are comparatively high, which can be related to some deviation from the stoichiometric composition (with increasing of the deposition temperature the layers have a composition closer to the stoichiometric one). The obtained Ne values of amorphous CVD-WO_3 films are close to those reported by Deb [32] for amorphous evaporated in vacuum films. Also for evaporated WO_3 films, a similar effect of decreasing of the N_e with increasing of deposition temperature has been observed by Miyake et al. [33].

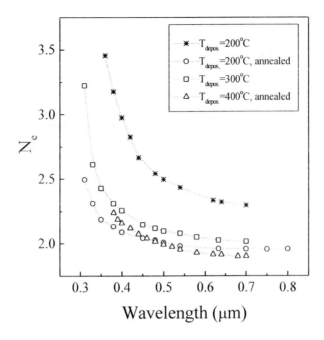

a

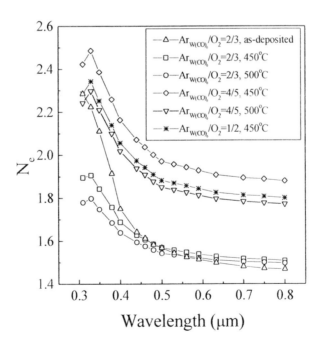

b

Figure 6 (continued)

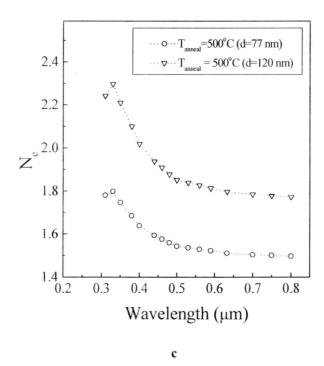

c

Figure 6. Effective refractive index vs. light wavelength for as-deposited and annealed tungsten oxide films. The technological conditions are given in the legend.

Additional annealing in oxygen-containing atmosphere leads to a decrease of the N_e, as well, because of the inclusions of oxygen molecules in the WO_3 film structure. In [34] the same experimental result was reported, i.e. decrease of N_e from 2.10 to 2.00 with increasing of the T_{anneal} from 500 to 600°C for WO_3 coatings obtained by oxidation of W and WO_2 layers. The other tendency observed was the occurrence of higher values of N_e for lower oxygen flow-rates, see figure 6b. This was expected since at higher oxygen flow rates, the films are fully oxidized and therefore stoichiometric. The same figure demonstrates also the tendency of decreasing of N as a result of thermal annealing in oxygen containing atmosphere. Even temperature differences of 50 K result in different N values. In figure 6c the influence of the film thickness on the refractive index for two WO_3 films, deposited under identical technological conditions is shown. Over the whole spectral range, the thinner film (d_1 = 77 nm) has smaller refractive index values compared to the thicker one (d_2 =120 nm). This experimental result confirms the fact that thinner films usually contain a higher amount of pores, which results in a lower density. A systematic investigation of the optical constants of tungsten oxide films up to 10 nm in thickness, formed by exposing tungsten thin layers to air at 250-350° C, was done by Thomas et al. [35]. In the case of real optical coatings the film deviates from perfect dielectric continuity due to a rough surface layer, which requires calculation of the dielectric function by applying multilayered analysis [36]. The real film was presented as consisting of homogeneous sublayers parallel to each other. The phase composition and the optical constants of our films were calculated on the basis of the effective medium theory of Bruggemann [37] using a multilayered model. This theory considers the film structure as a heterogeneous dielectric mixture of crystalline and amorphous phases, and voids. By positron lifetime annihilation spectroscopy in [15] it was proved that the fractions of all these three components vary with the preparation conditions. In the calculations using the effective medium approximation (EMA), fitting was made by minimizing the experimental and theoretical differences using the least-squares method. A two-layer model (surface rough layer and bulk layer) for the CVD-WO$_3$ films was accepted. The modeling results of coatings deposited on Si(111) substrates, as well as for those annealed for 1 h at 500°C are presented in table 4.

Table 4. Phase composition of CVD-WO$_3$ films as-deposited at Ar$_{W(CO)_6}$/O$_2$ = 1/6 and annealed (d$_1$ thickness of as-deposited film, d$^0_1$ film thickness after annealing)

T$_{depos}$ (°C)	film composition: as-deposited		d$_1$ (nm)	film composition: after annealing	d$^0_1$ (nm)
200	bulk layer	a-WO$_3$ +voids (96.8% +3.2%)	156.9	a-WO$_3$ + c-WO$_3$ +voids (0.16% +99.5% +0.34%)	212.0
	surface layer	a-WO$_3$ +voids (56.3% +43.7%)	36.4	a-WO$_3$ +c-WO$_3$ +voids (34.4%+12.0%+53.6%)	
300	bulk layer	a-WO$_3$ + c-WO$_3$ (82.4% + 17.6%)	60.0 17.5		
	surface layer	a-WO$_3$ + c-WO$_3$ +voids (18.5%+66.4%+15.1%)			
400	bulk layer	a-WO$_3$ + c-WO$_3$ +voids (0.4%+ 93.4%+6.2%)	221.5 22.3	100% c-WO$_3$	223.5 24.0
	surface layer	a-WO$_3$ +c-WO$_3$ +voids (6.0%+31.3%+62.7%)		c-WO$_3$ +voids (44.8% +55.2%)	

The film grown at 200°C is obviously amorphous with voids (inclusions of argon or oxygen). Although crystallization was observed for the film obtained at T$_{depos}$ = 300°C, the amorphous phase still dominates. Deposition at 400°C results in thoroughly crystallized films. Additional annealing leads to a small increase of the film thickness due to additional oxidation resulting in the incorporation of oxygen molecules in the crystal lattice of tungsten oxide. The ellipsometrically calculated phase composition of the CVD-WO$_3$ films is in good agreement with the RHEED study [38] and the Raman scattering results (above). The calculated values of the void fractions of the films were confirmed by density calculations, based on weight measurements. A considerable amount of voids was found to exist in the surface layer, which is only slightly influenced by the additional thermal treatment. The experimental results were successfully discussed in terms of EMA and the two-layer model. The good agreement achieved between the theoretical and the experimental data in a particular case: for samples deposited at T$_{depos}$=400°C is shown in figure 7.

In table 5 the modelling results for CVD-WO$_3$ deposited at 300°C and different ratios of the reactive gases are presented. The as-deposited films have predominant amorphous phase, although the crystalline phase is present too. A considerable amount of voids is also observed. This fact was again confirmed by independent weight measurements. The volume fractions of the amorphous and crystalline phases and the void fraction vary with the gas ratios. However, the influence of the deposition and annealing temperatures on the fractions of the amorphous and crystalline phase is much stronger. Annealing in oxygen slightly influences the void concentration and leads to an increase of the film thickness by about 8%. The surface roughness is influenced by the oxygen flow-rate, namely higher flow-rate results in rougher surface.

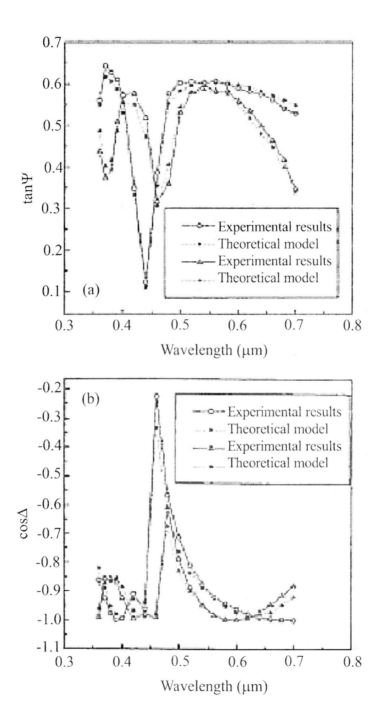

Figure 7. Comparison of the theoretical model with the experimental data for tan ψ and cos Δ of tungsten oxide films deposited at 400°C.

Table 5. Phase composition of as-deposited CVD-WO_3 films, obtained at a temperature of 300°C under different reaction gases and annealed at 450 or 500°C for 1 hour in oxygen-containing ambient

$Ar_{W(CO)_6}/O_2$ ratio	Annealing temperature	Film composition	Thickness d_1 (nm)
0.46	450°C	Bulk layer c-WO_3 + voids 58.6 % + 41.3 %	94.6
		surface layer a-WO_3 + c-WO_3 + voids 0.3% + 33 % + 66.7 %	26.4
0.62	450°C	bulk layer a-WO_3 + c-WO_3 + voids 5.2% + 35.3 % + 59.5 %	69.4
		surface layer a-WO_3 + c-WO_3 + voids 27.5% + 14 % + 58.5 %	6.8
0.62	500°C	bulk layer c-WO_3 + voids 39.4 % + 60.6 %	70.0
		surface layer a-WO_3 + c-WO_3 + voids 17.2% + 25.4% + 57.4 %	7.7
0.83	450°C	Bulk layer a-WO_3 + c-WO_3 + voids 0.2% + 63.5 % + 36.3 %	97.4
		surface layer a-WO_3 + c-WO_3 + voids 24.8% + 20 % + 55.2 %	2.5
0.83	500°C	bulk layer a-WO_3 + c-WO_3 + voids 0.3% + 56.7 % + 43 %	120.0
		surface layer a-WO_3 + c-WO_3 + voids 4% + 40 % + 56 %	2.0

From the modelling of the CVD-WO_3 films we calculated the values of refractive index (N) of the bulk layer. The dispersion of N in the VIS range is shown in figure 8. The refractive index N decreases with increasing of the deposition temperature.

The extinction coefficient (k) of all CVD-WO_3 films is lower than 0.1 up to 2 eV and rises towards the absorption edge. The small values of k indicate that the tungsten oxide films are transparent in a wide range of the visible spectrum.

Approaching the absorption edge, the extinction coefficient k, and correspondingly α, $\alpha = 4\pi k/\lambda$, increases due to transitions from the valence to conduction band. For all samples, below this spectral region the logarithm of the absorption coefficient α vs. photon energy E shows a linear dependence. This region exhibits the characteristic Urbach form. Assuming that the density of states for both valence and conduction bands is parabolic [39], then

$$\alpha = C[(E-E_g)^2/E], \qquad (7)$$

where C is a characteristic of the material. Plotting the dependence $(\alpha E)^{1/2}$ vs. E, the optical energy gap ($E_g^{(0)}$) can be determined as extrapolated value of $(\alpha E)^{1/2}$ vs. E to zero absorption. The plots of $(\alpha E)^{1/2}$ vs. E of samples deposited at 200, 300 and 400°C and additionally annealed at 450 and 500°C are shown in figure 9.

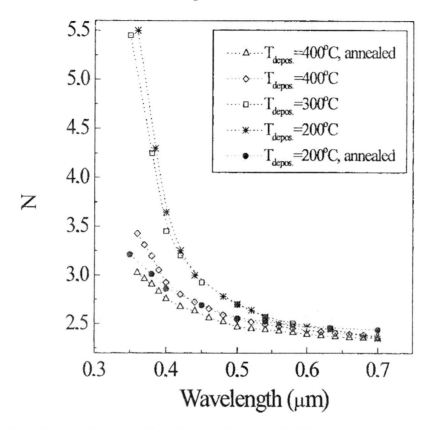

Figure 8. Refractive index dispersion of the bulk layer of tungsten oxide films.

The obtained room-temperature value of the optical band gap of amorphous WO_3 films is $E_g^{(0)} \approx 3.13$ eV. Analogous values are reported in literature for a-WO_3 films deposited by sputtering and evaporation [40]. In figure 10 the temperature dependence of $E_g^{(0)}$ is shown. One can see a decrease of $E_g^{(0)}$ with the crystallization of the material. A similar effect is reported by some other authors [33] for evaporated films. Possible explanations of this phenomenon, not typical of most semiconductors, are summarized by Granqvist [40] and they are related to structural transformations in WO_3 with increasing of the temperature or quantum confinement in the clusters.

The surface layer of tungsten oxide thin films, which is important in terms of device applications, was studied by means of atomic force microscopy (AFM), scanning electron microscopy (SEM) and spectroscopic ellipsometry in order to provide a reliable basis for comparison. Data from SE experiments and theoretical simulations showed that a layer forms at the surface of the WO_3 film that has a different structure and composition from the bulk film (see above). This surface layer becomes thicker with increasing oxygen flow rate during deposition.

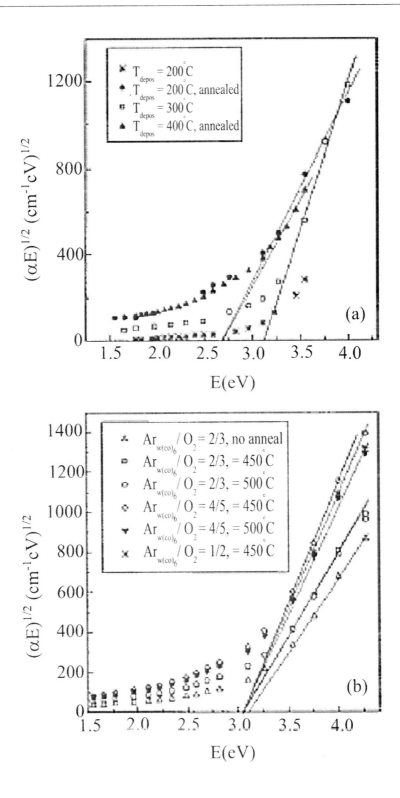

Figure 9. Plots of $(\alpha E)^{1/2}$ vs. photon energy E for as-deposited and annealed tungsten oxide films.

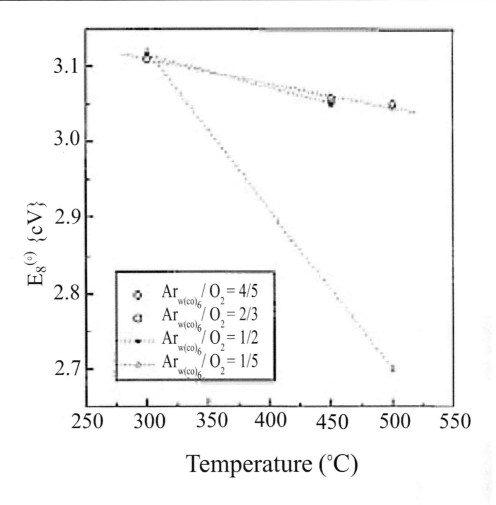

Figure 10. Temperature dependence of the optical energy gap of CVD-WO$_3$ films.

An AFM image of the film surface of a CVD-WO$_3$ film after annealing in air is presented in figure 11(a). Irregularly distributed domed crystallites form surface termination with a somewhat lumpy appearance [41]. Subjecting the samples to 180 min prolonged annealing at 470°C in 10^{-3} Torr of O$_2$ caused only minor qualitative modification of the surface (figure 11(b)). The general morphology of the surface was unchanged. This contrasts with the effect of annealing films prepared from a tungstic acid sol complex, where large straight-edged crystallites were created at the surface after annealing under the same conditions (Ref. 41 and references therein). On a smaller length-scale, however, linescans [42] and calculated rms surface roughness values show that the surface morphology has been altered by the annealing treatment: crystallisation in the surface layer is visible as a more jagged appearance in linescans and the AFM image.

The surface layer was predominantly amorphous for as-deposited films and predominantly crystalline after annealing. Root mean squared (rms) roughness values were calculated from AFM images of the surface layer. A high degree of surface roughness was revealed after deposition and annealing (~40 nm), and the roughness value increased after additional annealing to 470°C in 1x10^{-3} Torr of O$_2$.

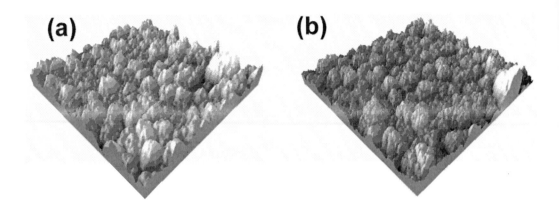

Figure 11. Pseudo-3D rendered AFM scans of the surface of a WO_3 film deposited on glass at 200°C and annealed. The images correspond to a 5 μm × 5 μm area. (a) Image of the film annealed at 400°C for 60 min. (b) AFM image of the surface following further annealing to 470°C in 1×10^{-3} Torr of O_2 for 180 min.

Figure 12. SEM micrographs of as-deposited WO_3 film on glass at 200°C and at a gas flow rate ratio of $Ar_{W(CO)_6}/O_2 = 1/12$.

Figure 12 shows a SEM micrographs of as-deposited WO_3 film on glass. Deposition was carried out at 200°C and at a gas flow rate ratio of $Ar_{W(CO)_6}/O_2 = 1/12$. The image reveal a typical of amorphous material featureless surface. In figure 13a-b. SEM micrographs of WO_3 films, (a) as-deposited on glass at 400°C and at a gas flow rate ratio of $Ar_{W(CO)_6}/O_2 = 1/12$, and (b) after additional annealing at 500°C in oxygen-containing environment for 60 min, are illustrated. The surface morphology demonstrated in figures 13 is typical of polycrystalline samples.

Figure 13. SEM micrographs of WO_3 film, (a) as-deposited on glass at 400°C and at a gas flow rate ratio of $Ar_{W(CO)_6}/O_2 = 1/12$, and (b) following subsequent annealing at 500°C in oxygen-containing environment for 60 min.

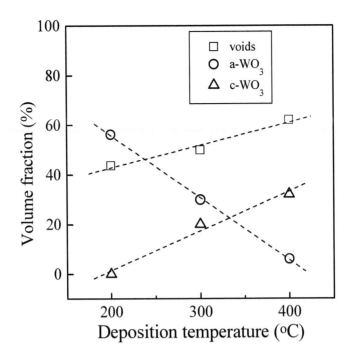

Figure 14. Volume fraction of the amorphous and crystalline phases of WO$_3$ and voids in the surface layer of WO$_3$ films as a function of deposition temperature. The films were deposited on Si substrates at Ar$_{W(CO)_6}$/O$_2$ = 1/6.

Our previous SEM investigation of the surface of CVD-WO$_3$ films has shown that the oxygen flow rate has a strong influence on the surface morphology [38]. Lower oxygen flow rates lead to smooth, homogeneous, small-grained film surfaces, while increased oxygen flow rates lead to the appearance of large crystallites on the smooth film surface.

Increasing the deposition temperature also increases the "grainy", crystalline appearance of the film, and for a deposition temperature of 400°C well-pronounced, randomly distributed crystallites develop on the film surface (figure 13(a)). Annealing at 500°C further enhances the crystalline structure of the as-deposited films. As an illustration, figure 13(b) shows a SEM micrograph of the WO$_3$ film after annealing at 500°C in air for 60 min. In addition to the larger average size of the crystallites, other changes in film morphology are visible, such as the appearance of elongated crystallites.

In figure 14 we present the results of simulated surface layer compositions for films deposited at different temperatures and at a gas flow rate ratio of Ar$_{W(CO)_6}$/O$_2$ = 1/6. With increasing of the temperature, the volume fraction of crystalline phase increases, indicating that higher deposition temperature promotes the crystallisation even during the film growth.

Digital processing of the AFM images was used to provide a quantitative guide to the roughness of the surface. Absolute values should be treated with caution due to the inherent effects of tip convolution with the surface topography, and of the dependence on the scan size. Despite this, it was possible to identify a clear trend in rms roughness by comparing images taken at the same magnification. Typical AFM image is shown in figure 15a, for films deposited at 200°C and at Ar$_{W(CO)_6}$/O$_2$=0.083. In this case, the rough surface texture was confirmed by the root mean square (rms) value of roughness of ~ 40 nm. Prolonged annealing at 470°C for 180 min in 10^{-3} Torr of O_2 did not change the general morphology, but increased the rms roughness to 50 nm. The change in the film surface by annealing is illustrated in figure 15b, where the two-dimensional (2D) linescans from the image in figure 15a are given. Obviously, annealing promotes film crystallization leading to increase of surface roughening. The AFM images show that the film surfaces are characterized by irregularly distributed domed crystallites forming a forest of cauliflower-like florets. Annealing leads to an increase of the surface roughness by 20–30%.

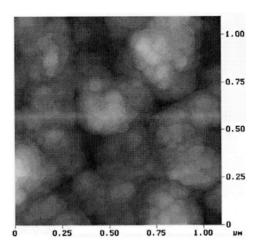

Figure 15a. AFM image of the surface of a WO_3 film deposited on glass at 200°C and annealed at 400°C in oxygen-containing environment.

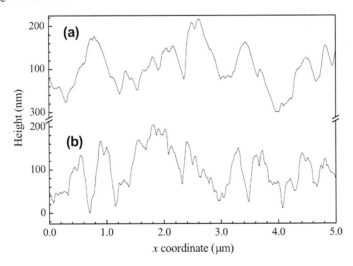

Figure 15b. 2D-linescans from the image in figure 15a before (a) and after (b) annealing at 470°C in O2 at 10–3 Torr for 180 min.

After annealing, the crystalline WO_3 and void fractions are dominant in the surface region. A typical dependence is demonstrated in figure 16, which shows results for films deposited at 300°C and annealed at 500°C in air. Each phase varies in an approximately linear fashion with gas flow rate ratio, with the proportion of crystalline material showing the strongest dependence on the $Ar_{W(CO)_6}/O_2$ ratio. At low O_2 partial pressures the void fraction in the surface layer was about 43%, with a high proportion of crystalline WO$_3$ and a negligible amount of amorphous material. However, the crystalline fraction decreased to 18% for higher O$_2$ pressures, with a concomitant increase in both the amount of amorphous material and void fraction. The large void fraction at high O$_2$ flow rate is confirmed by the high values of roughness calculated from the AFM images. The observed increase of the void fraction in the surface layer after annealing the films is an indirect indication that crystallisation leads to surface roughening, as observed in the AFM images.

The SE results and modeling revealed how the thickness of the surface layer of the WO$_3$ films changed as a function of gas flow rate ratio and deposition temperature. Figure 17 shows that the higher deposition temperature leads to a thinner surface layer in the WO$_3$ film. For both temperatures, however, the thickness dependence on oxygen flow rate is almost linear. The surface layer thickness decreased with increasing flow rate ratio of the reactive gases, *i.e.*, with decreasing oxygen content. There is good agreement with the surface roughness values obtained from AFM measurements, which is highlighted in figure 17 with a "star" data-point. In general, the annealing time (60 min in this case) had a weak influence on the surface layer thickness, but changes the proportion of the amorphous and crystalline phases of WO$_3$.

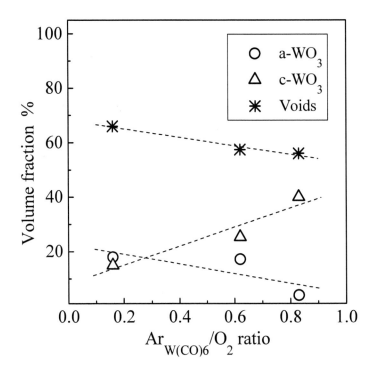

Figure 16. Volume fraction of the amorphous and crystalline phases of WO$_3$ and voids vs. ratio of gas flow rates Ar$_{W(CO)6}$/O$_2$ for the surface layers of films deposited at 300°C and annealed at 500°C.

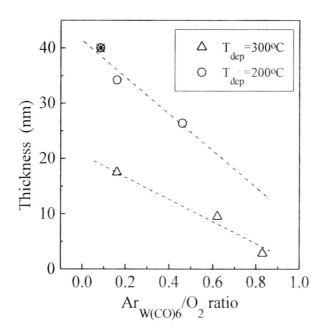

Figure 17. Surface layer thickness determined by SE vs. gas flow rates ratio for films deposited at 200 and 300°C, and annealed at 400 and 500°C. The starred point illustrates the AFM rms roughness value.

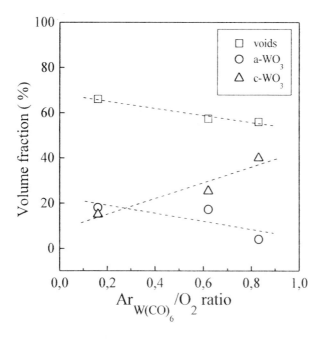

Figure 18. Volume fraction of the amorphous and crystalline phases of WO_3 and voids versus ratio of gas flow rates $Ar_{W(CO)_6}/O_2$ for the surface layer of WO_3 films deposited on glass at 300°C and annealed at 500°C for 60 min.

In figure 18, the volume fraction of the amorphous and crystalline phases of WO_3 and voids versus ratio of gas flow rates $Ar_{W(CO)_6}/O_2$ for the surface layer of WO_3 films deposited on glass at 300°C and annealed at 500°C for 60 min is illustrated.

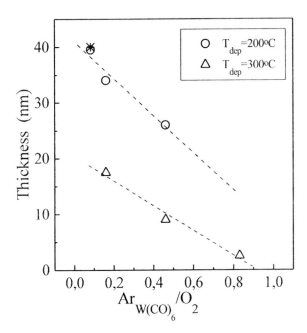

Figure 19. Thickness of the surface layer determined by SE as a function of oxygen content in the deposition ambient ($Ar_{W(CO)_6}/O_2$) for WO_3 films deposited at 200°C and 300°C and annealed at 400°C and above (450 or 500°C). The point with star illustrates the AFM rms roughness value of the corresponding film.

Figure 19 shows the thickness of the surface layer determined by SE as a function of oxygen content in the deposition ambient defined as $Ar_{W(CO)_6}/O_2$ for WO_3 films deposited at 200°C and 300°C and annealed. The point with star illustrates the AFM rms roughness value of the corresponding film. Obviously, the different characterization techniques demonstrate rather good coincidence.

In order to obtain dispersion parameters we used a single-oscillator description of the frequency-dependent dielectric constant [44]. For this purpose the single-oscillator energy E_0 and the dispersion energy E_d were obtained from the experimental data by plotting $1/(N^2-1)$ versus $(h\upsilon)^2$. The plots are given in figure 20. In the region of transparency, i. e. for $h\upsilon=E < E_g^{(o)}$, where $E_g^{(o)}$ is the optical bandgap, the dependences are linear. In the proximity of $E_g^{(o)}$ and above this energy, however, a deviation from linearity is observed. In this region the absorption sharply rises due to the electronic transitions occurring from the valence band to the conduction band, and gives a contribution to the frequency-dependence of the dielectric constant $\varepsilon_1(\omega)$ [44].

The linearity of the plots $1/(N^2-1)$ vs. E^2 in a considerably large energy region allows us to determine experimentally the dispersion parameters E_0 and E_d, as the resulting straight line yields values of E_0 and E_d. The results are given in figure 20.

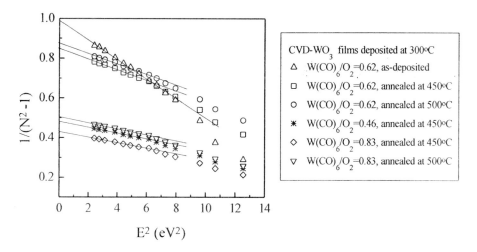

Figure 20. Plots of refractive index factor $1/(N^2-1)$ versus E^2. The treatment conditions are given in the legend.

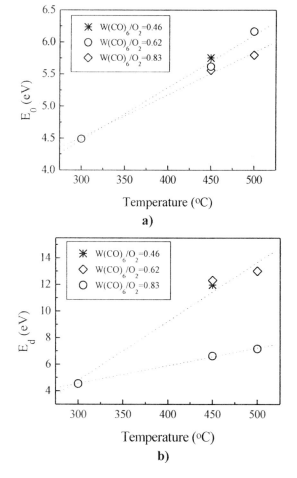

Figure 21. Oscillator energy E_0 (a) and dispersion energy E_d (b) as a function of treatment temperature for as-deposited and annealed CVD-WO$_3$ films at different gas ratios, described in the legend. The point of 300°C corresponds to the deposition temperature and the other ones represent the annealing temperatures [43].

As it is seen, both energies become higher with increasing of the treatment temperature (figure 21) and, therefore, with the change of the film structure from amorphous to crystalline one. A slight increase of E_0 values is observed for larger oxygen flow corresponding to smaller reactive gas ratios. However, the influence of oxygen flow on E_d values cannot be clearly seen due to their strong dependence on the film structure and on the amount of voids in the material.

It has been established [45] that the values E_0 and E_d of crystalline WO_3 are about 5 and 21 eV, respectively. For all noncrystalline films E_d has values smaller than that corresponding of the single crystal. This effect is due to their density deficit relative to the crystalline material. In contrast to this, E_0 seems to have values higher or lower than that of crystalline material [45]. The behavior of E_0 strongly depends on preparation methods and a large scatter in E_0 data is observed in the literature [45]. In our case, the considerably low values of E_d are a result of high porosity of the CVD-WO_3 films.

The oscillator energy E_0 can be related empirically to the lowest direct bandgap E_t by $E_0 \approx 1.5\ E_t$. E_t is the absorption threshold energy and is approximately equal to the bandgap energy E_g. The inequality of energy gaps can be written as $E_t < E_g^{(o)} < E_0$. For the as-deposited CVD-WO_3 films, the structure of which is amorphous, E_g is ~3 eV. It has been discussed above that the bandgap energy of tungsten oxide decreases with the degree of crystallization in contrary to the observed for other semiconductor materials. This tendency is also observed here, since the linear region of the plot $1/(N^2-1)$ vs. E^2 becomes narrower with increasing the treatment temperature. For crystallized films, however, the relation $E_0 \approx 1.5\ E_t$ cannot be used. Therefore, the $E_g^{(o)}$ values were calculated from the absorption coefficient, evaluated from the complex refractive indices. Considerable dependence of the optical bandgap energy on the deposition parameters was not observed. The value of $E_g^{(o)}$ of the as-deposited films was 3.11 eV. Annealing at 450 and 500°C resulted in a narrowing of $E_g^{(o)}$, which values were 3.06 eV and 3.04 eV, respectively. The observed low values of the refractive indices and the dispersion energies are a result of the high porosity of the material. The latter is an advantage for electrochromic device application.

In this paragraph compositional, structural and optical property changes that accompany changes in the substrate temperature were identified and studied.

2.A.3. Electrochromic Behavior of CVD Grown Tungsten Oxide films

The following paragraph deals with the preparation and performance characterization of electrochromic windows based on APCVD carbonyl tungsten oxide thin films.

The electrochromic (EC) material (usually tungsten oxide) is the most important layer in an electrochromic device and therefore its properties are the subject of a number of investigations ([1, 2, 46-52] and references therein). Thin WO_3 films deposited by evaporation, sputtering and electrochemical processes are very intensively studied nowdays [46-51]. Physical properties of CVD polycrystalline WO_3 layers obtained by oxidizing of carbonyl metal or black tungsten films were extensively studied [52], but the EC properties of amorphous CVD-WO_3 was done in Ref. 11 only.

Titanium oxide (TiO_2) coatings are attractive materials for use as photocatalysts, photoelectrodes and solar cells. Structural modification, quantum size effects and large surface to volume make the coatings interesting. The choice of a good counter electrode is important and often problematic. TiO_2 films have potential advantages for that purpose, because of their microstructural features which are favorable for the ion transport [53] and because of their high mechanical and chemical resistance [54]. However, their electrochromic characteristics are less known. For a passive electrode in an electrochromic device with a CVD-WO_3 basic layer, we used sol-gel produced TiO_2 films, because the sol-gel processing requires less capital to deposit coatings over large areas than the conventional vacuum methods, and hence offers a possible solution for lowering the cost of the EC devices. The sol-gel processing also offers advantages in controlling the microstucture or depositing films containing multiple cations. These parameters can influence the kinetics, durability, coloring efficiency and charge storage in the electrochromic electrodes [55-57].

The solution for TiO_2 films deposition was prepared by introducing of 15 mmol of acetylacetone peptized to a gel, which was previously formed from 30 mmol in situ made titanium-diethoxy-diacetate complex, 600 mmol of water in ethanol up to 100 ml total volume. A yellowish colloidal solution was formed for 24 h at room temperature [58,59]. Nanometric anatase titania films were deposited on conductive glass substrates by dipping, drying and heating at 560°C for 30 min.

The electrochromic properties of the EC active single layer — WO_3 or TiO_2 were examined by a standard three electrode system in 1 M $LiClO_3$ - PC electrolyte solution. The modulation of transmission and reflection of the window-structures in the VIS-NIR range were measured by Perkin Elmer spectrophotometers.

Authors in Ref. 11 discussed that carbonyl APCVD films, prepared at 300°C required 30 min for full bleaching from colored state. This interval is very long, because a microcrystalline phase exists in the layers. Therefore, we studied amorphous CVD-WO_3 films prepared at 200°C, despite of the low growth rate at this temperature, and long growth time, respectively.

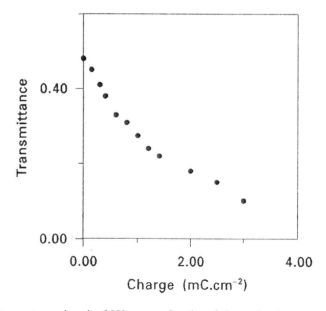

Figure 22. Transmittance at wavelength of 550 nm as a function of charge density.

In figure 22 the transmittance at a wavelength λ = 550 nm as a function of the injected charge is shown. The transmittance decreases monotonically as the injected charge density increases. If we assume that all injected charges participate in WO_3 transformation to $Li WO_3$, then we can write:

$$CE = \Delta OD/\Delta Q, \qquad (8)$$

where OD is the optical density, ΔQ - charge density and here CE is the coloration efficiency.

Figure 23 represents the change in the optical density at λ=550 nm (the middle of the visible spectrum) as a function of the injected charge density. The value of the coloration efficiency can be determined from the slope at the linear part of the curve in figure 23

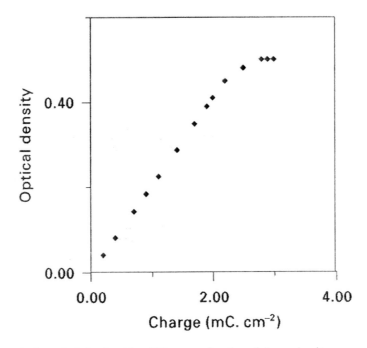

Figure 23. Change in the optical density at λ = 550 nm as a function of charge density.

The coloration efficiency of a 170-nm-thick amorphous APCVD-WO_3 film, selected as the best one, was estimated to be 205 $cm^{-1}C^{-1}$, which approximates the value of 230 $cm^{-1}C^{-1}$, reported in Ref. [11] and it is larger than not only the one reported for CVD polycrystalline WO_3 films (41 $cm^{-1}.C^{-1}$), but also for amorphous ones prepared by the PVD methods [2].

The WO_3 films have an intensive broad-band absorption throughout the visible spectral region which yields a deep blue coloration and a high coloration efficiency. Several mechanisms have been suggested to explain the coloration of the electrochromic films: formation of color centers at the oxygen vacancies, electronic transitions between W^{6+} and W^{5+} ions, small polaron absorption, intraband transitions, etc. [1, 2, 60].

Sol-gel prepared TiO_2 films have an anatase structure, so, their building blocks form a number of ideal vacant tunnels in which small ions can be easily intercalated/deintercalated. The optical modulation (ΔT) of TiO_2 sol-gel film with a thickness of about 100 nm, measured at 632.8 nm as a function of lithium charge capacity is shown in table 6. The average cycling durability of TiO_2 films is above 10^4., i.e. they demonstrate a long-term stability.

Table 6. The optical modulation as a function of lithium charge capacity for sol-gel TiO_2 film

Q (mC.Cm^{-1})	$\Delta T = T_{bleach} - T_{color}$
2.0	2.8
3.2	5.6
6.0	7.2
9.0	9.8

Solid-state Electrochromic devices containing solid polymer electrolytes can be conveniently constructed since the primary and secondary electrochromes are prepared separately on their conducting substrates (ITO). Then the two parts of the device are joined at room temperature using the polymeric electrolyte as an adhesive layer between them.

For a counter electrode in an electrochromic device with a CVD-WO_3 basic layer, we made a nanoporous sol-gel TiO_2 film. The last successively plays the role of a passive electrode because its transparency does not change substantially with the charge intercalation/deintercalation.

Figure 24 views an entirely solid state electrochromic window (device) - glass/ITO/WO_3/PVA+$LiClO_3$/TiO_2/ITO/glass - in bleached and colored state. The sign of the applied voltage is referred to the WO_3 electrode. The optimal modulation achieved is about 60% at 800 nm and the time response from transparent to blue state is 15 s. The above described device, when working in the absence of a counter electrode (TiO_2) has a 30 s. switching time (transparent→ blue state) and 12 min for transition in bleached state. Obviously, the APCVD-WO_3 thin films are capable to satisfy the requirements of "smart window" and may find possible applications in architecture and automobile industry for controlling the incident solar energy and thus allowing saving of energy for cooling or heating.

For CVD polycrystalline WO_3 films the deeply coloured state reflectivity of 56% has been achieved at 2.5 μm (not shown here) and it approximates the value, reported by Goldner in [61]. It is greater than the value of 35%, reported for thermally evaporated and crystallized by annealing WO_3 films [40, 52]. Polycrystalline WO_3 films have shown a typical semiconductor behavior in uncolored state, but a free-electron (Drude) behavior in colored state [40].

Durability tests demonstrated that the the carbonyl APCVD-WO_3 samples can stand more than 6 000 EC cycles without considerable changes in their performance. More details on other electrochromic parameters, such as memory effect, etc. of these films can be found in Ref. 62. Different types of polymer electrolytes [46] and anodic type PVA-pollyaniline composite were successfully implemented in combination with APCVD-WO_3 coatings [63]. Refs. 64-66 present reviews on a number of recently discussed EC devices for modulation of optical transmittance. Environmental assessment of electrochromic glazing production was done by Yianoulis et al. [67]

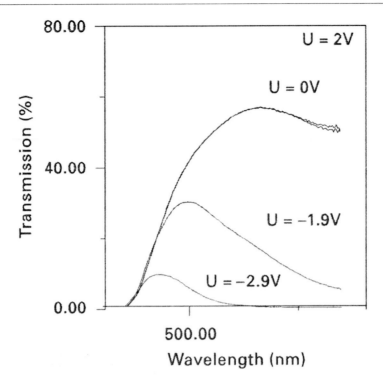

Figure 24. Modulation of spectral transmission of electrochromic device: glass/ ITO/WO$_3$/PVA+LiClO$_4$ /TiO$_2$/ITO/glass — in dependence of the voltage applied.

Photochromic behaviour of WO$_3$ films, deposited by Aerosol-assisted CVD using WCl$_6$ and phenol as precursors was shown recently in Ref. 68. Alternative wet chemical methods for deposition of WO$_3$ films possessing very good EC properties were also approached last years [69,70].

In conclusion, results on technology development of a low-temperature carbonyl process for direct deposition of amorphous and polycrystalline WO$_3$ films were presented and discussed. The relationship technological parameters-structure-optical properties of these films was revealed. Thus, an existing gap of knowledge was fulfilled. The capability of the APCVD technique to deposit WO$_3$ films with properties highly advantageous for "smart windows" application was demonstrated.

REFERENCES

[1] A.I. Gavrilyuk, N.A. Sekushin, Electrochromism and Photochromism in Tungsten and Molybdenum Oxides, *Nauka,* Leningrad, 1990 (in Russian);

[2] C.G. Granqvist, Handbook of Inorganic Electrochromic Materials, Elsevier, Amsterdam 1995;

[3] D.Davazoglou, A.Donadieu, R.Fourcade, A.Hugot-Le Goff, P.Delicher and A.Peres, *Rev. Phys. Appl.* 23 (1988)265;

[4] A. Donadieu, in Large-Area Chromogenics: Materials and Devices for Transmittance Control, Eds. C.Lampert and C.G.Granqvist, (SPIE Opt.Engr. Press, Bellingham, 1990), vol.IS4, pp.191;

[5] D.Davazoglou, and A.Donnadieu *Thin Solid Films.* 147 (1987) 131;
[6] G. Peev and I. Tsibranska, *Bulg. Chem. Commun.* 36 (2) (2004) 89;
[7] V.G.Surkin and A.A.Uelskij, *Journal of Physical Chemistry,* XLVI (1972) 2229;
[8] J.-P. Randin, *J. Electron. Mater.* 7 (1978) 47;
[9] K.A.Gesheva and D.Gogova, *J.de Phys.* IV, Coll. C3, *suppl. Au J.de Phys.* II, v.3 (1993) 475;
[10] S.S. Olevskii, M.S. Sergeev, A.L. Tolstikhina, A.S. Avilov, S.M. Shkornyakov, and S.A. Semiletov, *Dokl. Akad. Nauk SSSR.* 257 (1984) 1415 [*Soviet Phys. Doklady.* 29, 260 (1984)];
[11] T. Maruyama and S.Arai, *J. Electrochem. Soc.* 141 (1994) 1021;
[12] M.Rothschild and A.Forte, *Appl. Phys. Lett.* 59 (1991) 1790;
[13] J.J. Lander and L.H.Germer, *Trans. Met. Soc. AIME.* 175 (1948) 648.;
[14] D.Gogova, A.Iosifova, T.Ivanova, Zl.Dimitrova and K.Gesheva, *J. Crystal Growth.* 198-199 (1999) 1230;
[15] N.Djurelov, D. Gogova and M.Misheva, *Thin Solid Films.* 347 (1999) 302;
[16] C.G. Granqvist, Handbook of Inorganic Electrochromic Materials, Elsevier, Amsterdam 1995 (Ch. 7, p. 133);
[17] Y.Singesato, A.Murayama, T.Kamimori and K.Matsuhiro, *Rept. Res. Lab. Asahi Glass.* Co.Ltd. 38 (1988) 39;
[18] C.D.Wagner (Ed.), Handbook of X-ray Photoelectron Spectroscopy, Perkin Elmer Corp., *Physical Electronic Division,* 1994;
[19] B. Henley and G.J. Sacks, *J. Electrochem. Soc.* 144 (1997) 1045;
[20] O.Yu. Khyzun, *J. Alloys and Compounds.* 305 (1-2) (2000) 1;
[21] C. Kaito, T. Shimizu, Y. Nakata, and Y. Sait, *Jpn. J. Appl. Phys.* 24 (1985) 117;
[22] T.A. Taylor and H.H. Patterson, *Appl. Spectrosc.* 48 (1994) 674;
[23] JCPDS Powder Diffraction File, 1980, cart 24-747;
[24] Y. Shigesato, A. Murayama, T. Kamimori, and K. Matsuhiro, *Jpn. J. Appl. Phys.* 30 (1991) 814;
[25] C.G. Granqvist, M.H. Francombe, and J.L. Vossen (Eds.), *Physics of Thin Films: Mechanic and Dielectric Properties,* Vol. 17, Academic Press, Inc., San Diego 1993 (p. 307).
[26] G.M. Ramans, J.V. Garbusenoks, and A.A. Veispals, *phys. stat. sol.* (a) 74, K41 (1982;
[27] J. Garbusenoks, G.M. Ramans et al., *Solid State Ionics.* 14 (1984) 25;
[28] M.F. Daniel, B. Desbat, and J.C. Lassegues, *J. Solid State Chem.* 67 (1987) 235;
[29] C.G. Granqvist, see [1] (Ch. 5, p. 69 and Ch. 3, p. 38-41);
[30] P. Tagstrom, "Vapor Phase Deposition of WO3 and WC", PhD Thesis No398, *Faculty of Science and Technology,* Uppsala University, 1998;
[31] T.C. Arnoldussen, *J. Electrochem. Soc.* 128 (1981) 117;
[32] S.K. Deb, *Phil. Mag.* 27 (1973) 801;
[33] K. Miyake, H. Kaneko, M. Sano, and N. Suedomi, *J. Appl. Phys.* 55 (1984) 2747;
[34] J. Villachon-Renard, G. Leveque, A. Abdellaoi, and A. Donnadieu, *Thin Solid Films.* 203 (1991) 33;.
[35] S. Sacre and L.K. Thomas, *Thin Solid Films.* 203 (1991) 221;
[36] Ronnow, S.K. Anderson, and G.A. Niklasson, *Opt. Mater.* 4 (1995) 815;
[37] D. Bruggemann, *Ann. Phys.* (Leipzig) 24 (1935) 636;

[38] D. Gogova, PhD Thesis "Deposition and Investigation of the Structure and Optical Properties of CVD Thin Tungsten Oxide Films", *Bulg. Acad. Sci.*, Sofia, 1999;
[39] H. Demiryont and K.E. Nietering, *Appl. Opt.* 28 (1989) 1494;
[40] C.G. Granqvist, see [1] (Ch. 8).
[41] R. Tanner, A. Szekeres, D. Gogova, and K. Gesheva, *Applied Surface Science.* 218 (1-4) (2003) 163;
[42] R. Tanner, A. Szekeres, D. Gogova, and K. Gesheva, *Journal of Materials Science: Materials in Electronics.* 14 (2003) 769;
[43] A.Szekeres and D. Gogova, Proc.10th Int. School Condensed Matter Physics "Thin Film Materials and Devices - Developments in Science and Technology", *World Sci. Co.*, Singapore, eds. J. Marshal, N. Kirov and A. Vavrek, 1999, p. 257;
[44] S. H. Wemple and M. DiDomenico, *Phys. Rev.* B 3 (1971) 1338;
[45] D. Davazoglou and A. Donnadieu, *J. Appl. Phys.* 72 (1992) 1502;
[46] P.M.S. Monk, R.J. Mortimer, D.R. Rosseinsky, *Electrochromism: Fundamentals and Applications, VCH,* Weinheim, 1995;
[47] J.-G Zhang, D.K. Benson, C.E. Tracy, S.K. Deb, *J. Mater. Res.* 8 (10) (1993) 2657;
[48] A. Georg, W. Graf, R. Neumann, V. Wittwer, *Solid State Ionics.* 127 (2000) 319
[49] Anna-Lena Larsson, Gunnar A. Niklasson, *Solar Energy Mater. Solar Cells.* 84 (2004) 351
[50] M. Stolze, D. Gogova, L.-K. Thomas, *Thin Solid Films.* 476 (2005) 185
[51] M. Gillet, K. Aguir, C. Lemire, E. Gillet, K. Schierbaum, *Thin Solid Films.* 467 (2004) 239
[52] D. Davazoglou, G. Leveque, A. Donnadieu, *Sol. Energy Mater.* 17 (1988) 379;
[53] Sopyan, S. Murasawa, K. Hashimoto, A. Fujishima, *Chem. Lett.* (1994) 723;
[54] K. Kato, A. Tsuzuki, Y. Torii, H. Taoda, Y. Butsugan, *J. Mater. Sci.* 30 (1995) 837;
[55] Y. Paz, Z. Luo, L. Rabenberg, A. Haller, *J. Mater. Res.* 10 (1995) 2842;
[56] A. Agrawal, H.R. Habibi, R.K. Agrawal, J.P. Cronin, C.M. Lampert, *Thin Solid Films.* 221 (1992) 239;
[57] Ye Yonghong, Z. Jiayu, Gu Peifu, L. Xu, T. Jinfa, *Thin Solid Films.* 298 (1997) 197;
[58] O. Harizanov, A. Iossifova, P. Stefchev, *World Renewable Energy Congress, Denver, CO,* 1996, p. 1320;
[59] O. Harizanov, in I. Mojzes, B. Kovacs, *Nanotechnology a Dedicated Tool for the Future,*
[60] Yu. S. Krasnov and G. Ya. Kolbasov, *Electrochimica Acta.* 49 (2004) 2425.
[61] R.B. Goldner et al., *Sol. Energy Mater.* 12 (1985) 403;
[62] D. Gogova, N. Gospodinova, P. Mokreva, K. Gesheva, and L. Terlemezyan. *Bulgarian Chem. Commun.* 33 (2) (2001): 189 ;
[63] D. Gogova, and K. Gesheva, EUROCVD 11, Paris, France, 1997, Eds.: M. D. Allendorf and C. Bernard (*The Electrochem. Soc., Pennington, NJ), Proc. vol.* 97-25, 1482;
[64] C. -G. Granqvist, *Nature Materials.* 5 (2) (2006) 89;
[65] C. -G. Granqvist , *J. Europ. Ceramic Soc.* 25 (12) (2005) 2907;
[66] C. -G. Granqvist, E. Avendaño, A. Azens, *Thin Solid Films.* 442 (1-2) (2003) 20;
[67] E. Syrrakou, S. Papaefthimiou and P. Yianoulis, *Solar Energy Mater. Solar Cells.* 85 (2) (2005) 205;
[68] R. G. Palgrave and I. P. Parkin, *J. Mater. Chem.* 14 (19): (2004) 2864;

[69] V. V. Abramova, A. S. Sinitskii, T. V. Laptinskaya, A. G. Veresov, E. A. Goodilin, and Yu. D. Tretyakov, *Doklady Chemistry.* 407 Part 1 (2006) 31;

[70] G.A. Tsirlina, K. Miecznikowski, P. J. Kulesza, M. I. Borzenko, A. N. Gavrilov, L. M. Plyasova, I. Yu. Molina, *Solid State Ionics.* 176 (2005) 1681;

Chapter 2B

CVD MOLYBDENUM OXIDE THIN FILMS AS ELECTROCHROMIC MATERIAL

T. M. Ivanova and K. A. Gesheva
Central Laboratory of Solar Energy and New Energy Sources,
Bulgarian Academy of Sciences, Sofia, Bulgaria

2.B.1. INTRODUCTION

Molybdenum oxide compounds focus the scientific interest due to their important technological applications [1]. Currently, there exist several approaches available for preparing MoO_3 films with either an amorphous or a polycrystalline structure, e.g., sputtering [2], chemical vapour deposition (CVD) [3, 4], electro-deposition [5] and flash and thermal evaporation [6-8]. To fabricate nanoscopic MoO_3, methods like template-directed reaction of molybdic acid and the subsequent leaching process [9], and templating against carbon nanotubes [10] have previously been attempted. Electrochromic and sensing thin films of MoO_3 have been produced by sol-gel method [11] and reactively ion beam deposition [12].

MoO_3 is widely used as initial and/or intercalation material for producing charge - density wave conductors [13] and for optical coatings [14]. Molybdenum oxide under a thin film form is investigated as a capacitor and as a gas sensor [15]. The operation of the most gas sensors is based on the reversible changes of the resistivity of specific materials, caused by the presence of a certain gas in the environment [16]. MoO_3 has been found as having appreciable sensing capabilities for gases such as CO and NH_3 [17]. Mo oxide films allow easy insertion of organic compounds in its crystal lattice and this leads to their applications in electronic devices and heterogeneous catalysts [18]. Molybdenum trioxide is a material with a vast potential into the developing positive alpha-numeric display devices and solid-state microbatteries, due to unique layered structure of its orthorhombic phase, a high chemical stability and an electrochemical activity [12, 19].

MoO_3 has been also an intriguing intercalated material for ambient temperature solid state Li batteries. Molybdenum oxides become particularly attractive technologically due to their high electronic mobility and high lithium-ion mobility as well as a low-energy barrier for Li insertion and extraction reactions [20, 21]. Amorphous MoO_3 films attract attention as promising candidates for use in memory devices [22].

MoO_3 films become interesting in a respect to their chromogenic properties as they possess electrochromic, thermochromic and photochromic effects [23, 24]. Most of the research considerations according to electrochromic materials has been paid to the study of tungsten trioxide (WO_3), which has been found to be an excellent candidate exhibiting an efficient reflectance and absorption modulation depending on the film structure and phase. The molybdenum oxide (MoO_3) thin films, on the other hand, have received a comparatively insignificant amount of research attention, despite their electrochromic performance paralleling that of WO_3 films. Molybdenum oxide (MoO_3) thin films, exhibiting a cathodic coloration under lithium ion/electron double injection, have fairly high mixed ion-electron conductivity.

The electrochromic effect in the case of molybdenum oxide is suggested to be explained as follows:

$$MoO_3 + xe^- + xM^+ \leftrightarrow M_xMoO_3 \tag{1}$$

where M^+ is H^+, Li^+, Na^+, etc., which are small ions. The injected electrons are trapped by some Mo^{6+}, forming Mo^{5+} sites. The coloration is attributed to the intervalence charge transfer transitions between Mo^{6+} and newly formed Mo^{5+}. The intercalated compound M_xMoO_3 is known as a molybdenum bronze [25].

It is proposed that the electrochromic response in MoO_3 might be superior to many other electrochromic materials because it shows a stronger and more uniform absorption of light in its colored state. Molybdenum bronzes exhibit an improved open circuit memory compared to the more popular tungsten bronzes [26]. Furthermore, since the wavelength maximum of the intervalence band of the molybdenum bronze is more akin to the sensitivity of the human eye than the tungsten ones, so such materials show greater apparent color efficiency. It must also be noted that the MoO_3 films in proper deposition represent more neutral color than the blue color of tungsten oxides.

In summary, the attractive features of MoO_3 films and their applications arise from:

- The interesting layered structure of MoO_3 inducing two-dimensional properties – orthorhombic phase, $\alpha - MoO_3$.
- These materials exhibit always higher electrochemical activity.
- MoO_3 under the thin film form can be easily produced by numerous deposition techniques varying structural and optical properties.
- MoO_3 is one of the oxide compounds with highest stability.
- MoO_3 as well as the other transition metal oxides can easily form substoichiometric mixtures, leading to interesting optical behavior and enhancing the electrochromic effect.

2.B.2. CRYSTAL STRUCTURES OF MoO₃

Transition metal oxides exist in many different crystallographic forms, with stoichiometries differing only slightly from each other and transition metal ions exhibiting various oxidation states. This makes their experimental and theoretical study quite difficult and interesting at the same time.

Molybdenum can easily produced substochiometric structures. The two basic crystal phases are a unique layered structure (α - MoO_3 with orthorhombic symmetry) [27] and metastable monoclinic β - MoO_3 [28]. The schematic presentation of the two modifications is given at figure 1. These phases differ on their vibrational and optical properties they exhibit different refractive indices and optical band gap energy values [29-31].

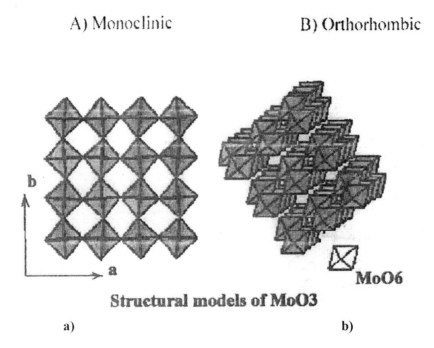

Figure 1. Schematic presentation of: a) monoclinic, β phase and b) orthorhombic, α - phase of molybdenum oxide.

Among its many polymorph phases, the thermodynamically stable phase of MoO_3 has an interesting layered structure, often marked as α – phase with an orthorhombic symmetry. The orthorhombic MoO_3 possesses a unique two-dimensional space layered structure, consisting of distorted MoO_6 octahedra with common edges and corners. Each octahedron possess one connected oxygen atom, two O atoms are common for two octahedra and three O atoms are with common corners and are shared for three octahedra, approaching the complete stoichiometry - $MoO_1(O_{1/2})_2(O_{1/3})_3$. The double layers are characterized with sublayers from periodically arranged MoO_6, where the oxygen atoms participate inside the adjacent octahedra into and among the sublayers. The internal interaction among atoms is dominated by ionic and covalent forces, but the sub layers pair has been interacted by the weak van der Vaals forces [27]. There are five different types of Mo-O bond lengths: 1.67, 1.73, 1.95

(twice), 2.25 and 2.33 Å. This structure gives an opportunity for injection of different donor ions into interlayered spaces and favors EC properties. From other hand, the metastable monoclinic β - phase possesses a cubic ReO_3 - like structure, consisting of three-dimensional rows from shared MoO_6 octahedra [32]. The β - phase has been found to be far more active for catalysts and would be expected to be superior for gas sensing applications [33]. Spontaneous transformation β→α phase is set at the temperatures of 370-400°C.

Molybdenum oxide has been known as a cathodic electrochromic material at least since 1974 [34]. From the point of view as electrochromic material, the layered structure of orthorhombic MoO_3 dictates highly anisotropic transport through the material and in some directions the inserted ions can move easily. Furthermore, there is also the monoclinic β – phase, which can be viewed as a metastable analogue of perovskite WO_3. As it was mentioned previously, tungsten oxide shows excellent electrochromic behavior. So, it can be expected that the both crystal modifications of molybdenum oxide will exhibit good electrochromic properties.

2.B.3. CVD Preparation of MoO_3 Thin Films

Generally, there exist two types of thin film depositions: Physical Vapor Deposition, which includes vacuum evaporation, sputtering, molecular beam epitaxy (MBE) and Chemical Vapor Deposition - atmospheric pressure chemical vapor deposition (APCVD), low pressure chemical vapor deposition (LPCVD), plasma assisted (enhanced) chemical vapor deposition (PACVD, PECVD), laser chemical vapor deposition (LCVD), metal-organic chemical vapor deposition (MOCVD), and chemical beam epitaxy (CBE), chemical vapor infiltration (CVI) and atomic layer CVD. Other thin film growth techniques are those, which are from liquid and solution phase: spray coating; spin coating; electroplating; liquid phase epitaxy.

Thin films of molybdenum oxide have been deposited using numerous methods, including spray pyrolisis, electrodeposition, evaporation, sputtering etc. Among these deposition techniques are the different modifications of CVD. Previous CVD research has focused primarily on producing molybdenum oxide films from chlorides, fluorides and carbonyls as vapor precursors. Lander and Germer [35] first deposited MoO_3 by low pressure CVD (LPCVD) by using $Mo(CO)_6$. Thin films of polycrystalline orthorhombic MoO_3 films were prepared and studied by double-step CVD at atmospheric pressure [36]. A. Abdellaoui et al. [36] obtained α – MoO_3 films with good optical and electrochromic properties by APCVD in two steps: first they deposited black Mo (MoO_2: Mo) or reflective Mo films. $Mo(CO)_6$ precursor is sublimated in a source vessel, which is heated in oil bath up to 75°C. The substrate temperature is kept 300°C. After obtaining the black Mo and reflective Mo films, the second step is the annealings at temperatures of 500 and 600°C in air enriched with oxygen. The deposited polycrystalline MoO_3 films have been characterized by studying their optical, structural and electrochromic properties. They crystallize in orthorhombic phase with a marked texture. The optical band gap varies in the range of 2.75 to 2.9 eV and the estimated color efficiency values are 45-51 cm^2/C^{-1}.

Cross et al. [37] proceeded a thermodynamical study of the low pressure chemical vapor deposition (LPCVD) of MoO_3, where the precursor used is Mo $(CO)_6$. The substrate temperature is varied in the range of 300-450°C. The films have been obtained in enriched O_2 ambient and the gas flow ratio $Mo(CO)_6/O_2$ has been changed from 1/25 - 1/65. Thermodynamic equilibrium calculations of the system showed that solid α-MoO_3 was the most stable phase over these ranges of the temperatures, the pressures and the gas compositions.

T. Maryama et al. [38] proposed CVD method at atmospheric pressure by using $Mo(CO)_6$. The substrate temperature has been changed from 150 to 400°C and polycrystalline MoO_3 have been obtained at 350°C. The deposition rate has been found to increase with temperature to 250°C, exhibited peak at 300°C and then a decrease with temperatures above 300°C. The determined color efficiency declared is 25.8 cm^2C^{-1} at λ=550 nm. Other CVD modifications were found as used for preparation of molybdenum oxide films with molybdenum carbonyl as precursor – laser ablated CVD [39, 40] and plasma-enhanced CVD [41, 42].

A relatively new technique Combustion CVD (CCVD) for MoO_3 films has been introduced recently [43, 44]. The precursors (liquids, often molybdates) have been dissolved in a flammable solvent and the solution delivered to a burner where it is ignited to give a flame. One advantage is that the deposition is performed at low substrate temperatures since the energy of decomposition of the precursor is provided by the flame. The grown films are amorphous with stoichiometry close to those of the molybdenum trioxides.

In our studies, the molybdenum oxide films have been obtained by atmospheric pressure chemical vapor deposition (APCVD) method, using the molybdenum carbonyl as a vapor precursor in an oxygen rich ambient. There were experimented different types of substrates and varying technological conditions [45-47].

The substrates used are ordinary glass, Si wafers and conductive glass substrates - Donnelly type - glass, covered with thin film of SnO_2:Sb with sheet resistance of 8 Ω/cm.

The structure and film properties are strongly influenced by the technological process parameters, a large number of experiments were done for obtaining the films with optimally good optical quality. The chosen experimental method is fixing one parameter and varying the others. The altered technological parameters of APCVD process are the substrate temperature ($T_{deposition}$), the sublimator temperature ($T_{sublimator}$) and the gas flow ratio ($Mo(CO)_6/Ar$).

It has been established that MoO_3 films are successfully obtained in the temperature range of 150-200°C. Experiments with increasing the deposition temperatures above 200°C (300 and 400°C) showed that there was no film deposits nevertheless there were enough carbonyl vapors in the reactor, because powder and deposits on the reactor walls were registered. This was a sign that this particular CVD process and the chosen construction of CVD equipment for MoO_3 film deposition requires lower substrate temperatures in comparison with the analogical process for WO_3 films, which were successfully deposited in the temperature range from 200 to 400°C [48-50].

Molybdenum hexacarbonyl is a white powder which at room temperature does not react with air, with very low temperatures of sublimation and its powder density is $Mo(CO)_6$ = 1.96 g/cm^3. V. Syijrkin [51] presents a formula for determination of vapor pressure at given temperature T, for the case of transition metal carbonyls:

$$\lg p = A - B/T \qquad (2)$$

where p – vapor pressure is in units кPa, and A and B are constants.

	A	B	P (vapor pressure) for $T=90°C$	$P(T=80°C)$	$P(T=70°C)$
$Mo(CO)_6$	1.17	373.8	0.00104	0.00314	0.000067

The chemical reaction that most probable takes place in the CVD reactor can be written as:

$$Mo(CO)_6 + O_2 \rightarrow MoO_3 + CO_2 \uparrow \qquad (3)$$

Into the CVD reactor, under the temperature influence a pyrolitical decomposition of molybdenum carbonyl molecules begins near the heated substrate. The Mo then gets oxidized in the oxygen ambient and the film growth starts.

The thicknesses obtained are in the range of 320-400 nm for the low-temperature set of samples and 120-240 nm for the 200°C-set of samples [45]. The two sets were grown for one and the same deposition time. It was interesting that the lower-temperature set process was proceeding with higher deposition rate (10 nm/min compare to 6 nm/min for the higher temperature samples), resulting in the thicker films. The lower temperatures obviously favor higher deposition rates for the Mo carbonyl CVD-process. This behavior might be associated with the thermal decomposition of the molybdenum carbonyl, because this chemical available for oxidation decreases in concentration with increasing deposition temperature, through thermal decomposition in the vapor phase. The two sets of samples differ in their color. The samples of the lower-temperature look bluish, when the 200°C-set of samples are yellowish. In the text below we will refer to these definitions.

The advantages of the chosen CVD process for deposition of MoO_3 electrochromic films are as follows:

- The possibility to obtain the molybdenum oxide thin films by in-situ CVD method.
- The deposition temperatures approached, $T_{deposition}$ = 150 and 200°C, are ones of the lowest reported CVD temperatures for obtaining MoO_3 films with CVD process.
- The capability for variations of the gas flow ratio of Ar, carrying the carbonyls vapors and oxygen – $Ar(Mo(CO)_6)/O_2$ leading to different oxygen environments in the CVD reactor.
- To change the precursor vapor pressure by increasing the sublimator temperatures - $T_{sublimator}$ = 70, 80 and 90°C.
- To obtain amorphous films (in their as-deposited state) and to crystallize these films after additional annealing.
- The CVD MoO_3 films can be crystallized either in orthorhombic structure, monoclinic or mixed ($\alpha + \beta$) phases.
- The preparation of two sets of samples (depending on the substrate temperatures) with different structural and optical properties.

2.B.4. CHARACTERIZATION OF CVD MoO₃ FILMS WITH SEM, TEM AND AFM METHODS

A. SEM Investigation of MoO₃ Film Morphology

Scanning Electron Microscopy (SEM) was imployed for studying the surface morphology of molybdenum oxide films. SEM micrographs, shown on figure 2, are taken for MoO₃ film deposited on the glass substrates at the following technological conditions: the substrate temperature is 200°C, the gas flow ratio is 1/4, the sublimator temperature is 90°C and the deposition time is 40 minutes. The as-deposited film possesses a smooth surface without any structural features (figure 2a). Annealing at temperature of 300°C, leads to film surface transformation in small-grained and polycrystalline structure. Increasing the annealing temperature leads to bigger crystallites with a random distribution. SEM study reveals that the higher oxygen flow causes rougher film surface.

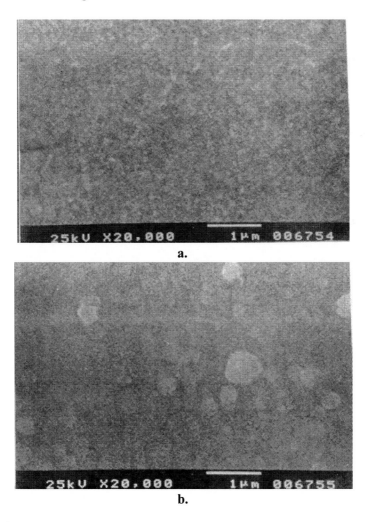

Figure 2. Continued.

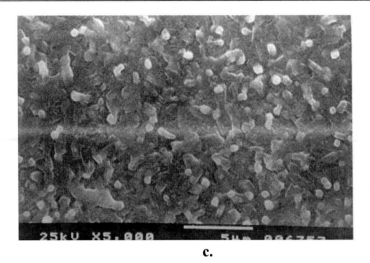

c.

Figure 2. SEM micrographs of MoO$_3$ films on glass substrates.

B. TEM Observation of MoO$_3$ Thin Films

Transmission electron microscopy (TEM) was performed by means of a TEM EM - 400 Philips equipment with a magnification of 14 000 times. The sample preparation has been done by using a double stage replica method. The substrates used were glasses coved with thin conductive layer. The technological conditions were: the two substrate temperatures – 150 and 200°C, the sublimator heated at 90°C, and the gas flow ratio was 1/40 for 40 minutes deposition. The as-deposited films of the two sets look like very similar and the film morphologies have changed after the additional thermal treatments [47].

The films obtained at the substrate temperature of 200°C showed increasing of the grain size with annealing at 300°C compared with as-deposited state, where the structure consists of close - packed grains (see figure 3). The anneling at 500°C leads to film cracking and a two-dimensional net formation.

Figure 4 exhibits the TEM observations of CVD MoO$_3$ films obtained at the lower substrate temperature of 150°C [47]. The small grained structure of as-deposited film transforms in heterogeneous after the 300°C treatment for 1 hour in air. It may be suggested that there a kind of mass transformation begins. The higher annealing temperature 400°C leads to bigger grains with a close ordering on the film surface.

CVD Molybdenum Oxide Thin Films as Electrochromic Material 173

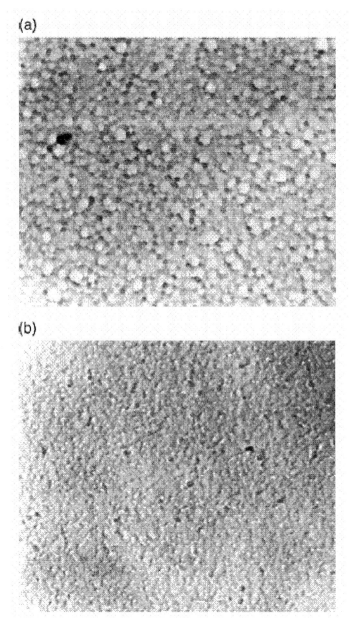

Figure 3. TEM micrograps of CVD MoO$_3$ films prepared at the substrate temperature of 200°C, obtained on the conductive glass substrates, where a) represents as-deposited film and b) after annealing at 300°C for 1 hour in air.

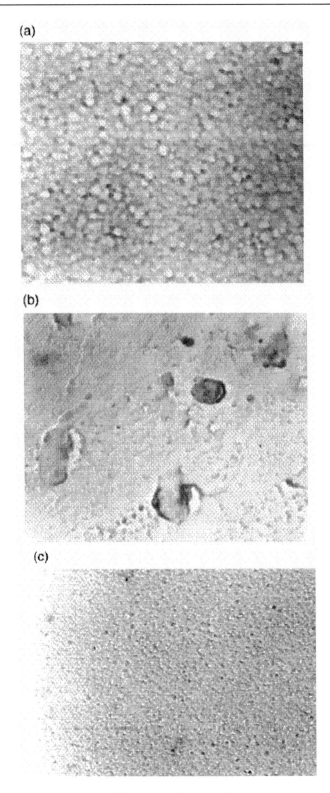

Figure 4. TEM micrograps of CVD MoO$_3$ films prepared at the substrate temperature of 150°C, obtained on the conductive glass substrates, where a) represents as-deposited film, b) after annealing at 300°C and c) after annealing at 400°C for 1 hour in air.

C. AFM Investigation of MoO$_3$ films, Obtained by APCVD

The atomic force microscopy (AFM) technique provides a way to analyse the surfaces of materials on an atomic scale due to its exceptionally high vertical resolution. The small probe-sample separation of the atomic force microscope (on the order of the instrument's resolution) makes it possible to carry out measurements over a small area. The resulting image resembles an image on a computer screen that consists of many rows or lines of information placed one above the other. Unlike traditional microscopes, the scanned-probe systems do not use lenses, so the sizes of the samples rather than the diffraction effects generally limit their resolution. The atomic force microscope measures topography with a force probe. AFM operates by measuring the attractive or the repulsive forces between a tip and the sample. In its repulsive "contact" mode, the instrument lightly touches the tip at the end of a leaf spring or "cantilever" to the sample. As a raster-scan drags the tip over the sample, some sort of a detection apparatus measures the vertical deflection of the cantilever, which indicates the local sample height. Thus, in the contact mode the atomic force microscope measures the hard-sphere repulsion forces between the tip and the sample [52]. In a non-contact mode, the AFM derives topographic images from measuring the attractive forces; the tip does not touch the sample [52, 53]. AFMs can achieve a resolution of 10 pm, and unlike electron microscopes, can image the samples in air and under liquids. In principle, AFM resembles the record player as well as the stylus profilometer. However, AFM incorporates a number of refinements that enable it to achieve the atomic-scale resolution.

The morphology and the surface roughness of CVD MoO$_3$ film, annealed at temperature of 300°C in air was determined by Atomic Force Microscopy (AFM). The images presented on figure 5, showing one film surface under different magnifications. As it could be seen, there is a beginning of the film crystallization. The observed film structure consists of small grains, almost uniformly distributed. This kind of morphology appears to be very typical for chemical vapored deposited films of transition metal oxides [54]. The magnification of the surface reveals that the separate grain is internal structured. The grains consist of smaller species gathering in agglomerates, which look like as cluster-like structures. It becomes obvious that the annealing temperature of 300°C is high enough for the crystallization to begin. The average grain size is 334 nm as shown on the AFM picture (figure 5). This value is taken from the picture made at the lowest magnification, where the accuracy of size determination is highest, due to the contribution of the biggest statistical number of grains per unit area.

Increasing the magnification, the films surfaces appear with their complex morphology. From the three dimensional view (figure 5a) of the AFM picture, the estimated average roughness was S_a - 21.5 nm, and the average cluster size was confirmed to be 334 nm.

Figure 5. AFM images of MoO$_3$ film, annealed at the temperature of 300°C with different magnifications.

Figure 5a. AFM images of MoO$_3$ film, annealed at the temperature of 300°C with smallest magnification in 2D and 3D view.

2.B.5. X-Ray Structural Study of APCVD MoO$_3$ Films

A. RHEED Investigation of CVD MoO$_3$ Thin Film

RHEED (Reflection High Energy Electron Diffraction) is a method for investigation of the sample crystal structure. A high energy beam (3-100 keV) is directed at the sample surface at a grazing angle, where the electrons diffracted by the crystal structure of the sample and then blow on a phosphor screen mounted opposite to the electron gun. The resulting pattern is a series of streaks, the distance between the streaks being an indication of the surface lattice unit cell size. The grazing incidence angle ensures surface specificity despite the high energy of the incident electrons. If a surface is atomically flat, then sharp RHEED patterns are seen. If the surface has a rougher surface, the RHEED pattern is more diffuse.

RHEED measurements were done by Carl Zeiss Jena EF-5 RHEED diffractometer. The RHEED were done for MoO$_3$ film, as deposited at the two (150 and 200°C) substrate temperatures. The RHEED patterns of APCVD as-deposited at 200°C MoO$_3$ films, obtained on glass substrates are diffuse and typical hallos, which proposed preferably amorphous structure (see figure 6a).

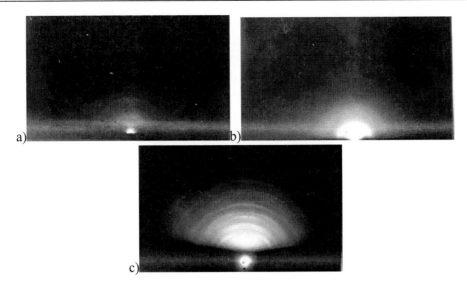

Figure 6. RHEED patterns for APCVD MoO₃ films, where a) presents the as-deposited film on glass substrate, b) as-deposited film on conductive glass substrate and c) the same film as b) but annealed additionally at 300°C.

Meanwhile, the Mo oxide films obtained at these technological conditions but the used substrates were conductive glass (a glass covered by a crystalline conductive layer) has a certain degree of crystallization, see figure 6b, although not clear, the RHEED pattern of 200°C-deposited film shows weak reflexes. The film deposited on conductive glass substrate reveals start of crystallization after annealing at 300°C, on figure 6c is presented a characteristic micrograph for polycrystalline structure with well defined reflexes. As it is seen, after the annealing at 300°C, a strong texturing occurs, revealing a well ordered crystalline structure. The presented below XRD analysis confirms this result. The determined d-spacings for Mo oxide film (namely the figure 6c, annealed at 300°C film) are d=3.78, 3.4, 2.86, 2.5, 2.295, 1.84, 1.55, 1.458 Å, which could be related to orthorhombic MoO_3. Similar results have been reported by other authors [6]. Due to the fact that the first three main d-spacings values are very similar for the orthorhombic and monoclinic modifications of MoO_3, in the literature these d-spacings are often found as attributed to the both crystal phases.

A. XRD Study of CVD MoO₃ Thin Film

XRD (X-Ray Diffraction) is an efficient and very used analytical technique for identification and characterization of the crystalline state of the materials. When a monochromatic X-ray beam with a proper wavelength is projected onto a sample at an angle θ, diffraction occurs only when the distance travelled by the rays reflected from successive planes differs by a complete number n of wavelengths. By varying the angle, the Bragg's Law conditions are satisfied. Plotting the angular positions and intensities of the resultant diffracted peaks of radiation produces a pattern, which is characteristic of the sample. Where a mixture of different phases is present, the resultant diffractogram is formed by addition of the individual patterns. Using X-ray diffraction, a great amount of structural, physical and chemical information about the material investigated can be obtained. XRD analysis can identify the crystalline phases into solid materials. From the XRD data can be determined the lattice parameters, the residual tensions, texture and the crystallite sizes.

First, the studied MoO_3 films were deposited at the substrate temperature of 200°C, the sublimator temperature 90°C, the gas flow ratio 1/12 on the conductive glass substrates. The samples were additionally treat at the temperatures of 200, 300, 400 and 500°C in air for 1 hour. XRD spectra were measured by a X-ray Philips spectrophotometer, using Cu K_α radiation. The characteristic peaks due to the crystalline substrate (a layer of SnO_2) have been eliminated from the spectra shown in figure 7. The observed strong reflection in (060) orientation is connected with existing of crystalline phase of α - MoO_3. MoO_x films obtained by other authors [2] have shown (0к0) and (1к0) orientations, indicative for a mixture $\alpha+\beta$ phases. On transformation, from mixed phase to orthorhombic α - phase, MoO_3 displays a very strong (0к0) preferred orientation. The presence of α - MoO_3 is proved by the (020), (040) and (060) planes [37]. The shown diffraction patterns are typical for polycrystalline material.

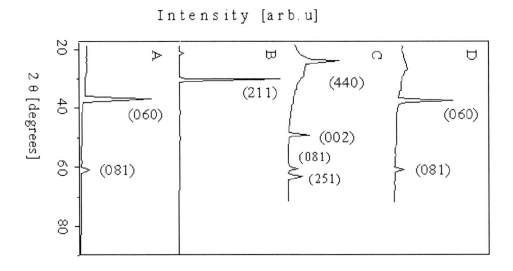

Figure 7. XRD spectra of CVD MoO_3 thin films, annealed at 200, 300, 400 and 500°C.

The XRD spectra of annealed at 200°C film does not differ from the as-deposited (not shown here). Annealing at 300°C resulted in XRD spectrum, showing a strong texture in (211) orientation. The observed d-spacings are d = 4.133, 2.959, 2.181, 1.797Å (according to card N 5-0513 JSPDS 1985) and they are not coincided with RHEED data of the same film. XRD data are related to Mo suboxides such as Mo_8O_{23} (d=2.959 Å) and Mo_4O_{11} (d=4.00, 2.207, 1.810 Å). One explanation may be the disposition of the sample towards the X-Ray beam, which leads to appearance of different XRD lines. Anyway, the molybdenum oxide film deposited at 200°C and annealed at 300°C represents a mixture of stoichiometric MoO_3 with suboxides, probably the duration of the annealing is not enough for fully oxidation of the oxygen vacancies in the film. Some authors (V. Sabapathi et.al. [6] and Cross et al. [37]) reported the presence of MoO_2 fraction in the Mo oxide films, annealed at 300°C. The annealed films at higher temperatures 400 and 500°C, revealed only XRD peaks due to stoichiometric MoO$_3$.

In order to study the MoO$_3$ structure for the films obtained at higher oxygen amount, new XRD investigation has been performed. The thin films were additionally annealed at the temperatures of 200, 300, 400 and 500°C in air and a small set was treated at 300°C in O$_2$ ambient. XRD study was made with URD diffractometer with secondary graphite monochramator using Cu Kα radiation. Two different configurations of the experiments have been applied - the standard Bragg-Brenato and the grazing-incidence asymmetric X-Ray diffraction (GIAXRD) with angle 3°. The investigated thin films are deposited on conductive glass at: $T_{deposition}$=200°C, $T_{sublimator}$=90°C, Mo(CO)$_6$/O$_2$=1/40.

The previous XRD and RHEED studies showed that when the Mo(CO)$_6$/O$_2$ ratio is smaller (1/15), the samples consisted of a mixture of Mo suboxides and traces of stoichiometric MoO$_3$. The XRD analysis shows that films which were additionally annealed at 200°C are amorphous in structure and the spectra exhibited peaks only due to the crystalline substrate (SnO$_2$:Sb-coated glass). The higher annealing temperatures favor the crystallization of the films (see Fig. 8). The XRD spectra present lines due to the orthorhombic modification of MoO$_3$ (according to the JCPDS card N 35-0609). Experiments were made in the standard Bragg-Brenato diffractometer arrangement in order to find if there was a preferential orientation in the oxide crystallization. It seems to have such preferential orientation most possibly in the (010) direction.

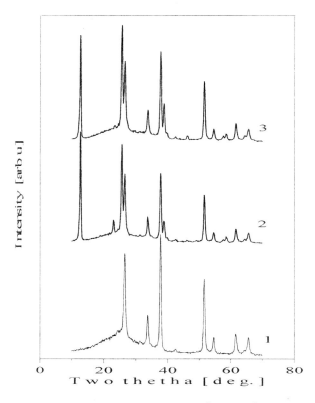

Figure 8. XRD spectra of annealed MoO$_3$ films at 1) 300°C, 2) 400°C and 3) 500°C in air.

The estimated parameters of the crystal lattice are shown in table 1. It can be observed that increasing the annealing temperatures leads to a slight decrease of the values of parameters **a** and **c**, meanwhile the parameter **b** increases.

Table 1. The lattice parameters of CVD MoO_3 films, derived from XRD analysis. The substrate temperature is 150°C, the sublimator temperature 90°C, and gas flow ratio $Mo(CO)_6/O_2 = 1/40$.

Additional annealings	a (Å)	b (Å)	c (Å)
300°C, in oxygen	3.955 ± 0.005	13.900 ± 0.004	3.710 ± 0.002
400°C, in air	3.932 ± 0.008	13.927 ± 0.006	3.705 ± 0.002
500°C, in air	3.939 ± 0.004	13.934 ± 0.003	3.697 ± 0.001
JCPDS card N 35 – 0609	3.9630	13.856	3.6966

2.B.6. IR Spectroscopy Investigation of MoO_3 Thin Films

Infrared spectroscopy is one useful undistructive analytic tool for characterization of the chemical bonds and the various vibrations of the molecules building the material. The IR radiation is absorbed by the sample leading to excitation of the vibrational, rotational and bending modes. At the same time the molecule remains in its initial ground state. The frequencies at which IR radiation is absorbed are characteristical and are directly related to certain chemical bonds.

In MoO_x derivatives, Mo^{6+} possesses an octahedral oxygen environment. The previously described unique thermodynamically stable orthorhombic structure of MoO_3 has been specified by the fact that in each octahedron, there is one oxygen unshared (Mo=O), two axial oxygens, which are common to the two octahedral and three equatorial oxygens jointed to the three octahedral. On the other hand, the metastable monoclinic modification of molybdenum trioxide involves corner sharing, distorted octahedral in a rhenium oxide, ReO_3 – type structure. As the molybdenum atom is about six times heavier than the oxygen atom, the expected vibrations are supposed to involve mainly oxygen atoms.

The orthorhombic cell of molybdenum trioxide extends over two layers and there 45 optical modes expected, where 17 modes of them are IR active and 24 are Raman active. The IR frequencies reported in the literature vary depending on the size and morphology of the crystallites. Also, there can be an overlapping of different vibrational modes split by transverse optic and longitudinal optic effects [55]. The main region, where characteristic absorption bands for all the transition metal oxides appear is in the range 200 – 1200 cm^{-1}. The lower wavenumbers beneath 400 cm^{-1} are assigned to the deformation bands. The stretching vibrations occur in the region of 1050 – 400 cm^{-1}. It worth mentioning that the absorption bands exhibited in 920-990 cm^{-1} are due to the terminal bonds Mo=O, which are a prove for appearance of the orthorhombic crystalline MoO_3 [56].

The studied films are prepared on silicon wafers for IR and Raman measurements. The cleaning procedure for Si substrates prior to film deposition has been a thermal treatment in a solution containing H_2SO_4 + 30% H_2O_2. The substrates are carefully washed with distilled water and dried. The wafer cleaning must be performed just before the beginning of the deposition process.

The transmittance spectra in IR region were recorded with the help of the double beam IR spectrophotometer Perkin–Elmer 1430 in the spectral range of 200 – 1200 cm^{-1}. The analysis has been carried out in order to determine the structure changes induced by the CVD technological parameters and the additional annealings at temperatures of 200 to 500°C.

Influence of the Substrate Temperature on the Vibrational Properties of CVD MoO$_3$ Thin Films

For IR study, CVD MoO$_3$ films have been chosen, obtained at the both deposition temperatures and all other technological parameters were kept the same. The sublimator temperature was 90°C, the gas flow ratio (Mo(CO)$_6$/O$_2$) was 1/20 and the deposition proceeded for 40 min. The additional annealings had been followed at the temperatures of 300, 400 and 500°C in air for 1 hour. The measured IR spectra are shown at figure 9.

The as-deposited at the deposition temperature of 150°C molybdenum oxide film reveals two strong peaks located at 578 и 728 cm^{-1} in its IR spectrum, meanwhile for the 200°C as-deposited MoO$_3$ a featureless spectrum can be observed with very weak broad bands and no peaks distinguished. The absorption band at 728 cm^{-1}, appearing in all IR spectra of 150°C deposited MoO$_3$ films independently on the other technological conditions, is characteristic for the stretching mode of the monoclinic β – phase of MoO$_3$ [57].

After the thermal treatment at 300°C, IR spectra changed. The 150°C MoO$_3$ film shows a strong peak at 269 cm^{-1} meanwhile its main absorption band, situated at 728 cm^{-1} in the as-deposited state, now has been splitted into two absorption bands at 555 and 835 cm^{-1}. The higher substrate temperature set reveals an increase of the IR bands intensities, located in the spectral range of 316 – 403 cm^{-1}. The characteristical band seems very broad, covering the spectral area of 779 – 926 cm^{-1} and centered at 866 cm^{-1}. As already mentioned, the 728 cm^{-1} line is due to the stretching Mo – O modes, connected with β – MoO$_3$, as the new absorption band at 555 cm^{-1} is attributed to the stretching vibrations of ν(OMo$_3$) groups, characteristical for the unique layered orthorhombic modification of MoO$_3$ [58]. The band at 835 cm^{-1} can be associated to the vibrations of Mo – O – Mo chemical bonds, which are a sign for the presence of β – phase [45-47]. The IR analysis leads to the conclusion that after the annealing at 300°C, the film structure of the 150°C obtained MoO$_3$ is a mixture of the two crystal phases α + β. For the 200°C films in the spectral range 316 – 403 cm^{-1}, the observed IR bands are overlapped and possess contributions of many absorption bands related to the deformation modes of different chemical groups specific for the two phases. The very strong intense and broad band at 779 – 926 cm^{-1} can also be an overlap of different absorption bands. The band at 866 cm^{-1} is related with ν(OMo$_2$) group vibrations and it is attributed to orthorhombic α – MoO$_3$ [58]. The other part of this broad band is probably due to the stretching vibrations of ν$_s$O – Mo – O groups (at 810 cm^{-1} located and associated with за α –MoO$_3$) and vibrations of νOMo$_2$ (about 819 cm^{-1}, observed in the orthorhombic oxide film as well as in the MoO$_3$ crystal) [59, 60]. Other possible contributions can be the stretching vibrations of Mo – O – Mo groups, specific for the β monoclinic phase. Probably, MoO$_3$ films deposited at 200°C and annealed at 300°C exhibited the unique orthorhombic modification with a small fraction of the monoclinic phase. After 400°C annealing, the IR bands appeared at the same positions with decreased intensities.

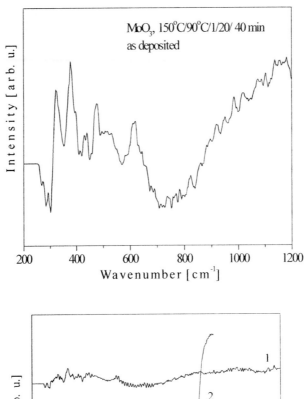

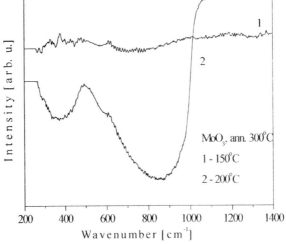

Figure 9 (continued)

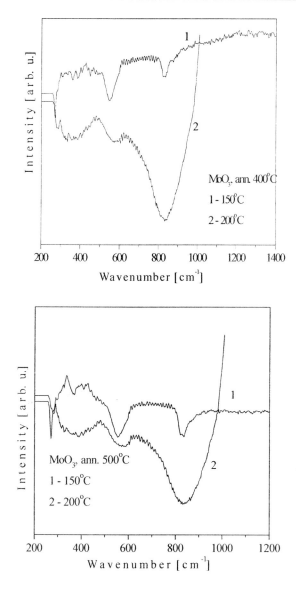

Figure 9. IR spectra of CVD MoO_3 films in dependence of the deposition and the annealings temperatures.

The highest annealing temperature (500°C) leads to the appearance of more features in the IR spectra. The 150°C deposited MoO_3 film reveals two lines at 555 and 836 cm^{-1}, which are similar to the 200°C films with IR bands at 570 and 833 cm^{-1}. The main difference can be observed at the lower frequencies for the strong peaks at 369 cm^{-1} (for the 150°C film) and 268 cm^{-1} (for the 200°C). The absorption line at 369 cm^{-1} characterizes the deformation modes of O - Mo bonds of the layered structure α-MoO_3 [55].

In order to determine if the tendency and differences remain, IR investigations have been performed for the molybdenum oxide films obtained at the enriched oxygen ambient in the CVD reactor during the film growth. The spectra were recorded for the MoO_3 films, prepared at the gas flow ratios (carbonyl vapor flow to oxygen) 1/28 and 1/40. The obtained results are summarized in table 2. The as-deposited films at the gas flow ratio 1/28 show an absorption peak at 729 cm^{-1} (for the 150°C film). For the 200°C film, there are seen two strong bands at 574 and 819 cm^{-1}. The IR lines at 574 and 819 cm^{-1} are proposed to be connected with the alternative bond lengths in the MoO$_6$ octaedra [55, 56].

Table 2. IR data for the MoO$_3$ films deposited at the two substrate temperatures and at the gas flow ratios of 1/28 and 1/40

Mo(CO)$_6$/O$_2$	Annealing temperature	IR bands [cm^{-1}]	
		T$_{deposition}$=200°C	T$_{deposition}$=150°C
1/28	Asdeposited	312 (intense) 570 (intense) 812 (very strong.)	281 (intense) 574 (weak) 778 (very weak)
	300°C	322 (weak) 356 (weak) 384 (weak) 572 824	324 454 562 (широка)
	400°C	280 (intense) 321 (intense) 387 (intense) 556 (intense) 803 (very strong)	268 (very strong) 370 550 (very strong) 668 (weak) 827 (very strong) 865 932 (слаба), 1042
	500°C	316 (very strong) 357 384 570 665 811 (very strong.)	268 (very strong.) 358 (very strong) 447 553 (very strong) 830 936, 986 (weak)
1/40	Asdeposited	Very broad 708 - 856 770 (center) 955 (weak)	717 (strong)
	300°C	248, 278 325 365 415 574 829 (broad)	303 371 420 482 571 778 888 638 - 940 (broad) 824 (center)
	400°C	837 (strong)	271 (very strong) 360 (very strong) 550 (very strong) 821 (very strong)

IR spectrum of the 150°C film, treated additionally at 300°C has no characteristic lines, which can be explained with start of some kind of transformation (TEM micrographs provoke a similar suggestion). IR spectrum of the 200°C deposited film presents two peaks at 570 and 824 cm^{-1}. Increasing the annealing temperature up to 400°C, the spectra of the two sets show weak bands, as the absorption band at 807 cm^{-1} observed for 200°C film is attributed to ν$_s$(O - Mo - O) vibrations – characteristic for α - MoO$_3$. For the 150°C film the splitting of the main band into two bands with annealing has been detected again. The peaks are positioned at 832 and 550 cm^{-1}. The most considerable fact is the appeareance of the weak IR peaks at 935 and 966 cm^{-1}. They are determined as vibrations of the terminal oxygen atoms Mo = O, which are specific and responsible for the layered orthorhombic structure of MoO$_3$. After the annealing at 500°C, IR spectrum of the 200°C film shows main band at 836 cm^{-1}. For MoO$_3$ film obtained at the lower substrate temperature of 150°C, IR bands are located at 553, 830 cm^{-1}. There, two well defined peaks at 966 and 1042 cm^{-1} exist, which are proof for the thermodynamically stable orthorhombic MoO$_3$ structure.

When the entering oxygen flow into the CVD reactor has the highest value used (Mo(CO)$_6$/O$_2$=1/40), IR analysis uncovers the following interesting features. For the as-deposited sample from the 150°C - set, the main absorption peak appears at 717 cm^{-1} and for the respective 200°C - deposited film there is a broad band in the range 708 - 856 cm^{-1}, centered at 770 cm^{-1}. Meanwhile, a well-pronounced peak can be seen at 995 cm^{-1}, suggesting start of crystallization in orthorhombic phase. After the 300°C annealing, the spectrum presents weak bands for 150°C film at 303, 371, 420, 482, 571, 778 and 888 cm^{-1}. Again the main band is very broad (638 - 940 cm^{-1}) with a center at 824 cm^{-1}. The IR spectrum of the 200°C film exhibits low intense bands at 248, 270, 325, 365, 415, 574 cm^{-1}, as the main peak is situated at 829 cm^{-1} and it is very strong and broad. After annealing at the temperature of 400°C, the strong peaks have been revealed at 271, 360, 550 and 821 cm^{-1} for the MoO$_3$ film obtained at the substrate temperature of 150°C, meanwhile the 200°C deposited film has only a specific, very intense band, located at 837 cm^{-1}.

It can be summarized that the molybdenum oxide films obtained at lower substrate temperature of 150°C possess higher fraction of layered α – phase than the films, which are deposited at 200°C. The other conclusion derived from IR spectroscopy is that the higher oxygen content also favors formation of the unique orthorhombic modification of MoO$_3$.

The Influence of the Sublimator Temperature on the Vibrational Properties of MoO$_3$ Thin Films

The IR investigations showed [citat] that the sublimator temperature can not significantly influence the vibrational properties of the CVD MoO$_3$ films, although the lowest sublimator temperature (70°C) leads to deposition of smoother and more uniform thin films as it was proved by the optical study. The followed investigations were performed predominantly for the thin films, deposited at the sublimator temperature of 90°C, where the highest film growth rate is achieved.

2.B.7. RAMAN SPECTROSCOPY INVESTIGATION OF MOLYBDENUM OXIDE THIN FILMS

Raman Spectra of MoO$_3$ Films after Annealings

Raman spectra were recorded for numerous APCVD MoO$_3$ films, prepared at various technological parameters [46, 57]. The main purpose is to determine the type of chemical bonds and the crystalline states of the studied CVD films. Raman scattering spectroscopy is a powerfull tool for material research and it is widely used for characterization of the transition metal oxide films. Raman method is a non-destructive and uncontact method for studying the structure of specific chemical bonds in materials [61]. Raman spectroscopy measures the wavelength and the intensity of the non-elastic scattering light from the sample molecules. The scattered light reveals at wavelength λ, shifted from the incident light exactly with the energies of the molecular vibrations. This can be imagined by a shift (Raman shift) towards the laser beam frequency over the frequency axis either to the left or to the right side of it, depending on the direction in which the exchange of energy between the vibrating crystal lattice atoms and the laser beam proceeds. In result of the energy exchange, the frequency of the crystal lattice atoms vibration comes in resonance with the frequency of the laser beam. The effect is a Raman shift (a peak, Raman band), which is seen on the X - axis. The difference with IR spectroscopy lies in the physical effects included: for IR spectroscopy there is absorption and a change of the dipole moment of the material molecules, while for Raman spectroscopy the scattering and the polarity changes are basic.

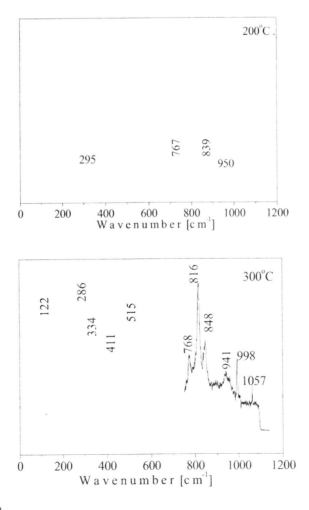

Figure 10. Continued.

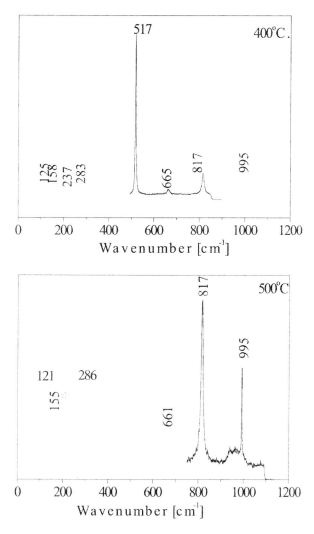

Figure 10. Raman spectra of the CVD MoO₃ films, deposited at 200°C, the sublimator temperature – 90°C and gas flow rate – 1/32 and additionally annealed at temperatures in the range 200, 300, 400 and 500°C.

The Raman scattering spectra were measured by a Dilor XY triple monochromator with a multichannel optical detector. During the recording of Raman spectra in the range of 100-1100 cm^{-1} the intensity of He-Ne laser (λ=632.8 nm) excitation was controlled with an accuracy of 1%. The excitation power on the sample surface was 0.5 mW and the excitation spot diameter was 3 μm. Almost all thin films deposited at the two substrate temperatures in their initial as deposited state and annealed at 200°C present smooth Raman spectra without characteristical peaks, which is a sign for amorphous structure of the films. Based on that experience, deeper analysing have been accomplished for Raman spectra of CVD MoO₃ films, annealed at the temperatures above 300°C. The intense peak appearing at 520 cm^{-1} presented in all Raman spectra is due to the optical phonon vibration of Si substrate [62]. The active Raman modes of MoO₃ are expected to appear in the spectral range of 100 - 1000 cm^{-1}. The spectral range beneath 200 cm^{-1} represents the lattice vibrations.

The spectra shown on figure 12 are recorded for MoO₃ thin films, prepared at the following technological parameters: the substrate temperature - 200°C, the sublimator temperature - 90°C, Mo(CO)$_6$:O$_2$=1/32 and the additional annealings at 200, 300, 400 and 500°C.

Table 3. Raman spectra data for CVD MoO$_3$ films, deposited at 200°C, the sublimator temperature – 90°C and gas flow rate – 1/32

Annealing temperature (°C)	Wavenumber (cm^{-1})	Attributed to	Crystal Modification
200°C	295	δ* O – Mo	Orthorhombic α- MoO$_3$
	767	Mo – O – Mo	Monoclinic β-MoO$_3$
	839	ν*O – Mo	
	950	ν O – Mo	
300°C	122	Deformation modes	Orthorhombic α- MoO$_3$
	151		
	190		
	241	δ OMo$_2$	
	286	δ O – Mo	
	334	δ OMo$_3$	
	354		
	373	δ O = Mo	
	411		
	670	ν O Mo$_3$	
	709	ν O Mo$_2$	
	768	Mo – O – Mo	Monoclinic β-MoO$_3$
	816	Mo – O – Mo	
	848	Mo – O – Mo	Monoclinic β-MoO$_3$
	941	ν O – Mo	Orthorhombic α- MoO$_3$
	998	ν O = Mo$_1$	Orthorhombic α- MoO$_3$
400°C	125	Deformation modes	Orthorhombic α- MoO$_3$
	158		
	237	O – Mo	Monoclinic β-MoO$_3$
	283	δ O – Mo	Orthorhombic α- MoO$_3$
	665	O $_{(2,2')}$Mo$_3$	
	817	Mo$_3$ – O – Mo$_3$	
	995	ν O = Mo$_1$	
500°C	121	Deformation modes	Orthorhombic α- MoO$_3$
	155		
	196	δ O Mo$_2$	
	243	δ O Mo$_2$	
	286	δ O – Mo	
	336	δ O Mo$_3$	
	377	δ O = Mo	
	661	ν O Mo$_3$	
	817	ν O Mo$_2$	
	958	ν O – Mo$_1$	
	995	ν O = Mo$_1$	

* By the letters δ and ν are presented the bending and stretching vibrations, respectively.

The spectra of the as-deposited at 200°C and additionally annealed at 200°C films have a big similarity, presenting Raman peaks, assigned to the stretching vibrations of Mo - O bonds in the spectral range of 750 - 960 cm^{-1}.

The broad shapes of the Raman lines suggest an amorphous-like structure with traces of crystallization as for the annealed at 200°C MoO_3 film, the crystallized fraction is bigger and the better pronounced peaks are located at 767, 839 and 950 cm^{-1}.

The Raman lines at 767, 839 cm^{-1} indicate the presence of the monoclinic modification (β-MoO_3). At the same time, the peak situated at 950 cm^{-1} and the broad band at 295 cm^{-1} (which is connected with deformation Mo – O modes) are attributed to the orthorhombic α-MoO_3 [67]. The most probable suggestion is the presence of a mixture of the two MoO_3 phases in the film structure. Increasing the annealing temperatures, new additional lines appeared in Raman spectra which for the most cases can be assigned to the orthorhombic phase. The description together with the vibrational modes and the corresponding crystal modifications are shown in table 3. As it can be seen from figure 10 and table 3, the treatment at 500°C causes full crystallization of MoO_3 film structure in orthorhombic modification.

As it is known, the orthorhombic crystal modification is layered structure, consisting of MoO_6 octahedra in the two dimensional arrangements in separate double layers, where the free spaces beyong them favor the intercalation/deintercalation of small alkali ions into film structrure. Thin CVD MoO_3 films with α – phase crystal structure are expected to show better electrochromic properties.

The IR and Raman analysis are based on the fact that MoO_3 structure is constructed from MoO_6 octahedra, for which the Raman and IR active modes are detected in the range of 100 - 1000 cm^{-1}. The terminal double bonds Mo = O are characterized by sharp IR and Raman bands in the range 920 - 1000 cm^{-1} [55]. For CVD MoO_3 films, the corresponding Raman peak is located predominantly at 947 cm^{-1}, but additional clearly seen lines are exhibited at 995 cm^{-1}. S. Lee et al. [64] employed the two terminal oxygen bonds to explane the origin of the observed at 955 and 995 cm^{-1} Raman bands – the first one as due to the cutting of the chemical bond at the corner common oxygens, meanwhile the 995 cm^{-1} line is originated from single coordinated oxygen in the layered MoO_3 structure.

Raman Spectra of MoO_3 films, Depending on Oxygen Gas Flow

The vibrational properties of MoO_3 thin films deposited at the 150°C substrate temperature by CVD method show significant variations depending on the different technological parameters and the annealing procedures. The studied films by IR and Raman spectroscopies revealed that the film structures are mainly mixtures of the orthorhombic and monoclinic crystal modifications of MoO_3. The influence of the reactive gas flow of oxygen (O_2) becomes more noticeable after the film annealings.

Table 4 presents the IR and Raman data for MoO_3 films deposited at the various gas flow ratios: carbonyl vapors to oxygen. The strong Raman peak at 847 cm^{-1} observed in the spectrum of the metal oxide film prepared at the highest oxygen content $Mo(CO)_6$:O_2=1/40, is related to the stretching vibrations of the bridge oxygen atoms connected with two metal atoms in the three dimensional arrangement of the monoclinic MoO_3 structure [60]. The reference [65] also stated that the intense Raman lines situated at 848 and 776 cm^{-1} must be attributed to the monoclinic MoO_3. Raman bands at 828 cm^{-1} for the film, deposited at 1/28 ratio and the band at 816 cm^{-1} for the obtained film at 1/20 ratio are probably related to the orthorhombic α - MoO_3. The bands at 548, 558 and 564 cm^{-1} can be suggested to be due to the stretching IR vibrations of Mo–O–Mo bonds in the orthorhombic modification of molybdenum trioxide [66].

The band at 564 cm^{-1} disappears in the spectra of the MoO_3 film, deposited at the highest oxygen quantity. Another band around 768 - 772 cm^{-1} can be observed and it is related to the monoclinic β phase of MoO_3 [60]. It seems that the higher oxygen content favors the formation of the monoclinic crystalline phase. The sharp line at 663 cm^{-1} in the spectrum of the MoO_3 film deposited at 1/40 ratio is connected with α – MoO_3. According to L. Seguin et al. [55] the Raman bands in the range of 660 - 700 cm^{-1} are attributed to the stretching vibrations of the bridge oxygen atoms connected with three Mo atoms in the orthorhombic α - MoO_3.

Table 4. Raman and IR data for MoO_3 films, deposited at 150°C and $Mo(CO)_6:O_2$ = 1/20, 1/28 and 1/40. The studied MoO_3 films are additionally annealed at 300 and 400°C for 1 hour in air

$Mo(CO)_6/O_2$ ratio	IR absorption bands [cm^{-1}]		Raman lines [cm^{-1}]	
	300°C annealed	400°C annealed	300°C annealed	400°C annealed
1/20	268 cm^{-1} 564 cm^{-1} 710 cm^{-1}	270 cm^{-1} 356 cm^{-1} 558 cm^{-1} 828 cm^{-1} (s)	303 cm^{-1} (m) 849 cm^{-1} (w) 952 cm^{-1} (s, broad)	155 cm^{-1} (w) 303 cm^{-1} (m) 428 cm^{-1} (w) 816 cm^{-1} (m) 936 cm^{-1} (s)
1/28	564 cm^{-1} 710 cm^{-1} (very broad)	270 cm^{-1} 356 cm^{-1} 558 cm^{-1} (s) 828 cm^{-1} (s)	150 cm^{-1}(w) 303 cm^{-1}(w) 680 cm^{-1}(w) 930 cm^{-1} (m, broad)	150 cm^{-1}(w) 303 cm^{-1} 820 cm^{-1} (m) 930 cm^{-1} (s,broad)
1/40	229 cm^{-1} 380 cm^{-1} 823 cm^{-1}	270 cm^{-1} 356 cm^{-1} 548 cm^{-1} 558 cm^{-1} 828 cm^{-1} (s)	303 cm^{-1}(w) 772 cm^{-1} (s) 841 cm^{-1} (s) 961 cm^{-1} (s, broad)	198 cm^{-1} (v. s.) 303 cm^{-1} (m) 663 cm^{-1} (s) 768 cm^{-1} (stronger) 847 cm^{-1} (s) 947 cm^{-1} (s, broad)

w – weak band, m – clearly seen band, s – strong band, v. s. – very intense band

The relative intense band at 303 cm^{-1} can be ascribed to the monoclinic phase. Raman spectra show a line at 820 cm^{-1}, characteristic for the orthorhombic Mo trioxide, which becomes especially intense for the MoO_3 films, annealed at 400°C. In the studied spectra, the bands at 564 cm^{-1} (for the annealed samples at 300°C) and at 558 cm^{-1} (after annealing at 400°C) can be associated with the orthorhombic crystal modification. Raman bands in the range between 400 cm^{-1} and 200 cm^{-1} are in general attributed to the deformation modes and the bands positioned beneath 200 cm^{-1} are connected with the vibrations of the crystal lattice.

From the Raman characterization can be concluded that the molybdyl bonds are not considerably influence on the oxygen content, entering the CVD reactor during film growth.

On the other hand, the oxygen bonds with two or three metal atoms exhibit changes for the thin films, obtained at different oxygen contents.

Vibrational Properties Depended on the Substrate Temperature

The substrate temperature is the most significant factor, affected the film structure and properties. Figure 13 presents Raman spectra of MoO_3 films, deposited at the two substrate temperatures - 150 and 200°C. The other technological parameters are: the sublimator temperature is 90°C, the gas flow ratio is 1/40 and the annealing fulfilled at 400°C for 1 hour in air.

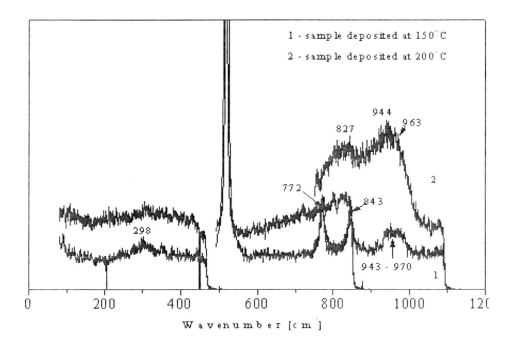

Figure 11. Raman spectra of 400°C - annealed MoO_3 films, prepared at the two substrate temperature of 150 and 200°C.

A higher degree of crystallization has been determined for MoO_3 films deposited at the lower substrate temperature of 150°C. Their spectra exhibited the sharper and intense Raman lines. For MoO_3 films, obtained at the substrate temperature of 200°C, the bands are broader, an indication for the presence of a considerable fraction of amorphous phase into film structure. The Raman spectrum of 150°C deposited film shows weak bands at 298, 353 and 443 cm^{-1}. Strong peaks appeared at 772 and 843 cm^{-1}, as well as a broad band located at 943-970 cm^{-1}, this broad band is due to vibrations of the terminal oxygen in the double bond Mo=O, specific for the orthorhombic structure. The higher substrate temperature causes the appearance of Raman lines situated at 756, 827, 944 and 963 cm^{-1}, which are broader and not intensive, but also assuming a presence of α - MoO_3.

Raman results suggest that the crystallization begins at the lower annealing temperatures for the molybdenum oxide films deposited at the substrate temperature of 150°C. MoO_3 films obtained at the substrate temperature of 200°C show a certain amount of the amorphous fraction even after the annealing at 400°C.

2.B.8. OPTICAL PROPERTIES OF CVD MoO$_3$ THIN FILMS

Spectrophotometry and spectral ellipsometry have been applied to characterize the optical properties of the metal oxide films. UV - VIS spectroscopy is a quick, precise and a relatively low cost optical method. The spectra in transmission and reflectance modes had been measured in the spectral range 200 - 900 nm and the obtained data are not difficult to interpretate [67]. Spectrophotometry is the quantitative measurement of the reflective and transmissive properties of a material as a function of wavelength. While it is relatively simple in concept, the determination of the reflectance and the transmittance involves a careful consideration of the geometrical and the spectral conditions for the measurements.

The transmission and reflection spectra are analyzed and the energy dispersion of the refractive index is considered. The optical absorption coefficient and the optical band gap energy E_g are determined.

On the other side, ellipsometry is known as a very sensitive optical technique for determining properties of surfaces and thin films. If a linearly polarized light of a known orientation is reflected at oblique incidence from a surface then the reflected light is elliptically polarized. The shape and the orientation of the ellipse depend on the angle of incidence, the direction of the polarization of the incident light, and the reflection properties of the studied surface. It can be measured the polarization of the reflected light with a quarter-wave plate followed by an analyzer; the orientations of the quarter-wave plate and the analyzer are varied until no light passes though the analyzer. From these orientations and the direction of polarization of the incident light, it can be determined the relative phase change, and the relative amplitude change, introduced by reflection from the surface [68, 69].

An ellipsometer measures the changes in the polarization state of light when it is reflected from a sample. If the sample undergoes a change, for example a thin film on the surface varies its thickness then its reflection properties will also differ. Measuring these changes in the reflection properties allow to deduce the actual change in the film's thickness.

The most essential application of ellipsometry is studying thin films. In the context of ellipsometry a thin film ranges from essentially zero thickness to several thousand Angstroms, although this range can be extended in some cases. If a film is thin enough that it shows an interference color then it will probably be a good ellipsometric sample. The sensitivity of an ellipsometer is so high that a change of a few Angstroms in the film thickness is usually easy to detect. Ellipsometry measures the change in the polarization, as well as the intensity, upon a reflection from an electrode.

The ellipsometric measurements were performed on a Rudolph Research ellipsometer in the spectral region of 300-820 nm at the incidence angle of 50°. The accuracy of the angles of the polarizer, the analyzer and the light incidence is within +0.01°. The transmittance and reflectance spectra of the films were registered in the range of 300-900 nm on the double-beam spectrophotometer Shimadzu UV-190. The measurements of the full reflectance (diffusive plus normal) were carried out in a 100-mm integrating sphere. As reference 100 % response was taken the magnesium oxide reflectance.

For improving the film structure and the optical properties, post-deposition annealing was applied in the temperature range of 200-500°C [45, 46, 54, 70]. Annealing at 500°C was a technological limitation due to the film cracking effects. Moreover, at this temperature possible softening and impurities diffusion of glass substrates occur.

UV - VIS Spectrophotometric Study of MoO₃ Films

UV-VIS spectra of MoO_3 films are recorded by UV – VIS Perkin – Elmer 330 spectrophotometer as a function of annealing temperature and the oxygen quantity entering the CVD reactor during the film deposition. First, there will be presented the results of the optical properties for MoO_3 thin films deposited at the lower substrate temperature of 150°C. The films of molybdenum oxide obtained at two substrate temperatures differ in their structural, vibrational properties as well as visually in color. The Mo oxide films, grown at 150°C look yellowish, while MoO_3 films at 200°C are bluish. The bluish color of the molybdenum oxide films, deposited by various methods is observed by other authors [71, 72]. The bluish coloring is characteristic for substoichiometric MoO_x and it is often related with the presence of oxygen vacancies in the structural arrangement of the layer.

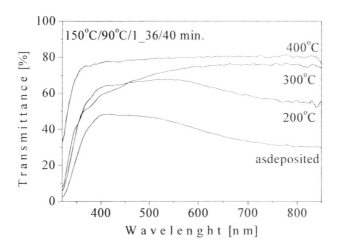

Figure 12. VIS-UV spectra of CVD MoO_3 films deposited at substrate temperature of 150°C, sublimator temperature of 90°C, gas-flow ratio of 1/36 and deposition time duration - 40 min. The films have been thermally treated after the deposition process.

Figure 12 shows the visible transmission spectra of the thin films deposited at the substrate temperature of 150°C, the sublimator temperature - 90°C and the gas flow ratio 1/36. As can be seen, the as – deposited film possesses transparency around 45% in the visible spectral range. The improvement of the optical transmission occurs after the annealing and the value approaches 78% for MoO_3 film, which is annealed at 400°C. It is proposed that the increase of the optical transparency is due to the additional oxidation of the molybdenum oxide films and correspondingly this causes a decrease of the oxygen vacancies in the material.

UV – VIS spectra of the 150°C deposited films, obtained for the different oxygen contents (Mo(CO)$_6$/O$_2$ = 1/15, 1/24 and 1/40) are presented on figure 13. The samples are annealed at the temperature of 300°C in air for 1 hour. The measurements reveal that the optical transparency is improved increasing the oxygen amount during the film growth.

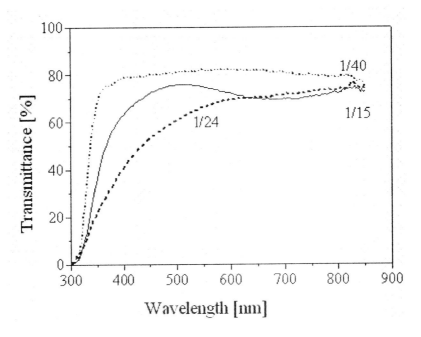

Figure 13. UV-VIS spectra of the MoO$_3$ thin films. obtained at 150°C/90°C and Mo(CO)$_6$/O$_2$ = 1/15, 1/24 and 1/40. They are additionally annealed at 300°C in air for 1 hour.

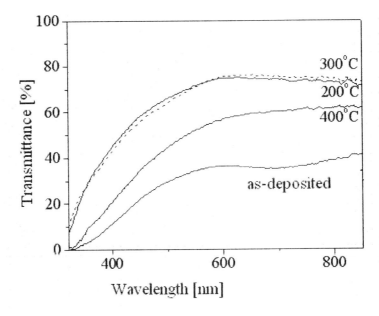

Figure 14. UV-VIS spectra of the MoO$_3$ thin films obtained at 200°C/90°C/1/15 and annealed at the different temperatures for 1 hour in air.

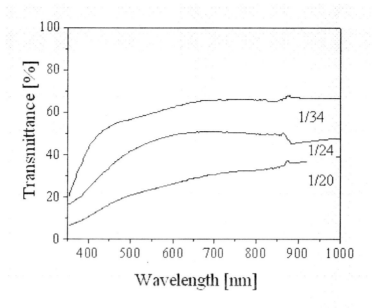

Figure 15. UV-VIS spectra of as deposited MoO_3 films, obtained at 200°C/90°C and varying oxygen gas flows. The used substrates are conductive glasses.

The chemically vapor deposited thin films of MoO_3 at the substrate temperature of 200°C are bluish colored. Figure 14 reveals MoO_3 films obtained at the following parameters: the sublimator temperature - 90°C and $Mo(CO)_6/O_2=1/15$. For these films, the optical transmittance increases from 35% for as-deposited state up to 70% for the annealed film at 300°C. When the gas flow ratio is higher (1/34), the transmittance of the annealed MoO_3 films reaches 80%.

On figure 15 a comparison of the UV – VIS spectra has been made for the as-deposited at 200°C films on conductive glass substrates, the sublimator temperature is 90°C, $Mo(CO)_6/O_2$ = 1/20, 1/24, 1/34. The optical transparency is lower due to the conductive layer of SnO_2:Sb beneath the MoO_3 films on the glass substrates. For the lowest oxygen flow, the transmittance reaches near 20%, the value becomes almost 40% for the increased oxygen content 1/24 and for the highest oxygen content the optical transmittance is 60% in the visible range. Again, the oxygen content influences the optical performance and more transparent films are obtained by increasing its amount during film deposition for the two substrate temperature.

MoO_3 thin films obtained at the substrate temperature of 200°C, $T_{sublimator}$=90°C and $Mo(CO)_6/O_2$=1/32 on the conductive glass substrates show 60% transperancy in the visible spectral range for the as-deposited state and slightly increases after the annealings. The as-deposited films, deposited at the same technological parameters on the ordinary glass substrates, possess transparency around 66 - 72% in the spectral area 400 – 800 nm (see figure 18). After the thermal treatments, the optical properties of MoO_3 films improved, as it is observed that the annealing at 400°C causes an increase of the transmittance to 75 – 80%, meanwhile the highest annealing temperature of 500°C approaches the values closed to 85% for the CVD MoO_3 thin films.

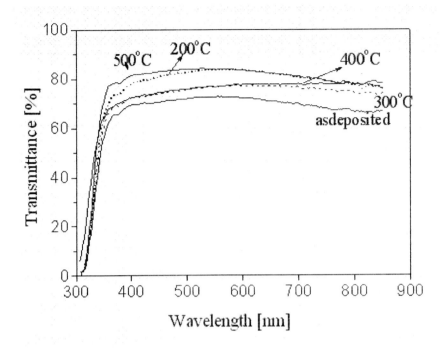

Figure 16. UV - VIS spectra of the MoO$_3$ thin films, prepared at 200°C/90°C/1/32 in the as-deposited state and after additional anneallings.

For MoO$_3$ thin films, deposited at the similar technological parameters and the substrate temperature 150°C, the optical measurements showed that the films transmittance is around 75% in the visible spectrum, but it considerably decreases if the oxygen content during chemical vapor deposition of MoO$_3$ films is lower.

Optical Transparency Depending on the Sublimator Temperatures

The studied films of molybdenum oxide have been obtained at the technological regime, where the only variable parameter is the sublimator temperature (varying 70, 80 and 90°C), which causes a change of the carbonyl vapor pressures entering the CVD reactor [45]. There it must be noted the interesting experimental result that the MoO$_3$ films deposited at the higher substrate temperature manifest no dependence of the optical transmittance in the visible spectral range on the sublimator temperatures.

The MoO$_3$ films obtained at the lower substrate temperature (150°C), behave differently as can be seen from figure 17. The increase of the sublimator temperature changes the optical properties. The transmittance in the range 300 - 850 nm significantly varies with the sublimator temperature. The films, prepared at 70°C show the best optical performance, meanwhile the transmittance gets worse for the higher sumblimator temperature. This optical behaviour can be explained with a higher growth rate and respectively the thicker films. But this possible explanation could be only partial, as the thickness of 200°C films also increases with the sublimator temperatures but this does not affect the optical transmittance.

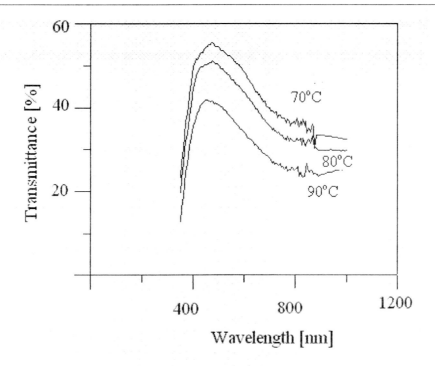

Figure 17. UV-VIS transmittance spectra of MoO₃ films, deposited on the conductive glass at $T_{deposition}=150°C$ and varying the sublimator temperature ($T_{sublimator}$ - 70, 80, 90°C).

The absorption coefficient α as well as the optical band gap E_g for thin films can be determined from the spectrophotometric data [73]:

$$\alpha d = \ln(I_o/I), \qquad (12)$$

I_o and I are the intensities of the incident and passed light, d is the film thickness. The optical band gap can be estimated using the equation of band-to-band transitions:

$$\alpha(\nu) = A(h\nu - E_g)^n \qquad (13)$$

where E_g is the semiconductor energy band gap, and A is a constant. The exponent n has the value 1 for the direct transition band gap and 4 for the indirect transition band gap. Analysing the dependence $\alpha^{1/2}$ versus the photon energy hν, E_g can be determined by extrapolation of the linear part of the curve at $\alpha^{1/2} = 0$.

The estmited values for E_g for the MoO₃ films deposited at the higher substrate temperature of 200°C are as follows: for the as-deposited film the band gap is 3.37 eV, the E_g value increases after the 400°C annealing to 3.51 eV and for the 500°C treated MoO₃ film the optical band gap is close to 3.45 eV.

The 150°C deposited films for the different oxygen content Mo(CO)₆/O₂ and after the annealing at 300°C are 2.7 eV (Mo(CO)₆/O₂=1/24) and 3.4 eV (Mo(CO)₆/O₂=1/40). The optical band gap value is significantly influenced by the oxygen content. The higher values of the optical bandgap for MoO₃ films, obtained at increased oxygen gas flow during the film deposition can be explained with the partial filling of the oxygen ion deffects [70]. Similar results are reported by other authors, who also observed higher values of E_g due to more oxygen during the film deposition. Scarminio et al. [74] presented E_g values varying from 2.8 to 3.2 eV for evaporated MoO_3 films with high oxygen content. Sabhapathi et al [75] also determined the increase of E_g from 2.98 to 3.25 eV, when the partial pressure of O_2 is changed from 6.7×10^{-6} to 2.7×10^{-4} mbar during film deposition process.

Spectral Ellipsometry Investgation of MoO_3 Thin Films

The Ellipsometry is a well known optical measurement technique, developed from P. Drude in the beginning of 20 century. The spectral ellipsometry is undistructive method for determination of the optical constants *n* and *k* (refractive index and extinction coefficient), film thickness, interface roughness and compositions of the surface thin films and of multilayer structures. This method is based on the measurements of the polarity change of light at a reflection from the sample surface by changing the wavelength and the incident angle. For each wavelength λ, the reflective coefficient (ratio of R_p and R_s) is measured and then *PSI (Ψ)* and phase difference *DELTA (Δ)* are estimated. The optical constants and the film thickness can be determined from the best fitting of the experimental and the theoretical values of *PSI* and *DELTA* characteristics. The fitting process is proceeding by using a database of the optical indices and having certain knowledge of the prehistory of the sample. For these estimations, the probabilities for modeling and provement of unknown optical constants are extremely important. Comparing with the other optical methods such as the spectrophotometries, where the spectra of transmittance and reflections are recorded, the ellipsometry gives more experimental data and based on the fact, that it has measured the ratio of two parameters, this technique possesses a very good precision and it is highly sensitive and reproducible measurement method [70].

Dependence of the Optical Constants on the Substrate and Sublimator Temperatures for MoO_3 Films, Deposited on Si Wafers

The optical parameters of the MoO_3 films had been deduced from the spectroscopic ellipsometric measurements carried out in the spectral region of 300 - 800 nm using Rudolph Research ellipsometer with PCSA configuration. The refractive index, *n*, and the extinction coefficient, *k*, were calculated for a single-layer/substrate optical system assuming that the films are a homogeneous medium. The optical band gap energy, E_g, had been determined by studying the absorption edge behavior and its dependence on the process temperatures. The E_g values were derived from a plot of $(\alpha h\nu)^{1/2}$ versus photon energy (hν), where $\alpha = 4\pi k/\lambda$ is the absorption coefficient.

The investigated MoO_3 films are deposited on Si substrates and at the technological parameters: $T_{deposition}$=150 and 200°C, $Mo(CO)_6/O_2$=1/12. Although the films, obtained at the two deposition temperatures differ in color and structure, they possess closed values of the optical constants. The effective refractive indices and the extinction coefficients are determined from the ellipsometric data and are presented on figure 18. For MoO_3 films on the crystalline Si substrate and the substrate temperature of 150°C, *n* is varied in the range of 1.89

- 2 (for the spectral region of 400 - 700 nm), meanwhile the extinction show the values between 0.35 – 0.34 in the same spectral region and the curve possess a minimum at 550 nm (the k value is 0.08) in the visible part of the spectrum. For the MoO_3 films deposited at 200°C, n changes from 1.8 at 400 nm to 1.97 at 700 nm, at the same time k is around 0.1 (400 nm) and 0.056 (at 700 nm). Other authors [76] reported similar values of the refractive index for MoO_3 films derived by the oxidation of the metallic Mo thin films where n is 1.96-2.08 at 500 nm. For MoO_3 thin films prepared by the two-stage CVD method (the deposition of black Mo (MoO_2:Mo) followed by the additional oxidation) shows n=2.0-2.1. The same authors stated that the extinction is near zero for the range 1-3 eV, which explains the good transparency of the molybdenum oxide films. Our CVD MoO_3 films have very low absorption, so the extinction coefficient in the visible spectrum is 0.06.

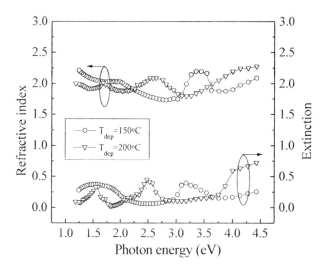

Figure 18. Effective values for refractive index and extinction of CVD MoO_3 films, obtained at the two substrate temperatures - 150°C and 200°C.

Table 5. Data for the CVD-MoO_3 films thickness in dependence on the deposition process parameters. The deposition time duration was 40 min for all runs

$T_{deposition}$ (°C)	$T_{sumlimator}$ (°C)	Ar flow rate carrying $Mo(CO)_6$ vapors (l. min^{-1})	Oxygen flow rate, (l. min^{-1})	Thickness (nm)
150	70	0.1	1.2	193
150	80	0.1	1.2	277
150	90	0.1	1.2	360
200	70	0.1	1.2	105
200	80	0.1	1.2	133
200	90	0.1	1.2	240

The deposition temperatures of the CVD process and the corresponding film thickness are summarized in table 5. As can be seen, the higher deposition temperature results in much smaller film thickness for all the sublimator temperature used. Moreover, the growth rate varies with the sublimator temperature, as the higher the temperature the thicker the film. The difference in the thickness almost reaches 90 nm for the 150°C set. The same tendency was observed for the 200°C set.

Typical spectral dependences of the refractive index for CVD MoO$_3$ films at the different process temperatures are given in figure 19. As can be seen, the n values increase with the deposition temperature in the 300 – 550 nm spectral range. The shape of the dependence $n(\lambda)$ is obviously affected by the sublimator temperature. For films deposited at the lowest (70°C) sublimator temperature the $n(\lambda)$ curves are comparatively featureless, suggesting more uniform surface and film morphology. When the sublimator temperature increases (80°C and 90°C), a minimum in the spectral dependence appears for both the deposition temperatures. For all APCVD MoO$_3$ films, the refractive index varies in the range of 1.7-2.3 in the studied spectral region.

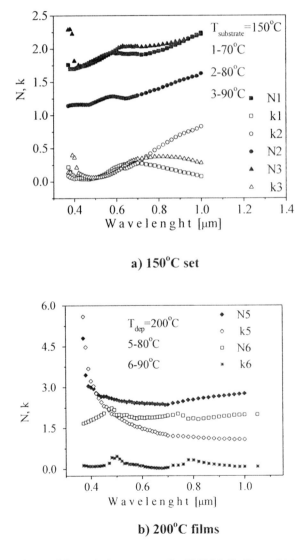

a) 150°C set

b) 200°C films

Figure 19. Spectral dependence of the optical parameters for CVD MoO$_3$ films, obtained at different substrate temperatures.

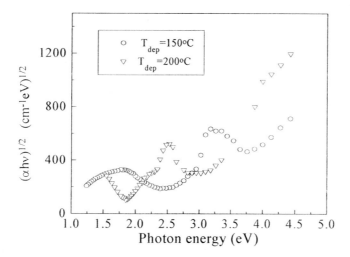

Figure 20. Absorption coefficient as a function of the photon energy for the MoO$_3$ films, obtained at T$_{deposition}$=150 и 200°C.

The absorption coefficient for the different process temperatures can be evaluated from the corresponding extinction coefficient. The spectral dependences are presented in figure 20. There can not be derived a clear dependence on the deposition temperatures below the absorption edge. The α values remain smaller than 5×10^4 cm^{-1} for the wavelengths below the absorption edge. Approaching the absorption edge, band to band electronic transitions take place and the absorption strongly increases, having higher values for the films of 200°C-sets. The shapes of the curves are found to be similar for T$_{sublimator}$=70 and 80°C and differs significantly only for highest sublimator temperature T$_{sublimator}$=90°C. For the latter two well-pronounced strong absorption bands appear. These peculiarities at energies below the absorption edge could be connected with a proper type of defects giving states in the forbidden band gap. The absorption bands in that energy range were observed also by Julien et al. [77] and these bands were associated with the defects of oxygen-ion vacancies in the MoO$_3$ film structure.

The plot of $(\alpha h\nu)^{1/2}$ versus $h\nu$ determines the E$_{og}$ values. The received values lead to the conclusion that the sublimator temperature has no influence on the E$_{og}$ values (2.91-3.04 eV) for the molybdenum oxide films, obtained at the substrate temperature of 150°C, since the difference is in the range of the measurement accuracy.

For the 200°C set, this distinction becomes more noticeable. The values of the optical bandgap energy for all the films studied are within 2.76 - 3.14 eV, which is in agreement with the literature data [1]. The authors in Ref. [76] show that the optical band gap energy for the polycrystalline MoO$_3$ films were in the range of 2.79 - 3.1 eV. The optical band gap energy values of the as-deposited and the annealed CVD MoO$_3$ films are summarized in table 6.

Some scientists have observed a thickness dependence of the optical properties and, therefore, of the optical band gap energy [79]. The observed E$_{og}$ variation for the different temperatures could be connected either with the structure or the thickness changes. In our experiments, however the deposition time was kept constant and different thicknesses were obtained due to the kinetics reasons. Because of the correlation between structure and film thickness, the influence of neither one on E$_{og}$ values can be determined as dominant. The structure of the films revealed by XRD, IR and Raman analysis is a mixture of amorphous and crystalline phases.

After annealing at 300 and 400°C, the n and k values turn to be closed for both sets, but their spectral dispersion curves reveal specific behaviors. The refractive index is within 1.8-2.2 and the extinction coefficient is near the value of 0.2. The curves character and maxima suggest the presence of some structural inhomogeneity. The crystal lattice defects contribute to the optical absorption and the maxima in the extinction spectral dependences indicate that the proper types of defects have been appeared. After annealing, the spectral dependence of the refractive index becomes featureless and this is an indication for the improved structural uniformity. The film thickness increases with ~10%, which can be explained with the additional oxidation of the unbounded metal atoms. The k values are lower in the whole studied range, manifesting more transparent molybdenum oxide films.

Correspondingly to the extinction, the absorption in the annealed film is smaller (see figure 21). For the 150°C films, the same absorption peaks exist as their intensity decreases after the annealing procedures. The peak situated at 1.75 eV is shifted towards the higher energies up to ~ 2.2 eV closing the eye-sensitivity maximum at 2.5 eV [78]. This is very important feature for the electrochromic properties of this material.

Table 6. Optical band gap energy values estimated for the MoO$_3$ films, depending on the process temperatures

Annealing temperature	E_{og} (eV)	
	$T_{substrate}$ 150°C	$T_{substrate}$ 200°C
As - deposited	2.75	2.89
300°C	3.07	3.09
400°C	3.16	3.16

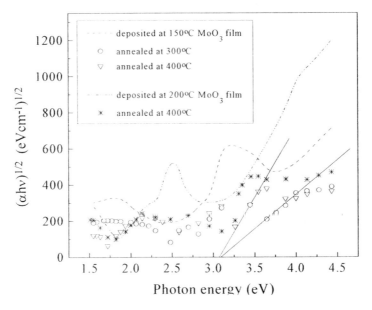

Figure 21. Plot of $(\alpha h\nu)^{1/2}$ vs. $h\nu$ of the annealed CVD MoO$_3$ films, deposited at 150 and 200°C.

The absorption peak at ~ 2.5 eV observed for the MoO₃ films deposited at 200°C, after annealing stays at the same position with its intensity is considerably decreased. Apparently, the oxidation in air decreases the amount of the defects due to the saturation with oxygen of the dangling Mo bonds. Further crystallization takes place due to the higher temperatures (300 and 400°C). The Raman spectra analysis of the MoO₃ films has shown that the annealings at these temperatures leads to the formation of the crystalline phases including the two - monoclinic and orthorhombic - modifications.

Optical Constants of CVD MoO₃ Thin Films Deposited on Glass Substrates

Thin films of APCVD MoO₃ on glass substrates were measured by the spectroscopic ellipsometer before and after annealing at 300, 400 and 500°C. An incidence angle of 50° was used to assure a separation of the signals reflected from the front and the backside of the samples (film-glass system). For a further minimization of the backside reflectance, 2 mm thick glass substrates were used. The film thickness preliminary determined by a Talystep profilometer was in the range of 250-290 nm.

In figure 22 the spectral dependences of the refractive index for the as-deposited and annealed at 400 and 500°C MoO₃ films are presented. The refractive index varies within the range of 1.9-2.5 in the studied spectral range of 280-820 nm. For the as-deposited film, the dispersion itself is not significant since the change is only within 2.4-2.6 for the whole spectrum. The refractive index values decrease with increasing the annealing temperature. Similar shape of the curves is bserved up to 400°C annealing. At 500°C, the curve becomes more featureless.

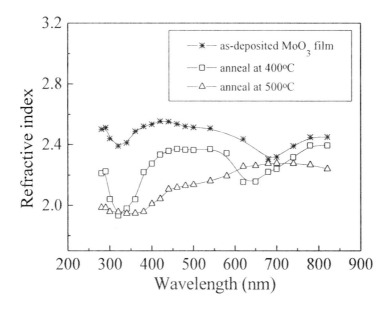

Figure 22. Refractive index vs. wavelength of CVD-MoO₃ films before and after annealing at 400 and 500°C.

The curves character suggests a presence of some structural inhomogeneity. Raman results prove that the annealing at temperatures up to 400°C leads to the MoO₃ film structure that is a mixture of the amorphous and crystalline phases. As it is discussed above, the monoclinic and orthorhombic modifications have been detected in these films. Such a structural inhomogeneity could define the observed dispersion behavior of the refractive indices. The featureless dispersion dependence for MoO_3 films annealed at 500°C, as seen in figure 22, can be a consequence from a more uniform structure of MoO_3 of thoroughly orthorhombic crystalline modification as given by the corresponding Raman spectrum.

The maxima in the spectra and the high value of absorption ($\alpha \sim 10^5$ cm^{-1}) in the range below the absorption edge indicate a high concentration of different types of defects, most probably oxygen related. After annealing at 400 and 500°C the absorption in the 1.5-2.2 eV region drops down, the corresponding refractive index values are minimal. These films in fact are visually more transparent. The extinction values demonstrate the MoO_3 film transparency as well as the improvement of the optical transmittance with the annealings and with the higher oxygen content.

Approaching the absorption edge, band to band electronic transitions take place and the absorption increases. For the as-deposited films the optical band gap energy is 2.51 eV. An increase in the E_{og} value is obtained after annealing, the higher the temperature, the larger the band gap E_{og}. After annealing, the E_{og} values are found in the range of 2.72-3.63 eV, typical for a polycrystalline material.

It can be concluded from all optical characterizations for a great number of samples that the CVD MoO_3 films exhibit properties suitable for electrochromic applications. They are transparent in the visible spectral region and knowing the correlation between the process temperatures and the film structure, it becomes possible to prepare thin films appropriate for application in electrochromic devices.

2.B.9. ELECTROCHROMIC BEHAVIOR OF MoO_3 THIN FILMS

The electrochromic effect is defined as an absorption change induced by electric field or voltage. The MoO_3 films are almost transparent in their usual state, after injection of small ions (such as Li^+, H^+ or K^+) and electrons electrochemically into these films, there occurs a color change to dark blue. The physical mechanism of this optical effect is still not fully understood. The color change is supposed to be directly related to the double intercalation/deintercalation of electrons and ions in the films, it can be written as:

$$xM^+ + xe^- + MoO_3 \leftrightarrow M_xMoO_3 \qquad (14)$$

In the equation M^+ is a small alkali ion. The current theories, as mentioned before, suggest that the optical absorption of the colored ion-injected MoO_3 film is caused by reduction of the Mo^{6+} states to Mo^{5+} with ion/electron injection and electron transitions between adjacent Mo^{5+} and Mo^{6+} states [1].

The electrochemical measurements were carried out for the MoO_3 thin films, deposited on conductive glass substrates. All the films show electrochromic reversible effect. Coloration as a result of Li intercalation is observed. After 5 cycles it was also found that the electrochemical and optical properties of the MoO_3 films are reversible.

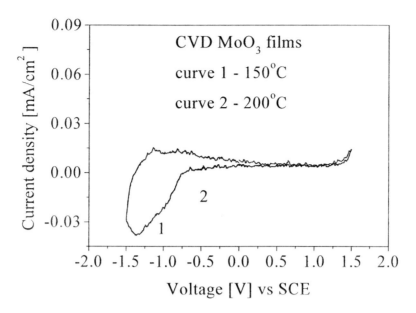

Figure 23. CV curves of CVD MoO$_3$ films, deposited at the substrate temperatures of 150°C (curve 1) and 200°C (curve 2).

Figure 23 shows cyclic voltamommetric (CV) curves of the as-deposited MoO$_3$ films from the 150 and 200°C sets. The curves reveal no significant features and peaks, which indicate the amorphous disordered structure of the as-deposited APCVD molybdenum oxide films. The peak due to the injection of Li$^+$ ions appears around -1.37 V and the corresponding peak caused by ion extraction is at -1.14 V. The differences between the two main sets of samples can be clearly seen. Higher current densities are observed for the 150°C - set.

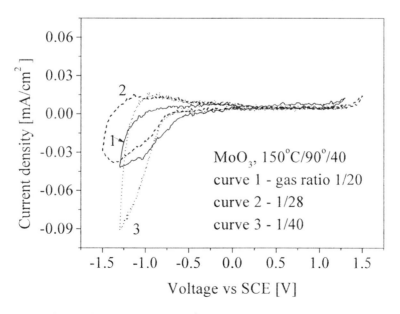

Figure 24. CV curves of MoO$_3$ films, deposited at 150°C and different gas flow ratios of 1/20, 1/28 and 1/40.

Figure 24 represents the CV curves for the MoO_3 thin films in dependence of oxygen contents entering the CVD reactor during the film growth. The differences in the peak positions are related to the ion intercalation. The peaks, corresponding to Li ions intercalated into the host lattice, are found at -1.37 and -1.13 V for the gas ratios 1/28 and 1/20, respectively. The CV curve for the higher oxygen content exhibits no noticeable cathodic peak. The anodic peaks can be observed at -0.82, -1.12, -0.96 V when the oxygen content varies from 1/20 to 1/40. After the annealings of these films, the current density considerably increases.

Table 7. Diffusion coefficient values D_{Li}, in the case of Li-containing electrolyte for CVD MoO_3 thin films.

$T_{deposition}$	$Mo(CO)_6/O_2$	$T_{annealing}$ [°C]	D_{Li} [cm²/s]
150°C	1/20	As deposited	2.6×10^{-12}
150°C	1/28	As deposited	1.6×10^{-11}
150°C	1/40	As deposited	1.16×10^{-11}
200°C	1/20	As deposited	1.5×10^{-13}
200°C	1/20	400°C	2.2×10^{-13}
200°C	1/28	As deposited	5.3×10^{-12}
200°C	1/28	400°C	6.76×10^{-12}
200°C	1/40	As deposited	4.9×10^{-12}

The diffusion coefficient D of Li ions, entering the film structure can be estimated from the obtained experimental CV data and the equation (16). The determined values of the diffusion coefficient D_{Li} are presented in table 7. It is found out that the diffusion coefficient is strongly dependent on the deposition process parameters of the molybdenum oxide thin films and its values can vary significantly. This difference comes from the fact that CVD MoO_3 films possess various surface and structural properties depending on the technological conditions of CVD preparation. Granqvist [1] reported the values for the diffusion coefficient of Li ions in the mixed structure of monoclinic and orthorhombic phases in the range of 10^{-12} - 5×10^{-12} cm²/s, correspondingly for the bulk orthorhombic MoO_3, the estimated values are 10^{-11} - 10^{-10} cm²/s. Our results show that the as-deposited films obtained at the substrate temperature of 150°C and the oxygen contents ($Mo(CO)_6/O_2$) - 1/28 and 1/40, define values of the diffusion coefficient close to that of the bulk material. As it was revealed from IR and Raman studies at that temperature the orthorhombic modification is favorable.

The color efficiency can be determined from the cyclic voltammometric data and equation (17) and it is found to be 39 cm²/C for as deposited MoO_3 films at the substrate temperature of 200°C and highest oxygen content - 1/40.

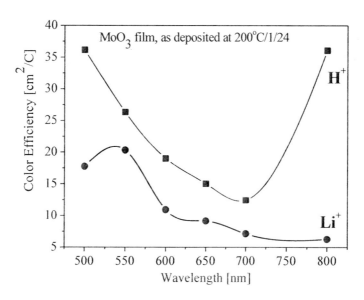

Figure 25. Spectral dependence of the coloration efficiency values, determined for the as-deposited MoO_3 films, under Li^+ and H^+ ions intercalation.

Figure 25 presents the coloration efficiency of molybdenum oxide films in lithium and hydrogen containing electrolytes. The values of CE for hydrogen electrolytes are larger. Glycerin is added to the electrolyte in order to reduce the damage to the oxide film. However, the MoO_3 film was damaged after 50 cycles.

In conclusion we may say that molybdenum oxide thin films can be successfully deposited by atmospheric pressure CVD techniques, using carbonyl as precursor. The structural, optical and vibrational properties of MoO_3 films are investigated in details. The MoO_3 films show about 80% transmittance in UV-VIS for samples, deposited on ordinary glass and the transparency is about 60% for the films on conductive glass substrates. It has been determined that with an appropriate selection of the technological parameters we can obtain thin MoO_3 films crystallized either in the layered orthorhombic or in the monoclinic structure. The unique orthorhombic layered structure allows easy intercalation/deintercalation of small ions into the intermediate layer spaces in case of application of the MoO_3 films in electrochromic cells. Varying the technological conditions, thin films can be obtained with variety of structural, optical and electrochromic properties. The research done proved that these thin films could be applied in electrochromic cells.

REFERENCES

[1] M. Figlarz, *Progress in Solid State Chemistry.* 19 (1) (1989) 1.
[2] M. Kharrazi, L. Kullman, C.G. Granqvist, *Solar Energy Mater, Solar Cells.* 53 (1998) 349.
[3] W. Gulbinski, D. Pailhavey, T. Suszko, Y. Mathey, *Appl. Surf. Sci.* 475 (2001) 149.

[4] G.E. Carver, A. Divrechy, S. Karbal, J. Robin, A. Donnadieu. *Thin Solid Films.* 94 (1982) 269.
[5] J. Tang, Y. Lu, B. Liu, P. Yang, Y. Huang, J. Kong, *J. Solid State Electrochemistry.* 7 (4) (2003) 244.
[6] V.K. Sabhapathi, O.M. Hussain, S. Uthanna, B.S. Naidu, P.J. Reddy. *Materials Sci. Eng.* B 32 (1995) 9.
[7] C. Julien, G.A. Nazri, J.P. Guesdou, A. Gorenstein, A. Khelfa, O.M. Hussain. *Solid State Ionics.* 73 (1994) 319.
[8] T.S. Sian, G.B. Reddy, *Solar Energy Mater, Solar Cells.* 82 (2004) 375.
[9] V. Hornebecq, Y. Mastai, M. Antonietti, S. Polarz, *Chemistry of Materials.* 15 (2003) 3586.
[10] B.C.Satishkumar, A. Govindaraj, E.M. Vogl, L. Basumallick,C.N.R. Rao, *J. Materials Research.* 12 (3) (2000) 604.
[11] Gouma, P., Comini, E., Sberveglieri, G. ,Proceedings of SPIE - *The International Society for Optical Engineering.* 5275 (2004) 68.
[12] Hussain, K. Srimivasa Rao, K.V. Madhuri, C.V. Ramana, B.S. Naidu, S. Pai, J. John, R. Pinto, *Applied Physics.* A 75 (2002) 417.
[13] L. Degiorgo, B. Alavi, G. Mihaly, G. Gruner, *Physical Reviews.* B 44 (1991) 7808.
[14] M. Yahaya, M. Sallen, I. *Tabib Solid State Ionics.* 113-115 (1998) 421.
[15] M. Ferroni, V. Guidi, G. Martinelli, P. Nelli, M. Sacerdoti, G. Sberveglieri. *Thin Solid Films.* 307 (1997) 148.
[16] D. Davazoglou, T. Dritsas, *Sensors and Actuators.* B 77(2001) 359-362
[17] M. Ferroni, V. Guidi, G. Martinelli, P. Nelh, M. Sacerdoti, G. Servegheri, A.K. Prasad, P.I. Gouma, *J. Mater. Sci.* 38 (2003) 4343
[18] M.K. Dongare, V.V. Bhagwat, C.V. Ramana, M.K. Gurjar, *Tetrahedron Letters.* 45 (24), (2004) 4759.
[19] C. Julien, L. El-Farh, M. Balkanski, O.M. Hussain, G.A. Nazri: *Appl. Surf. Sci.* 65–66 (1993) 325.
[20] A. Martínez-De La Cruz, I.J. Ramírez, *J. Power Sources.* 133 (2) (2004) 268.
[21] J.W. Bullard III, R. L. Smith, R.L. *Solid State Ionics.* 160 (3-4) (2003) 335.
[22] Prasad, P.Sathya Sainath, Radhakrishna, S. *Solid State Ionics.* 28-30 (1988) 814
[23] J. N. Yao, K. Nashimoto, A. Fujishima, *Nature.* 355 (1992) 624.
[24] J. Scarminio, A. Lourenco, A. Gorenstein, *Thin Solid Films.* 302 (1997) 66
[25] X.A. Yang, J.W.Cao, B.H. Loo, J.N. Yao, *J. Phys. Chem.* B 1998, 102, 9392.
[26] G. Leftheriotis, S. Papaefthimiou, P. Yianoulis. *Solar Energy Mater. Solar Cells.* 83 (2004) 115.
[27] C. Fukushima, M. Nagano, T. Sumita, H. Kubota, M. Nagata, Y. Honda, T. Oku, j. Imahori, *Physica.* B 239 (1997) 56.
[28] Mizushima, T., Fukushima, K., Tran, M.H., Ohkita, H., Kakuta, N. *Chemistry Letters.* 34 (7) (2005) 986
[29] C.G. Granqvist, "Handbook of Inorganic Electrochromic Materials", Elsevier Science, 1995.
[30] P.F. Garsia, E. McCarron III. *Thin Solid Films.* 155 (1987) 53
[31] A. Donnadieu, D. Davazoglou, A. Abdellaoui, *Thin Solid Films.* 164 (1988) 333.
[32] W. Li, F. Cheng, Z. Tao, J. Chen, *J. Physical Chemistry.* B 110 (1) (2006) 119.
[33] M.R. Altman, *Thin Solid Films.* 414 (2002) 205.

[34] M.R. Tubbs, *Phys. Stat. Solidi.* (a) 21 (1974) 213.
[35] J.J. Lander, I.H. Germer, *Trans. AIME.* 175 (1948) 648
[36] A. Abdellaoui, G. Leveque, A. Donnadieu, A. Bath, B. Bouchikhi. *Thin Solid Films.* 304 (1997) 39.
[37] J.S. Cross, G.L. Schrader, *Thin Solid Films.* 259 (1997) 5.
[38] T. Maryama. T. Kanagawa, *J. Electroch. Soc.* 142 (1995) 1644.
[39] R.R. Tummala, R.B. Shaw, *Ceramics International.* (1987)
[40] P.M. Sousa, A.J. Silvestre, N. Popovici, M.L. Paramês and O. Conde, *Material Science Forum.* 455-456 (2004) 20
[41] W. Xu, J. Xu, N. Wu, J. Yan, Y. Zhu, Y. Huang, W. He, Y. Xie, *Surface and Interface Analysis.* 32 (2001) 301.
[42] C.E. Tracy, D.K. Benson, *J. Vac. Sci. Technol.* A 4 (5) (1986) 2377.
[43] G. Benito, M.J. Davis, S.J.Hurst, D.W. Sheel, M.E. Pemble, *Electrochemical Soc, Proceed. Vol.* 2003 -08, 2003, pp. 557.
[44] M.J. Davis, G. Benito, M.E. *Pemble, CVD* 10(1) (2004) 29
[45] K. A. Gesheva, T. Ivanova, A. Iosifova, D. Gogova, R. Porat, *J. de Physique IV France.* 9 (1999) Pr 8-453 – Pr8-459.
[46] K. Gesheva, T. Ivanova, A. Szekeres, A. Maksimov and S. Zaitzev, *J. Phys. IV France.* 11 (2001) Pr3 - 1023 - Pr3 - 1028.
[47] T. Ivanova, M. Surtchev, K. Gesheva, *Materials Letters.* 53 (4-5) (2002) 250.
[48] D. Gogova, A. Iossifova, T. Ivanova, Zl. Dimitrova, K.A. Gesheva, *J. Crystal Growth.* 198/199 (1999) 1230-1234.
[49] D. Gogova, K. Gesheva, A. Kakanakova-Georgieva, M. Surtchev, *EPJ Applied Physics.* 11 (3) (2000) 167-174
[50] D. Gogova, G. Stoyanov, K.A. Gesheva, *Renewable Energy.* 8 (1-5) (1996) 546-550.
[51] V. G. Syijrkin, *Metal Carbonyls, Chimii,* Moscow, 1984 (in Russian).
[52] T. R. Albrecht, S. Akamine, T.E. Carver, C.F., *J. Vac. Sci. Technol.* A 8(4) (1990) 3386-3396.
[53] S. Alexander, L. Hellemans, O. Marti, J. Schneir, V. Elings, P.K. Hansma, M. Longmiro, J. Gurley, *J. Appl. Phys.* 65(1) (1989) 164.
[54] K.A. Gesheva, T. Ivanova, *Chemical Vapor Deposition.* 12 (2006) 231.
[55] L. Sequin, M. Figlarz, R. Cavagnat, J. Lasseques, *Spectroch. Acta.* A 51 (1995) 1323.
[56] G. Nazri, C. Julien, *Solid State Ionics.* 80 (1995) 271.
[57] T. Ivanova, K. Gesheva, A. Szekere s, *J. Solid State Electroch.* 7 (2002) 21-24.
[58] C. Julien, B. Yebka, G.A. Nazri, *Mater. Sci. Engin.* B 38 (1996) 65.
[59] T.S. Sian, G.B. Reddy, *Solid State Ionics.* 167 (3-4) (2004) 399.
[60] W. Dong, A. Monsour, B. Dunn, *Solid State Ionics.* 144 (2001) 31.
[61] Lewis I.R., Edwards, H.G.M., Eds.; Marcel Dekker: "Handbook of Raman Spectroscopy: From the Research Laboratory to the Process Line;," New York, 2001
[62] Wu, D., Li, A.-D., Ge, C.-Z., Lu, P., Xu, C.-Y., Xu, J., Ming, N.-B. *Thin Solid Films.* 322 (1-2), (1998) 323.
[63] E. Haro-Poniatowski, C. Julien, B. Pecquenard, J. Livage, M.A. Camacho-López, *J. Materials Research.* 13 (4) (1998) 1033.
[64] S. Lee, M. Seong, C. Tracy, A. Mascarenhas, J. Pitts, S.Deb, *Solid State Ionics.* 147 (2002) 129.

[65] M. Dieterle, G. Weinberg, G. Mestl, *Physical Chemistry Chemical Physics.* 4 (5) (2002) 812.
[66] G.M. Ramans, J.V. Gabrusenoks, A.R. Lusis, A. Patmalnieks, *J. Non-Crystalline Solids.* 90 (1987) 637.
[67] Agrawal, Solar Energy Mater. *Solar Cells.* 31 (1993) 9.
[68] S. Bosch and F. Monzonis, *J. Opt. Soc. Am.* A 12, (1995) 1375.
[69] M. Ghezzo and K. M. Busen, *J. Phys. D: Appl. Phys.* 2 (5) (1969) 655.
[70] A. Szekers, T. Ivanova, K. Gesheva, *J. Solid State Electroch.* 7 (2002) 17.
[71] M. Anwar, C. A. Hogarth, R. Bulpett, *J. Materials Science.* 24 (1989) 3087.
[72] C. Julien, A. Khelfa, J. P. Guesdon, A. Gorenstein, *Applied Physics A: Materials Science and Processing.* 59 (1994) 173.
[73] G. Benno, K. Joachim, "Optical Properties of Thin Semiconductor Films", 2003, http://users.physik.tu-muenchen.de/jkopp/pdf/F/hl_spekt.pdf
[74] J. Scarminio, A. Lorenco, A. Gorenstein, *Thin Solid Films.* 302 (1997) 66.
[75] V. Sabhapathi, O.M. Hussain, P. Reddy, K. Reddy, B. Uthana, B. Naidu, *Phys. Stat. solidi (a).* 148 (1995) 167.
[76] A.Abdellaoui, G. Leveque, A. Donnadoeui, A. Bath, B. Bouchikhi, *Thin Solid Films.* 304 (1997) 39.
[77] C. Julien, B. Yebka, G.A. Nazri, *Materials Science Engineering.* B 38 (1-2) (1996) 65
[78] K.A. Gesheva, A. Scekeres, T. Ivanova, *Solar Energy Mater. Solar Cells.* 76 92003) 563.

Chapter 2C

INVESTIGATION OF CVD MIXED OXIDE FILMS BASED ON MOLYBDENUM AND TUNGSTEN

T. M. Ivanova and K. A. Gesheva
Central Laboratory of Solar Energy and New Energy Sources,
Bulgarian Academy of Sciences, Sofia, Bulgaria

2.C.1. APPLICATIONS OF MIXED OXIDE SYSTEM WO_3 - MoO_3

The group of inorganic materials possessing chromogenic properties is the oxides of the elements located in a well – defined region in the Periodic table [1, 2] and these elements are transition metals. Furthermore, it was discovered that different parts of that region pertain to oxides with cathodic and anodic coloration. So far, the tungsten oxide is the most intensively investigated electrochromic material [1], applied in the first prototypes of electrochromic windows. WO_3 possess a defected structure within the anion sublattice and oxygen vacancies being the predominant defects. Tungsten oxide crystals have perovskite-like atomic configuration based on the corner-sharing WO_6 octahedra. In general, tungsten oxide has a tendency to form sub-stoichiometric phases. Due to this structure, amorphous and polycrystalline WO_3 thin films possess a high capacity for reversible Li^+ insertion and highly reversible intercalation and transport properties [3-6]. The tungsten oxide films also show high color efficiency and relatively low prize as well it is non-toxic material. Tungsten oxide films show well expressed photochromic and gasochromic behaviour. Thin films of WO_3 are studied as gas sensor with respect of their microstructure and film morphology, as photosensitive semiconductor electrodes [7, 8] and as well as diodes.

Mixed oxide systems have been proposed in order to modify the physico-chemical properties. Experimental research over the past decades and the great interest that have been devoted to the electrochromic materials and their practical large-area applications require optimization of the electrochromic properties and the improvement of the existing electrochromic thin films. Possible beneficiary effects on the EC host materials are the increased color efficiency, the improved cycling lifrtime, the color neutrality and a larger switching potential range or the faster reaction kinetics. Mixed oxide films as electrochromic materials were reviewed by Grangvist [1] and Monk [2].

The advantages of the mixed oxide based on the transition metals for electrochromic-based technology can be summarized as follows:

- is challenging and of considerable interests for practical applications.
- The increase of the switching applied voltages range.
- The combination of two or three transition metal oxides can achieve a higher optical modulation that means a greater diff The mixed oxide films often exhibit favorable, usually neutral, colors.
- Improving of the device durability (the cycling lifetime).
- The ability to produce a higher luminous CE in the mixed films than for the pure constituents erence between the transmittance in bleached state and the transmittance in colored dark state.

The advantages of the pure components can be combined into mixed oxide system and in the same time attempting to avoid or to diminish the disadvantages as low as possible.

Mo trioxide has showed many advantages, one of which is the absorption band close to sensitivity of the human eye. It has to mention that molybdenum oxide films are still not very used as active electrochromic layer due to their unstability in acid and base media. That's why WO_3 mixed with molybdenum oxide system (with general formulae - $Mo_xW_{1-x}O_3$) has been studied aiming to improve the electrochromic properties and to unite the promising properties of the individual oxides. Mixed oxide films are investigated as well for gas sensitive materials. Electrochromic Mo-W oxide films can be deposited by electrodeposition, simultaneously evaporation of the two oxides in vacuum, sol-gel and CVD techniques [9-11].

Galatsis et al. [12] deposited sol-gel films of WO_3 - MoO_3 for usage as oxygen gas sensors. They found out that the mixed oxides are inhomogeneous in structure due to the differences in their vapor pressures of the pure materials - WO_3 and MoO_3. The obtained thin films had been analyzed and reported to possess improved gas sensing characteristics in comparison with the individual oxides.

Khawaja et al. [13] had focused their study on the optical properties of the evaporated Mo - W oxide films. They also found inhomogeneity of the mixed films.

Cao et al. [14] investigated the preparation of WO_3 - MoO_3 films by chemical vapor deposition using the corresponding hexacarbonyls as precursors. The carbonyls were physically mixed, then sublimated and carried out towards the CVD reactor, where the oxide films had been deposited in the oxygen rich ambient. The authors reported that the stoichiometry of the obtained films can be controlled according to the reaction:

$$2aA(CO)_6 + 2вB(CO)_6 + 9O_2 \rightarrow 2A aBвO_3 + 12(a + в)CO_2, \text{ where } a + в = 1$$

The mixed film structure and the stoichiometry had been studied by X-Ray diffraction analysis.

Papaefthimiou et al. [15] performed a study of the working characteristics for the experimental electrochromic cells incorporated WO_3 and MoO_3, and then compared with EC device using WO_3 - MoO_3 film as the working electrode.

Genin et al. [16] and Figlarz [17] described the dimensional space structure of WO_3 - MoO_3 system obtained by the method "soft chemie douce". WO_3 structure is ReO_3 type, a three-dimensional sequence from WO_6 octahedra with common corners, where the distortion depends on temperature. On the other hand, MoO_3 possesses the layered orthorhombic structure, builded from the connected octahedral sharing corners and edges. The crystallographic configurations formed in the mixed oxide system, $Mo_xW_{1-x}O_3$, can be characterized with a ReO_3-like structure for $x \leq 0.95$ (a greater content of the tungsten oxide) and increasing the Mo including the structural arrangements are more likely the orthorhombic α - MoO_3 modification.

Granqvist [1] showed that bulk $Mo_xW_{1-x}O_3$ crystals have hexagonal, orthorhombic or monoclinic structures in the range of the composite content and depending on the deposition method. WO_3 similar structure dominates up to $x \leq 0.96$, the layered MoO_3 like arrangement is found at $0.96 < x < 1$. It is supposed that the electrochromic effect is influenced from the value of x and the quantity of injected ions y in a very characteristical way. Taking in mind that the absorption is due to the charge transfer transitions, in the mixed oxide system are expected to take place electronic transitions from the types: $Mo^{5+} \rightarrow Mo^{6+}$, $W^{5+} \rightarrow W^{6+}$ and $Mo^{5+} \rightarrow W^{6+}$. Mo states are with a lower energy than the corresponding tungsten levels. When $x<y$, the inserted electrons are trapped at the Mo sites and the transition $Mo^{5+} \rightarrow Mo^{6+}$ and $Mo^{5+} \rightarrow W^{6+}$ are possible, whereas in the case $y<x$, the Mo^{6+} sites are saturated and the transitions $Mo^{5+} \rightarrow W^{6+}$ and $W^{5+} \rightarrow W^{6+}$ can take place. In that way to the absorption peak of the mixed oxide possesses two different contributions: depending on the values of x and y.

2.C.2. TECHNOLOGICAL PARAMETERS FOR CVD PREPARATION OF $W_xMo_{1-x}O_3$ THIN FILMS

The mixed oxide films based on tungsten and molybdenum are also prepared by CVD method using our previous experiences on the depositions of WO_3 and MoO_3 films. The precursors are the molybdenum and tungsten hexacarbonyls. The two chemicals are both powders slightly yellowish crystal compounds, inert to the air at room temperatures.

Metal oxide films have been deposited by the pyrolytical decomposition of the corresponding hexacarbonyls ($W(CO)_6$, $Mo(CO)_6$ or their physical mixture in a ratio of $Mo(CO)_6$:$W(CO)_6$=1:4) in argon-oxygen at the atmospheric pressure (APCVD process) in the horizontal cold walls CVD reactor [18-21]. The precursor powder placed in the sublimator immersed in the silicon oil bath has been heated at the temperature of 90°C and has been controlled with an accuracy of ±1°C. This sublimator temperature provides a sufficient vapor pressure of the hexacarbonyls. The gas lines, made of "Galtek" type teflon were heated up to at least the temperature of the sublimator, in order to assure successful transport of the vapors to the CVD reactor. Argon (99.995%) flow carries the precursor vapors to the reactor. The selected flow rate of argon through the sublimator assures a constant amount of the precursors vapor. Through a separate line oxygen (99.95%) enters the reactor. In the present study the ratio of flow rates of argon to oxygen is 1:32.

For the comparative study the deposition temperature was kept at 200°C because only at that temperature it was possible for the three kinds of the metal oxide films to be obtained. WO_3 can be prepared at higher temperatures (up to 400°C) and the growth rate increases with increasing the temperature. The growth rate of MoO_3 decreases with increasing the temperature due to the gas phase reaction [21, 22]. From that point of view, the temperature of 200°C is a cross point for the deposition of the mixed oxide films. The deposition time was kept constant (40 min) and because of the different growth rates the film thickness was 300 nm for MoO_3, 400 nm for WO_3 and 120 nm for mixed oxide films.

The sublimation temperature of the carbonyls is determined from their density – for $Mo(CO)_6$ is 1.96 g/cm^3 and for $W(CO)_6$ - 2.65 g/cm^3. From the formula (number 2 given in Chapter 2B) can be estimated the vapor pressures at the given temperatute:

Carbonyls	A	B	P (vapor pressure) at T=90°C [kPa]
$Mo(CO)_6$	1.17	373.8	1.04×10^{-3}
$W(CO)_6$	1.09	364.0	1.12×10^{-3}

A second set of the mixed oxide films had been deposited by using two separate vapor sources. The corresponding carbonyls were put in two different vessels and two silicon oil baths. The sublimator temperatures were 110°C ($W(CO)_6$) and 90°C for $Mo(CO)_6$. The substrate temperature was kept 200°C. The gas flow ratio of the two Ar flows carrying the carbonyls vapors to oxygen flow was 1/16 and deposition time was 40 minutes. The obtained films were compared with pure MoO_3 films deposited at similar deposition conditions.

2.C.3. AUGER ANALYSIS OF MIXED OXIDE FILMS ON THE BASIS OF MO AND W OBTAINED BY CVD METHOD

Auger spectroscopy has been used for the determination of the chemical compositions as well as the depth profile of the elements distributions in the mixed oxide thin films. Auger results show that APCVD mixed films deposited from the precursors situated in two different vapor source chambers, contain less than 10% molybdenum nevertheless our attempt to assure the equal vapor pressures by keeping the source chambers at the proper temperatures (90°C for the molybdenum precursor and 110°C for $W(CO)_6$. Obviously, at the substrate temperature used (200°C), some kinetic and thermodynamic reasons exist, which favor the deposition of tungsten oxide, and less - the molybdenum oxide. Although pure tungsten oxide used to grow at 200°C very slow [23] here, in presence of molybdenum vapors it grows with preference.

The molybdenum precursor somehow acts as a catalyst precursor for the growth of tungsten oxide. In result, as figure 1 presents, to the substrate about 35% W and less than 5% Mo are included in the film. For the annealed film, the results are similar: W-36%, Mo less than 3%, the rest is carbon and oxygen. For the other type mixed films, deposited from a precursor-physical mixture, the Auger results presented on figure 2 are as follows: near the film surface rich of Mo-23%, W less than 8%, deeper in the film the opposite situation is found out. There is the film rich of tungsten – 30%, Mo-less than 8%, the other is 9% carbon and 53% oxygen. It is seem that closed to the heated substrate, WO_3 in the presence of the molybdenum carbonyl vapors first starts growing and when the tungsten oxide film is formed, the molybdenum oxide in the film begins increasing favored probably by the solubility of the two oxides. After the annealing at 400°C, the Auger data reveal that near to the surface W is less than 4%, while Mo is 26%, and deeper in the film, close to the substrate, the corresponding values are W is 20%, and Mo is less than 10%. In this thickness level of the film, oxygen is 57% and carbon 7%. As the authors H. A. Wreidt (for Mo-O) and L. Brewer and R. H. Lamoreaux (for O-W) had been reported [24] the phase diagrams show as an evident that it is possible to form the two oxides in the temperature range 0-500° C for MO_x (up to 200°C would grow solid MoO_3, gas phase of it, as well, also Mo_8O_{23} suboxide) and a number of other suboxides at the end of the temperature range) and from 200-400°C for WO_3.

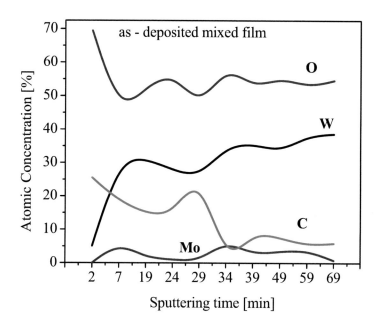

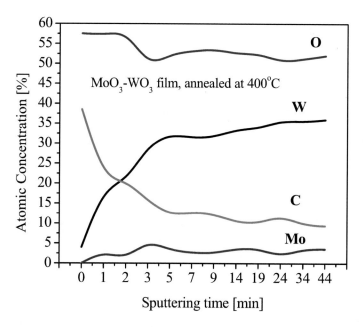

Figure 1. Auger spectra of the as-deposited and the annealed at 400°C mixed MoO_3-WO_3 films, obtained from Mo and W carbonyls situated in the separate vapor source chambers. The ratio between the carbonyl vapors and the oxygen is 1:16.

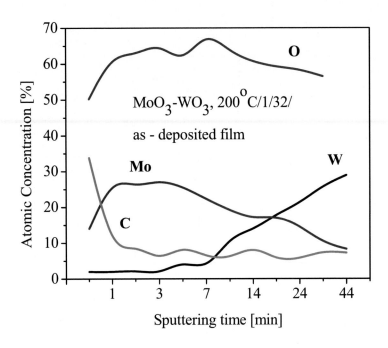

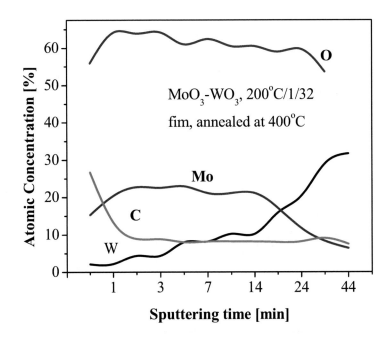

Figure 2. Auger spectra of the as-deposited and the annealed at 400°C MoO$_3$-WO$_3$ films, obtained from Mo/W precursor made as a mixture in proportion 1:4 in favor of W(CO)$_6$. The gas ratio between the precursor vapors and the oxygen gas flow-rate is 1:32.

Investigation of CVD Mixed Oxide Films Based on Molybdenum and Tungsten   217

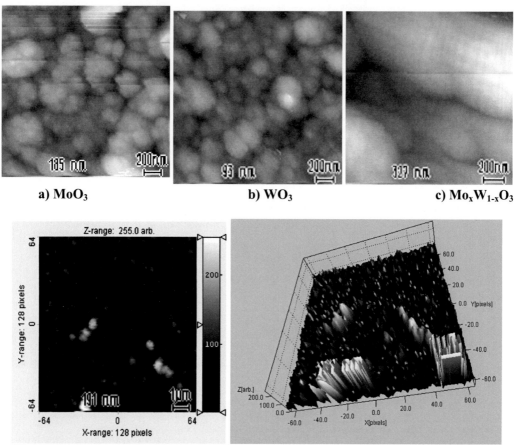

a) MoO_3 b) WO_3 c) $Mo_xW_{1-x}O_3$

D) AFM images of WO_3 films, annealed at 300°C in the smallest magnification in 2D and 3D views.

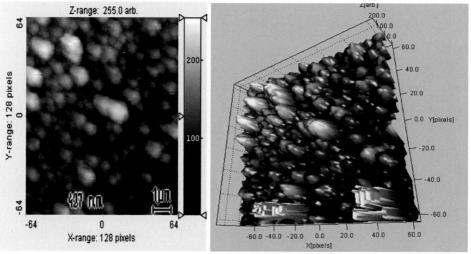

E) AFM images of as-deposited MoO_3 - WO_3 films in the smallest magnification, 2D and 3D views.

Figure 3. 2D and 3D AFM images of annealed at 300°C CVD MoO_3 and WO_3 and also as-deposited mixed Mo-W oxide films (the precursor is a mixture of Mo and W carbonyl powders).

2.C.4. SURFACE MORPHOLOGY OF APCVD MoO₃, WO₃ AND Mo/W OXIDE FILMS

The morphology was determined by AFM technique for CVD MoO_3-WO_3 films, deposited by applying the two approaches. Figure 3 presents the micrograph of the as-deposited state of mixed films compared with annealed at 300°C WO_3 and MoO_3.

The 3D images show that the mixed oxide film structure in as-deposited state consists of much larger grains, size 327 nm, compare to the smaller grains of MoO_3 even after annealing - average grain size of MoO_3 is 185 nm and for WO_3 (93 nm).

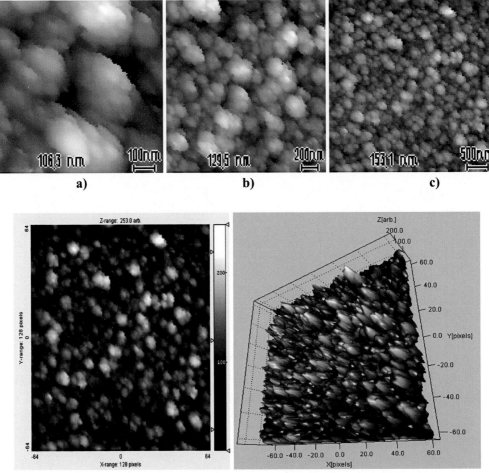

C1) 2 D view and 3D view of the annealed CVD-mixed oxide film (smallest magnification)

Figure 4. AFM images of CVD Mo-W oxide films, annealed at 400°C. Films are deposited by using of Mo and W carbonyl precursors situated in separate vapor source chambers.

MoO_3 film, obtained at 200°C and annealed at 300°C has an average surface roughness S_a - 31.4 nm and average cluster size –185 nm. Similarly obtained WO_3 film annealed at 300°C has an average surface roughness Sa - 24.8 nm and average cluster size –93 nm. The surface average roughness of the mixed films is S_a - 45.7 nm and the average cluster size –327 nm. It should be pointed out that the thickness of the films is different, although the technological conditions at which the films grow are equal. Because of the growth rate difference, the WO_3 film has the smallest thickness, the grains are smallest, and the film surface roughness has also the smallest value. In general, thin films surface roughness increases with increasing in their thickness. With respect to the pure component films (MoO_3 and WO_3 films), the AFM images of the film morphology confirm our previous conclusions that CVD MoO_3 films have higher degree of crystallization even at 300°C (Raman, TEM studies), while WO_3 and the mixed oxide films stay predominantly amorphous. As annealed at 400°C the mixed films stay still amorphous-like (figure 4).

Figure 5 presents the SEM micrographs of CVD mixed oxide films, obtained from mixed carbonyl mixture and after annealing. It can be clearly seen that the morphology of the CVD film consists of randomly distributed grains. The enclosed observation of the film surface reveals that the grains are not homogeneous; they looked as agglomerates with a different contrast through the whole grain. The same feature has been observed in studies by AFM analysis [10] of the similar obtained CVD oxide films. This effect of some kind of agglomeration or clustering has been seen for the pure MoO_3 and WO_3 films, deposited at the same technological conditions.

Figure 5. SEM micrographs of CVD MoO_3-WO_3 films, obtained from the carbonyl mixture annealed at 500°C and deposited on conductive glass substrate.

2.C.5. INVESTIGATION OF OBTAINED FROM MIXED PRECURSOR APCVD $Mo_xW_{1-x}O_3$ FILMS

X-Ray Diffraction Analysis of CVD Mixed Oxide Films

The development of crystalline phases of pure CVD Mo and W oxide films comparing with mixed oxide films, due to thermal annealing at 400°C, has been analyzed by large-angle XRD measurements [25]. The obtained spectra and XRD data are summarized in figure 6, where the XRD patterns for the MoO_3, WO_3 and MoO_3-WO_3 films are given.

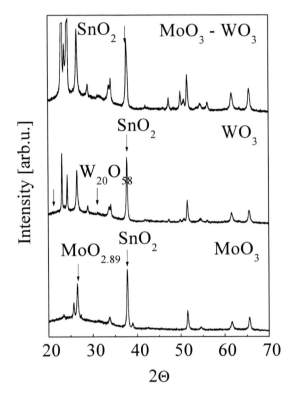

Figure 6. XRD spectra for MoO_3, WO_3 and mixed MoO_3-WO_3 oxide films deposited on the conductive glass substrates at the deposition temperature of 200°C and annealed at 400°C.

Most of the XRD peaks found are specific for the corresponding trioxides. The non-specific peaks are pointed out in the spectra, corresponding to the crystalline conductive layer beneath the studied films of MoO_3, WO_3 and MoO_3-WO_3. The appearance of the sharp XRD peaks above the broad amorphous one reveals that all the three kinds of oxide films contain crystalline phases. However, it is difficult to separate the crystalline part from the amorphous one, because the structure of the glass substrates is also amorphous.

In the case of the MoO_3 films, the orthorhombic molybdenum trioxide and the triclinic $MoO_{2.89}$ sub-oxide have been identified among the crystalline phases (see figure 6). Analysing the ratio of the peak intensities, it has been established that the orthorhombic MoO_3 crystallites prefer to grow in the <010> direction. This preferential direction was previous determined for CVD MoO_3 films by XRD study (see Chapter 2B). The sub-oxide percentage, however, could not be determined because the corresponding X-Ray lines coincided with the strongest XRD peak of the crystalline SnO_2 used as a conductive layer on the glass substrate. This SnO_2 peak appeared in all the XRD spectra. The larger part of the crystalline phases in the tungsten oxide films consisted of triclinic WO_3, and there was less than 2 % of $W_{20}O_{58}$ sub-oxide (figure 6). The XRD pattern of the mixed MoO_3-WO_3 film was very similar in shape to that of the WO_3 film. Although the peak positions were the same, the peak intensity was strongly enhanced, because it is suggested that W atoms are partly substituted by Mo atoms. Obviously, in the mixed oxides, (W, Mo) O_3 crystallites are present with the same triclinic modification found for the pure WO_3 film. It has been suggested that the Mo atoms replaced some of the W atoms in the triclinic WO_3 lattice.

IR Study of $Mo_xW_{1-x}O_3$ Films Compared to MoO_3 and WO_3

The three types of CVD metal oxide films are visually distinguished in color in their as-deposited state. The films look slightly colored either in blue (WO_3), in yellow (MoO_3) or as a bluish with brown suggestion (mixed oxides) tint [20]. The CVD thin films have been analyzed by IR spectroscopy in the spectral range 200 – 1200 cm^{-1}. The IR spectra reveal (as shown at figure 7) that there is a shifting of the main absorption band of the mixed $Mo_xW_{1-x}O_3$ films between the principal bands of the pure metal oxides. This effect could be considered as an indication of the existence of a new phase, expected to be a mixed MoO_3 - WO_3 phase. IR spectra of the as-deposited WO_3, MoO_3 and $Mo_xW_{1-x}O_3$ films manifest very broad and intense main absorption bands, supposing almost fully amorphous structures. For the molybdenum oxide film (figure 7) there exists an IR absorption line centered at 790 cm^{-1}, meanwhile for WO_3 the characteristical band is located at 690 cm^{-1}. It can be clearly observed that the IR band of the mixed oxide film is situated between the the main bands of the pure components and it is closed to 720 cm^{-1}. No other characteristic IR lines have appeared in the spectra of $W_xMo_{1-x}O_3$ and WO_3, at the same time the spectrum of MoO_3 reveals two additional bands at 572 and 998 cm^{-1}. The absorption band at 790 cm^{-1} is assigned to Mo - O stretching vibrations in the amorphous oxide network [26], 572 cm^{-1} is reported to be due to the bending vibrations of Mo-O-Mo bonds [27]. The interesting feature is the IR band placed at 998 cm^{-1}. It is connected with the existence of the double Mo=O bonds (called molybdyl group) and correspondingly to the vibrations of the terminal oxygen atoms. These molybdyl bonds are responsible for the unique layered arrangement of the orthorhombic molybdenum trioxide [26]. The appearance of these IR bands propose that the molybdenum oxide film structure has contained not only amorphous fraction, but as well a small crystallized presence, which is suggested to be a mixture of the orthorhombic and monoclinic ($\alpha+\beta$) phases. The IR band at 720 cm^{-1} of the WO_3 film is related with the stretching W-O vibrations [28]. For the case of the mixed oxide film, the observed IR band is most probably due to the overlapping of the stretching vibrations of different Mo-O and W-O groups [29].

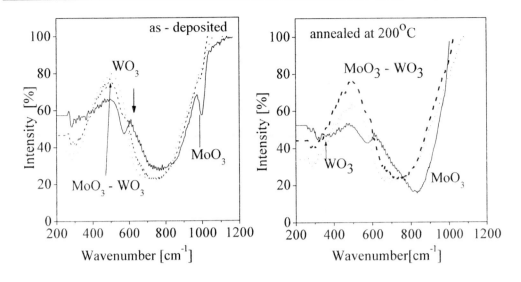

Figure 7 and figure 8. IR spectra of as deposited and 200°C annealed films of MoO_3, WO_3 and MoO_3-WO_3.

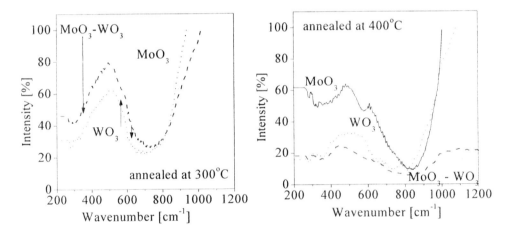

Figures 9 and 10. IR spectra of MoO_3, WO_3 and MoO_3-WO_3 annealed at 300 and 400°C.

After annealing at the temperature of 200°C, the IR spectra uncover that MoO_3 is definitely different from the other two oxide materials (see figure 8). The main band of the molybdenum oxide film is shifted to 834 cm^{-1} and the 570 cm^{-1} line appeared again. On the other hand, the spectra of WO_3 and $W_xMo_{1-x}O_3$ express their main bands at 700 and 747 cm^{-1}, respectively [30, 31]. The IR band at 834 cm^{-1} is related with the stretching vibrations of Mo - O - Mo bonds appearing in the three dimensional crystal structure of β - MoO_3 [27].

Increasing the annealing temperature to 300°C (figure 9), leads to the same tendency of the similarity of WO_3 and $W_xMo_{1-x}O_3$ IR spectra differing from the MoO_3 spectrum. The main band of $W_xMo_{1-x}O_3$ film can be again found between the main bands of the pure oxides, which are at 809 cm^{-1} for MoO_3, at 690 cm^{-1} for WO_3 and for the mixed system at 732 cm^{-1}, respectively. The new band at 280 cm^{-1}, is related to the deformation modes of the crystal lattices [25]. The annealing at 400°C causes broadening of the absorption bands of the tungsten oxide and the Mo/W oxide films as can be observed from figure 10. IR bands at 841, 572 cm^{-1} and weak lines around 328, 312 and 280 cm^{-1} have been exhibited for the MoO_3

film. The very broad band of WO_3 can be expressed three very weak maximums at 807, 721 and 638 cm^{-1}. Where the IR bands located at 638 and 807 cm^{-1} are supposed to be connected with the stretching W-O vibrations of the crystalline monoclinic structure of WO_3 [29] and the 721 cm^{-1} line is due to W=O bending vibrations [32]. $W_xMo_{1-x}O_3$ spectrum exhibits a featureless shape without well defined bands. The possible explanation for such behavior might be with a beginning of a certain structural transformation. Later on, Raman study reveals that the mixed oxide system is amorphous with small traces of crystallization appeared after the high temperature annealings at the temperatures above 400°C [33].

The IR spectra recorded for the three types of films after 500°C annealings (figure 11), have presented the different behavior once again of MoO_3 film with the main band located at 834 cm^{-1}. The IR bands of $W_xMo_{1-x}O_3$ and WO_3 are broad with the greater overlapping of the two spectra with three maximum observed at 810, 717 and 640 cm^{-1}.

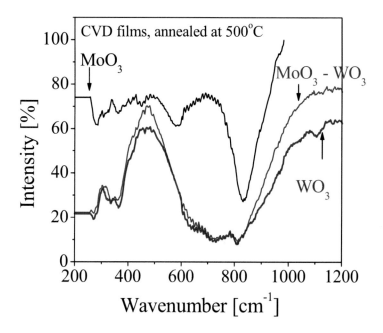

Figure 11. IR spectra of the 500°C annealed films of MoO_3, WO_3 and MoO_3-WO_3.

IR study leads to the conclusion that although the carbonyls mixture contains higher amount of $Mo(CO)_6$, into the CVD obtained films have the predominant presence of WO_3 fraction. That conclusion has been proved by Auger analysis (see chapter 2C 3). The shifting of the main absorption band of MoO_3-WO_3 films between the two absorption bands of the pure components can be the proof for the appearance of the mixed phase [33].

Raman Investigation of CVD Mixed Oxide Compared to MoO_3 and WO_3 Films

The sets of CVD thin films were investigated by Raman spectroscopy to reveal the bonding and their crystalline structure. An observation of the WO_3, MoO_3 and Mo/W oxide film surfaces under X100 magnification of the optical microscope mounted on the Raman equipment showed that the film surfaces undergo some changes depending on the annealing temperatures as the most common surface features observed are the big prolonged (up to 2-3 µm) or the shaped as cauliflower crystallites [34].

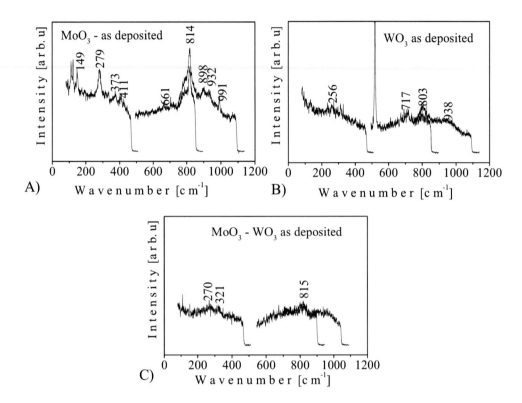

Figure 12. Raman spectra of as-deposited MoO_3, WO_3 and $Mo_{1-x}W_xO_3$ thin films.

On figure 12 are presented Raman spectra of the as-deposited films, where the band shapes suggests that the films are probable completely amorphous in structure. $Mo_xW_{1-x}O_3$ film exhibits a very broad Raman spectrum without expressed Raman lines, which are the signs for the amorphous-like material. Similar spectrum has been shown by the tungsten oxide film, where weak peaks have been detected at 256, 717, 803 and 938 cm^{-1}. This behaviour also assumed predominantly amorphous fraction. Raman lines at 717 and 803 cm^{-1} can be assigned to the stretching vibrations of W-O bonds in the crystalline monoclinic WO_3 [35]. The band, located around 938 cm^{-1} is related to the double W=O bond [32]. But the intensities of the observed Raman peaks are too weak to assume higher degree of a crystal phase. Only Raman spectrum of the molybdenum oxide film has revealed some intense lines, proposing the beginning of the crystallization with the greater percentage of amorphous phase. The peaks located at 149 and 279 cm^{-1} in the region beneath 300 cm^{-1} are due to the deformation modes of O-Mo groups [25]. Raman line at 373 cm^{-1} is observed by other authors and is connected to orthorhombic MoO_3. The line situated at 611 cm^{-1} can be related to stretching Mo-O vibrations, where oxygen is coordinated with three different Mo atoms. Raman line located at 814 cm^{-1} is assigned to the alternative bond lengths in MoO_6 octahedra and to the bridging O-Mo-O bond [36]. This Raman peak is characteristic for α - structure. 932 cm^{-1} line is a sign for Mo=O bonds in polymorph $Mo^{6+}O_6$ groups [37]. It is worth noting that for these technological parameters, MoO₃ film possesses orthorhombic structure without other phases. Even the as-deposited films exhibited Raman lines proving the presence of the layered orthorhombic arrangement. The characteristic double bond Mo=O is represented by 991 cm^{-1} peak, which is due to the terminal oxygen Mo^{6+}=O stretching mode as a result of unconnected oxygen.

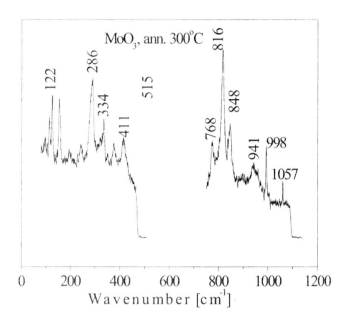

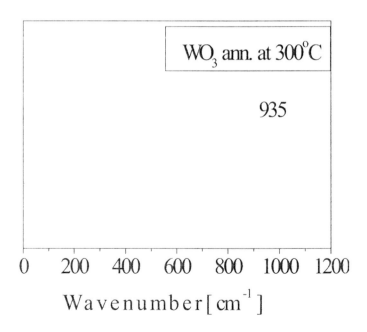

Figure 13 (continued)

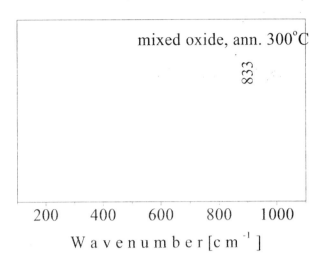

Figure 13. Raman spectra of MoO_3, WO_3, and $Mo_{1-x}W_xO_3$ films, annealed at 300°C.

After annealing at 300°C, Raman spectra of $Mo_xW_{1-x}O_3$ and WO_3 exhibit no characteristic features again (see figure 13), showing the amorphous structure of the films. For MoO_3, Raman spectrum reveals sharp intense lines, suggesting a complete crystallization. The strong peaks at 816 and 998 cm^{-1} are characteristical for α - MoO_3 as well as the other well defined peaks. Similar behavior has been observed after the films annealing at the temperature of 400°C. For WO_3, there are indications for a starting of the crystallization, marked by Raman peaks appeared at 109, 139, 171, 271, 321, 372, 717, 804, 930 and 971 cm^{-1} (see figure 14). Most of them are weak, while the two Raman bands at 717 and 804 cm^{-1} are of considerable intensity. Raman peaks found beneath the 200 cm^{-1} region are related to the lattice vibrations of the crystalline WO_3 [35]. The peaks at 271, 321 cm^{-1} are attributed to the bending vibrations of the crystalline oxide [32], as the bands at 717 and 803 cm^{-1} are characteristical for the stretching W-O vibrations of the monoclinic WO_3. Raman lines situated at 938 and 971 cm^{-1} can be assigned to the stretching vibrations of the doble bonds W=O. Raman spectrum of MoO_3, annealed at 400°C (figure 14) shows less peaks then the molybdenum oxide film, which have undergone annealing as well, at the lower temperature of 300°C. The two Raman main lines at 817 and 995 cm^{-1}, proving the existence of orthorhombic phase, are still strong. The mixed Mo/W oxide film presents predominantly amorphous features as can be seen from its spectrum. The weak Raman bands have appeared at 267, 317, 727, 817 and 942 cm^{-1}. The Raman line, located at 727 cm^{-1} can be due to the both Mo-O and W – O vibrations, as the broad band centered at 817 cm^{-1} has been most probably attributed to the Mo- O stretching mode of the orthorhombic MoO_3. The very broad absorption band at 942 cm^{-1} is an overlapping of the stretching vibrations of the Mo = O and W = O bondings, which is a result from the presence of the two oxides in the film structure.

The annealing at 500°C contributes for the crystallization process (see figure 15), although WO_3 and $Mo_xW_{1-x}O_3$ films still exhibit a high degree of amorphous phase. The main Raman lines at 661, 817 and 995 cm^{-1} relating to the stretching type of vibrations can be connected with the orthorhombic α phase of MoO_3. The best pronounced Raman peaks for CVD WO_3 films appear at 265, 713 and 802 cm^{-1}. The mixed oxide spectrum reveals Raman lines at 718 and 825 cm^{-1} as well as a broad band at 967 cm^{-1}.

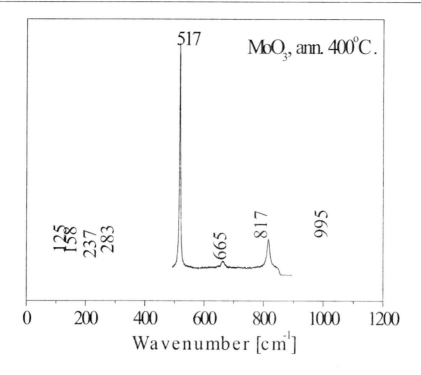

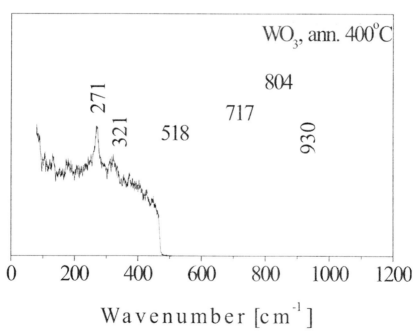

Figure 14 (continued)

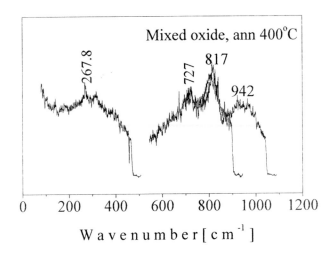

Figure 14. Raman spectra of CVD MoO_3, WO_3, and mixed $Mo_{1-x}W_xO_3$ films, annealed at 400°C.

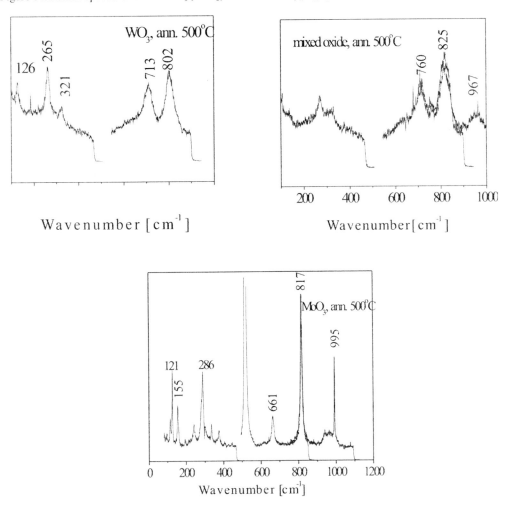

Figure 15. Raman spectra of CVD WO_3, mixed $Mo_{1-x}W_xO_3$ and MoO_3 films, annealed at 500°C.

The Raman study states that the MoO$_3$ films deposited at certain process parameters crystallize into the unique layered structure, favorable for expressing electrochromic properties. Therefore, the Raman data and the exhibited peaks indicate the presence of the crystalline MoO$_3$ phases in an amorphous WO$_3$ framework, where the fraction of the molybdenum oxide grows predominantly in α - phase. Raman data clearly shows that the pure MoO$_3$ films crystallize in the orthorhombic α – phase as even the as-deposited film shows the traces of the crystallization. From other hand, the mixed oxide and the pure WO$_3$ stay mainly amorphous, showing a slight influence of the thermal treatments as the amorphous features exist in all their Raman spectra even after the annealing at 500°C.

Optical Properties of CVD Mo$_x$W$_{1-x}$O$_3$ Films Compared with MoO$_3$ and WO$_3$

Spectrophotometry and spectral ellipsometry have been applied for the optical characterization of the metal oxide films, obtained by APCVD techniques. Their optical properties have been compared to study the impact of combining of two individual transition metal oxides in one system. On figures 16-18 are presented the transmittance and reflectance curves for the as-deposited (curves 1) and for the annealed films (curves 2). The transmittance in the visible wavelength range for all the films (MoO$_3$, WO$_3$ and MoO$_3$-WO$_3$) varies within 60-80 %, but a direct comparison of the film transparencies cannot be made due to the different film thicknesses. Correspondingly, the values of the reflectance range around 20 %. After the additional annealings, MoO$_3$ and WO$_3$ films become more transparent; while the transmittance of the mixed oxide films decreases in the spectral range of 400-750 nm, showing a slight increase towards the longer wavelengths (see figure 16 and 17). The thermal treatments do not affect considerebly the reflectance spectra of the pure oxides films, at the same time the annealings (especially at 400°C) result in a noticeable decrease in the reflectance, pertaing to the mixed MoO$_3$-WO$_3$ oxide films [33].

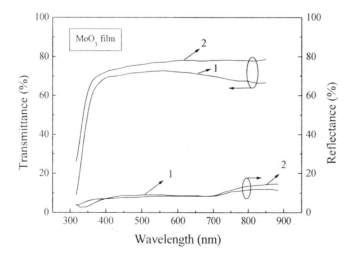

Figure 16. UV-VIS spectra of CVD MoO$_3$ films, recorded for the as-deposited (1) and the annealed at 400°C (2) states.

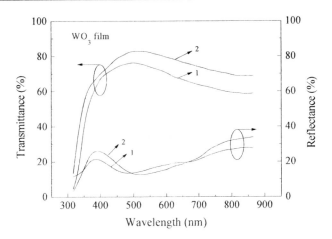

Figure 17. UV-VIS spectra of CVD WO_3 films, recorded for the as-deposited (1) and the annealed at 400°C (2) states.

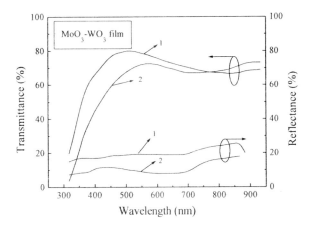

Figure 18. UV-VIS spectra of CVD WO_3 - MoO_3 films, recorded for the as-deposited (1) and the annealed at 400°C (2) states.

The different optical behaviour of the mixed oxide system towards the optical properties of the pure components is a very interesting and intriguing feature. From the technological point of view, this means that for CVD WO_3 - MoO_3 films no necessity of after-deposition temperature procedures is needed.

The absorption coefficient α has been determined by two independent methods: UV-VIS spectrophotometry and spectroscopic ellipsometry. Since the two dependences of α coincide very well, only the spectrophotometric results are shown in figure 19, showing the absorption spectra of the as-deposited and the annealed films. As it is seen, there is no maximum in the WO_3 optical absorption curve it looks like a plateau up to ~ 3 eV. The small maximum located at 1.7 eV in the spectrum of MoO_3 transforms in the mixed oxide spectrum into a broad and deep maximum at 1.8 eV. For the tungsten-molybdenum mixed oxide films had been found the absorption peak in the same energy range by Faughnan et al. [38]. After the annealing at 400°C, the absorption in the pure metal oxides decreases, while in the mixed oxide films significantly increases. It had been found out a certain decrease of α in these mixed oxide films with annealing, when they are deposited on the crystalline silicon substrates. This effect can be related to the different film structure. The amorphous glass substrate does not encourage the crystallization, while the crystalline Si substrate facilitates the crystallization into the films even during the deposition growth.

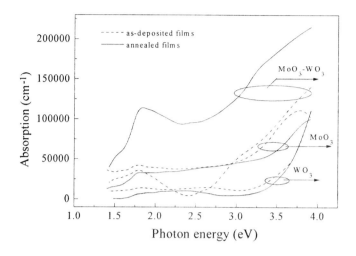

Figure 19. Spectral dependence of the absorption coefficient (α) for MoO_3, WO_3 and MoO_3-WO_3 for the as-deposited (dotted line) and after 400°C (solid line) annealed films.

For the MoO_3, WO_3 and MoO_3-WO_3 films in the as-deposited and annealed states, the optical band gap energies E_g have been derived from the spectrophotometry and the spectroscopic ellipsometry absorption coefficient data. At the absorption edge, the absorption rapidly increases due to the band-to band electron transitions. The absorption can be expressed by $\alpha h\nu \sim (h\nu-E_g)^\eta$, where $h\nu$ is the photon energy and $\eta=2$. Assuming indirect and allowed electron transitions in MoO_3 and WO_3 films [39], the optical band gap energies have been determined. However, for the mixed MoO_3-WO_3 films this assumption leads to difficulties of defining the value of the optical band gap, as it has been no sufficient linear part in the spectral dependence for making the extrapolation. Accepting the mechanism of the direct and allowed electron transitions as prevailing, the experimental data points defined a rather long linear part of the expression $\alpha h\nu \sim (h\nu-E_g)^\eta$, for the case of $\eta=1/2$. The extrapolation gives the real values of E_g.

The E_g values calculated are summarized in table 1. Any discrepancy in the results can be related to the different experimental accuracy and the systematical errors of each method. It is empirically established that the structural transformation from amorphous to crystalline phase causes an increase of the energy band gap values for MoO_3 [40] and a decrease for WO_3 [41]. The explanation of the optical band gap narrowing in a result of crystallization is still on a discussion level. For the annealed MoO_3 films, the E_g value increases to 2.7 eV, suggesting crystallization. For the as-deposited WO_3 films, the optical band gap value is 3.1 eV and remains unchanged after the annealing. This is an indication that 400°C annealing does not affect the film structure. Raman spectra of the tungsten oxide films show that WO_3 films stay amorphous after the annealing procedures.

Table 1. Optical band gap energy values estimated by the spectrophotometric (VIS) and by the spectroscopic ellipsometric (SE) data

CVD metal oxides		E_g (eV) VIS	SE
MoO_3	As-deposited	2.55	2.51
	Annealed at 400°C	2.7	2.92
WO_3	As-deposited	3.1	3.1
	Annealed at 400°C	3.1	3.07
MoO_3-WO_3	As-deposited	3.23	3.15
	Annealed at 400°C	2.93	3.0

For the as-deposited MoO_3-WO_3 films the estimated E_g value is 3.23 eV as after annealing, the optical band gap decreases to 2.93 eV. The change in the E_g value could be discussed by coexistence of the crystalline MoO_3 inclusions in the host amorphous WO_3 matrix. This suggestion is supported by the Raman and IR results, which reveal that the temperature annealings influence strongly the film structure of MoO_3, leading to complete crystalline films. These crystalline inclusions may probably define certain defects, the energy levels of which contribute to the band gap narrowing.

The refractive index of the studied metal oxide films had been determined from the ellipsometric measurements data. In general, the refractive index values vary within 1.8 - 2.6 in the spectral range of 350 - 850 nm. The film annealing leads to a slight decrease of the refractive index in the case of MoO_3 and MoO_3-WO_3, meanwhile the refractive index of WO_3 films remains almost the same. These results have been expected as they are in good agreement with the observed changes by the temperature of the film structures and the bond configurations [33-42]. The peculiarities of the refractive index dependences could be connected with some structural inhomogenities, such as amorphous and crystalline phases of different crystalline modifications. The lower index values of WO_3 films indicate that these films are probably more porous. For these films the relative density varies with the technological conditions and with the film thickness [43].

The values of the absorption (α) are determined from the corresponding extinction coefficient data. As it can be seen in figure 20, the absorption remains below 10^5 cm^{-1} for wavelengths up to the absorption edge, where it strongly increases due to band to band electronic transitions. The absorption of films before annealing is higher in the whole spectral region. In the lower energy-range there is a maximum in the optical absorption. The peak of that maximum for as-deposited films is situated at energy of ~ 2 eV and moves to ~1.8 eV for the annealed films. For MoO_3, as-deposited at the same conditions, absorption peak was observed at around 1.5 eV.

Our results are in good agreement with the observations that optical absorption peaks of all the mixed oxides are higher in energy than the peak of ether pure WO_3 (~1.4 eV) or MoO_3 (~1.56 eV) [44]. The optical absorption for as-deposited films has, in addition, a second maximum around 2.75 eV, which in the case of MoO_3 [45] appears at 2.5 eV. The enhanced absorption at energies below the absorption edge could be associated with proper type of defects in the metal oxides. Absorption bands in this energy range are also observed in [46] and they are associated with oxygen - ion vacancies in MoO_3 film.

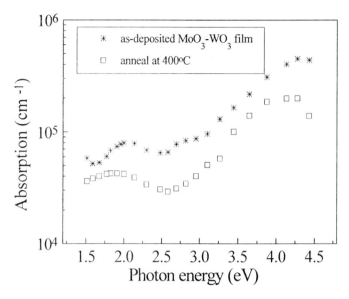

Figure 20. Absorption coefficient (α) determined for the as-deposited and the annealed at 400°C MoO_3-WO_3 films.

Electrochromic Behavior of MoO_3 - WO_3 Thin Films

CVD $Mo_xW_{1-x}O_3$ films were electrochemically characterized and compared to MoO_3 and WO_3. All the films show coloration as a result of Li intercalation. The expected reaction can be schematically written as:

$$Mo_xW_{1-x}O_3 + yLi^+ + ye^- \rightarrow Li_y Mo_xW_{1-x}O_3 \qquad (2)$$

To our knowledge, these are the first "in-situ" made APCVD MoO_3-WO_3 films with a strong electrochromic behavior. After changing the voltage polarity (extracting Li^+), the films bleach and become transparent.

The cyclovoltammetric curves show the most interesting feature: a rapid increasing of the current for MoO_3-WO_3 films as in the same time their transmittance decrease significantly as a result of the intercalation of Li ions. The mixed oxide films color in deep blue and the transmittance drops under 10% both for the as-deposited and the annealed films [42]. The most intensive color modulation and change has been found for the "in-situ" made APCVD mixed $Mo_xW_{1-x}O_3$ films [42, 47, 48].

The cyclovoltametric curves for the as-deposited films (figure 21) possess no distinct anodic and cathodic peaks, suggesting amorphous film structure. The cathodic wave for the MoO_3 film has no intense peak and a weak one can be found at - 0.42 V. For WO_3 film, there exist two separate peaks at - 0.46 and - 0.82 V. For the mixed oxide system, the cathodic maximum is situated at - 0.73 V, which corresponds to the Li^+ ions insertion into the film structure. The anodic peaks are situated at -0.2 and -0.25 V for the mixed oxide and the tungsten oxide, respectively. MoO_3 does not exhibit an anodic maximum. The anodic peaks are related to the extraction of lithium ions. As it can be observed, for the as-deposited as well as for the annealed MoO_3-WO_3 films, the current density is considerably larger than for pure metal oxide (MoO_3, WO_3) films.

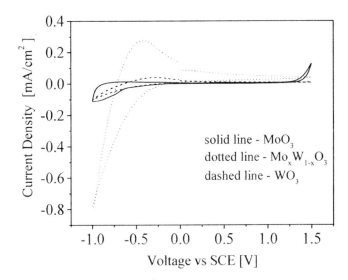

Figure 21. The CV curves of the as-deposited MoO_3, WO_3 and MoO_3-WO_3 films.

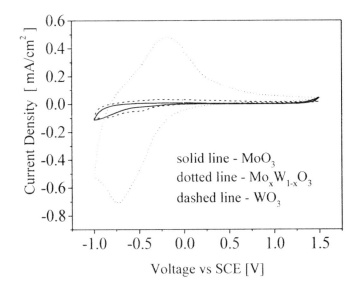

Figure 22. The cyclovoltametric curves of CVD MoO_3, WO_3 and MoO_3-WO_3 films, annealed at the temperature of 400°C in air for 1 hour.

Figure 22 presents the electrochemical behavior of the annealed films. They show a higher degree of the curve features with well expressed peaks for the mixed oxide films.

Table 2 shows the estimated values for the inserted and extracted charges for CVD oxide thin films. There is some improvement in the electrochromic properties with the annealings. As it was mentioned before, the interesting feature is the rapid charge increase for the mixed oxide films.

Table 2. The determined inserted and extracted during the electrochemical reaction charges for CVD thin films. Q is measured in [mC/cm^2].

Material	Annealing temperature	$Q_{inserted}$	$Q_{extracted}$
MoO$_3$	As deposited	31.6	30.4
	400°C/1 hour	16.6	41.2
WO$_3$	As deposited	23.6	20.5
	400°C/ 1 hour	39.2	31.4
Mo$_x$W$_{1-x}$O$_3$	As deposited	116.3	171
	400°C/ 1 hour	195.6	204.5

The diffusion coefficients for Li ions of the three types of EC films have been estimated from the experimental data and the obtained results are given in table 3.

Table 3. Diffusion coefficients values of Li ions inserted in thin films of MoO$_3$, WO$_3$ and Mo$_x$W$_{1-x}$O$_3$

Material	Annealing T$_{annealing}$	D [cm^2/s]
MoO$_3$	As-deposited	6.8x10^{-12}
	Annealed at 400°C	5.3x10^{-12}
WO$_3$	As-deposited	9.6x10^{-11}
	Annealed at 400°C	2.3x10^{-11}
Mo$_x$W$_{1-x}$O$_3$	As-deposited	4.4x10^{-10}
	Annealed at 400°C	1.4x10^{-11}

The smallest values are obtained for the mixed oxide films where lithium ions can fastest to move in and out the films comparing with the pure oxide films.

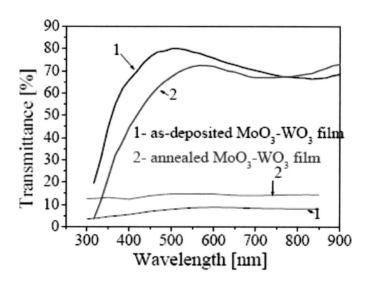

Figure 23. Change in the optical transmittance before and after Li ions inserting in CVD MoO$_3$-WO$_3$ films.

The transmittance of the mixed MoO_3-WO_3 films decreases rapidly after intercalation of Li ions into the film structure. Figure 23 shows the transmittance change before and after Li intercalation. The films were colored deep bluish and the transmittance values are under 10% for the as-deposited and annealed (400°C) mixed films. Prior to the electrochemical measurements the as-deposited and annealed films had high transmittance (around 80-60%). The optical modulation is found to be 60-70%.

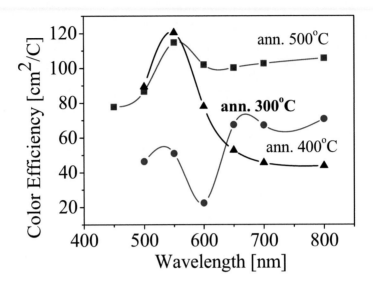

Figure 24. Color efficiency values of MoO_3-WO_3 films, annealed at 300, 400 and 500°C.

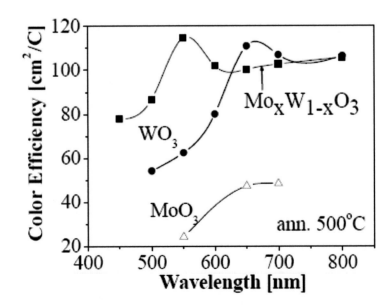

Figure 25. Color efficiency values for the three kinds of films, annealed at 500°C in air for one hour.

The coloration efficiency (CE) strongly depends on the wavelength and the changes of the optical density ΔOD. The CE values in the visible range for MoO_3-WO_3 films are better than those for WO_3 films (figure 25). Knowing that pure WO_3 films grow very slowly at 200°C, the mixed films are superior because the growth rate is much higher reaching that of MoO_3. Thus, the mixed films obtained by APCVD carbonyl process combine the positive features of the pure oxides: the WO_3 high color efficiency with the MoO_3 high growth rate. In that way, the most appropriate APCVD films for EC devices are MoO_3-WO_3. The optical modulation (transmission difference between colored state after Li insertion and bleached state) is above 50%. It is interesting experimental result that for the 400°C annealed MoO_3-WO_3 films the obtained CEs are not very good, while for the mixed films annealed at 300°C and 500°C, they exceed 110 mC/cm^2 (see figure 25).

Table 4. The optical modulation of $Mo_xW_{1-x}O_3$ thin films, deposited by CVD process using precursor made as a physical mixture of the two precursor powders, $W(CO)_6$ and $M_O(CO)_6$.

λ [nm]	ΔT [%] as deposited	ΔT [%] ann. 300°C	ΔT [%] ann. 400°C	ΔT [%] ann. 500°C
500	71.7	38.1	52.8	52.1
550	72.3	52.6	56.3	52.3
600	66.2	32.5	56.9	37.7
650	63.7	18.6	55.1	57.0

As it can be seen from table 4, the optical modulation of MoO_3-WO_3 films strongly depends on the wavelength. The highest modulation for the as-deposited films is around 72% for wavelength 550 nm (maximum of the solar spectrum). For 400°C - annealed film, ΔT is lower - 56%, which confirms the optical studies that after annealing, the optical transmittance decreases and the optical absorption increases. Table 5 reveals the comparison of two electrochromic characteristics for CVD obtained tungsten oxide films and Mo/W based oxide films.

Table 5. The color efficiency and optical modulation of WO_3 films and $Mo_xW_{1-x}O_3$ thin films (precursor is 1:4 ratio mixture of the two precursor powders - $M_O(CO)_6$) and $W(CO)_6$

λ [nm]	WO_3, Ann. 400°C		WO_3, Ann. 500°C		MoO_3-WO_3, Ann. 400°C		MoO_3-WO_3, Ann. 500°C	
	CE	ΔT [%]	CE	ΔT [%]	CE	ΔT [%]	CE	ΔT [%]
500	56.7	29	54	32	46.4	53	86.7	52
550	82.7	39	62	36	51	56	114.7	52
600	81	38.5	80	44	22	57	102	38
650	100.7	45	111	55	67	55	100	57
700	98	45	107	50	67.7	56	102.7	55
800	96	44	107	54	70.8	58	105.7	55

The influence of the temperature annealing can be also observed. For the both types of CVD films, the structural investigations (Raman and IR spectroscopy, as well as XRD analysis) have found out that their structural arrangements are slightly influenced by the additional thermal treatments as the tungsten oxide and Mo/W based oxide films remain with the high degree of amorphness independently on the annealing temperatures. But still there can be discovered some differences in their electrochromic responses due to the temperature effects. The best CE value for WO_3 has been determined for the wavelength at 650 nm (100.7 cm^2/C), meanwhile the mixed film based on WO_3 and MoO_3 has revealed the highest value at 550 nm (114.7 cm^2/C), which guessed by the absorption spectral dependences. The annealings have an impact on the electrochromic properties of CVD films, as their EC characteristics improve.

2.C.6. INVESTIGATION OF MoO_3-WO_3 FILMS, OBTAINED FROM THE SEPARATE PRECURSORS

A second approach for the preparation of the mixed Mo/W based oxide films by APCVD method has been tested, namely using of two separate sublimators. In that way, the sublimator temperatures of the two carbonyl precursors can differ allowing the achievement of the equal vapor pressures entering the CVD reactor. It has been expected a better control of the thin film growth. These mixed oxide films had been deposited at the substrate temperature of 200°C and the corresponding sublimator temperatures were 110°C ($W(CO)_6$) and 90°C for $Mo(CO)_6$. The gas flow ratio (the Ar flows carrying the carbonyls vapors/the oxygen flow) was 1/16 for the deposition time of 40 minutes. The parallel set of CVD pure MoO_3 films deposited at similar deposition conditions had been obtained. The thin films have been structurally and optically studied. Their electrochromic properties were also determined.

The FTIR studies were performed for CVD mixed Mo - W oxide films additionally annealed at 300, 400 and 500°C for 1 h. FTIR spectrum of 300°C annealed films (figure 26, curve 1) exhibits a very strong and broad band, centered at 668 cm^{-1}, which is obviously an overlapping of many contributing IR bands. A shoulder can be seen at 796 cm^{-1}. The 668 cm^{-1} band is associated to the Mo - O - Mo bending vibration of the orthorhombic MoO_3 film. In this spectral range (632 - 670 cm^{-1}) the infrared active modes of W - O - W stretching vibrations for amorphous tungsten trioxide is also located, so they could participate as well. The weak shoulder at 796 cm^{-1} can be associated to MoO_3 or to WO_3 vibrations. The interesting features appear in the spectral range 930 - 1066 cm^{-1}, where the stretching modes of the terminal double bonds of Mo = O and W = O give absorption bands [49, 50].

The main absorption band of the 400°C annealed MoO_3 - WO_3 film (figure 26, curve 2) is shifted towards 703 cm^{-1}, due to the vibrations of OMo_2 units, the stretching mode of MoO_3 [51]. Again, there is a very broad band (in the range of 627 - 824 cm^{-1}), where other modes can also contribute. The weak bands at 932 and 998 cm^{-1} are due to the terminal bonds of the both oxides. After the annealing at 500°C, the FTIR spectrum reveals two peaks at 724 and 803 cm^{-1}, superimposed over a broad and very strong absorption band, centered at 750 cm^{-1}. The two separate peaks are mostly reported to be related to the W - O stretching vibrations of the crystalline WO_3 films [52]. The weak bands, attributing to the terminal Mo=O and W=O bonds have also been registered.

The centers of the main absorption bands of the three spectra have different positions. The lowest temperature annealed film has IR peak centered at 717 cm^{-1}, and after annealing the center position is moved to 736 and 770 cm^{-1} for 400°C and 500°C, respectively. This shifting can be associated with the crystallization of the films. Previous IR study of CVD mixed oxides, prepared by physical mixture of carbonyls and higher amount of oxygen showed that IR spectra of MoO_3 film has main band at 770 cm^{-1}, WO_3 - at 730 cm^{-1} and mixed Mo - W oxide - at 750 cm^{-1}. We note that the main absorption band of the presently studied mixed Mo/W oxide films is also located at the same (750 cm^{-1}) position. In general we should stress that all the transition metal oxides studied have their basic absorption in one and the same range. In addition, depending on the structural modifications, characteristic absorption bands, due to different vibrational modes can lead to more specific IR spectra features.

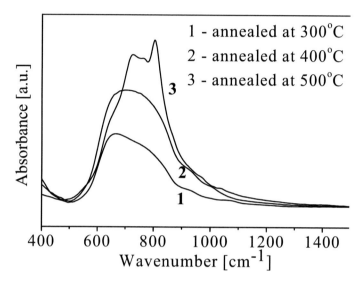

Figure 26. FTIR spectra of CVD MoO_3 - WO_3 films, annealed at different temperatures.

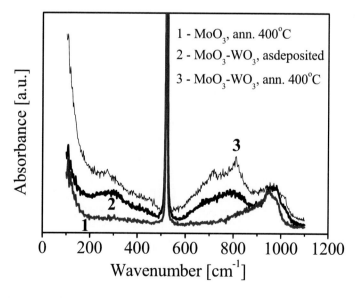

Figure 27. Raman spectra of (1) MoO_3 film annealed at 400°C, (2) as deposited MoO_3-WO_3 film and (3) MoO_3-WO_3 film, annealed at 400°C.

Figure 27 presents Raman spectra of CVD MoO_3-WO_3 films, obtained from the separate sources and comparing with the pure MoO_3 film deposited at the similar technological conditions. The Raman spectrum of molybdenum oxide exhibits two Raman lines at 948 and 988 cm^{-1}, characteristic for the orthorhombic modification. The as-deposited mixed oxide film exhibited broad and flattened bands, proving a great degree of amorphous fraction. The line at 289 cm^{-1} is attributed to the lattice vibrations and the Raman peak at 779 cm^{-1} can be related to the Mo-O-Mo bonds of the monoclinic MoO_3 [25]. The band centered at 975 cm^{-1} is due to the terminal bonds of both oxides MoO_3 and WO_3. After the 400°C annealing the Raman spectrum of the Mo/W oxide film show Raman bands at 268, 716, 806 and 963 cm^{-1}. The Raman lines at 716 and 806 cm^{-1} are the most reported and characteristic Raman bands for the crystalline WO_3 [35], suggesting that WO_3 begins to crystallize at that temperature as it was suggested by FTIR study.

CVD mixed films obtained by using separate precursors were also electrochemically characterized. As it was discussed above, the MoO_3 films are considerably more transparent than the mixed films. Their optical properties are improved by the thermal treatments. The mixed oxide films show a slightly decrease of the transmittance after the annealing. The coloration as a result of Li intercalation has been observed for the films. As an illustration, current density versus voltage - voltammograms of MoO_3-WO_3 films are presented (figure 28). As can be seen, the curves shapes of the mixed oxide films are considerably different than of the pure metal oxide (MoO_3) films. The CV curves of the pure metal oxide show some distinct peaks, which might be related to a certain degree of crystallization in the film. The CV curves of the mixed oxide films, on the hand, possess no pronounced peaks - a feature for amorphous-like structure. The existing of a second component obviously has an inhibiting influence on the process of crystallization. The CV curves of the Mo/W oxide films are smooth with no peaks seen neither in the anodic, nor in the cathodic regime. A considerable decrease in the magnitude of the anodic peak current with the increase in Mo content in MoO_3-WO_3 films had been reported by other authors [33]. This effect is related to a highly disordered structure. The authors recall that the disorder enhances the electrochromic effect, as postulated long ago. These CVD mixed films have better pronounced electrochromic effect.

In this case, the color efficiency has been determined by measuring the transmittance for a certain wavelength in the as-deposited and annealed states. The transmittance decreases significantly with the ion insertion. After Li ions intercalation, the mixed oxide films color in dark brownish-bluish and the transmittance at λ = 600 nm is 14% and 7.8% for the as-deposited and annealed MoO_3-WO_3 films, respectively. The estimated color efficiencies for the as-deposited film CVD MoO_3-WO_3 were 50.4cm^2/C and 50 cm^2/C – for the annealed mixed oxide film. These values are in a good agreement for the reported color efficiencies. The cited CE values for MoO_3 are 41cm^2/C [55] or even smaller than 12-16 cm^2/C [56] and for WO_3 films are in the range 24-50 cm^2/C [57-59] and Patil et al. [60] reported the improvement of the color efficiency for the mixed Mo – W oxide films – 63 cm^2/C. On the other hand, the color efficiencies estimated for CVD MoO_3-WO_3 films obtained from a physical mixture precursor are much higher, but direct comparison can not be done because the oxygen content during CVD preparation are different as well as the film thickness.

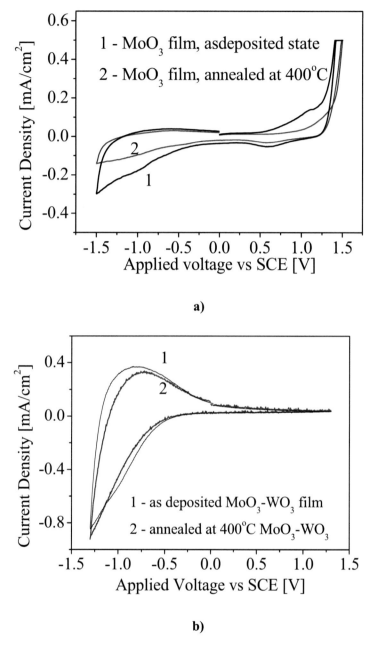

Figure 28. CV curves of CVD MoO_3 (a) and $MoO_3 - WO_3$ (b) films, deposited at 200°C and gas ratio 1/15 in as – deposited state and annealed at 400°C for 1 hour.

2.C.7. ELECTROCHROMIC CELL

EC devices with different design and electrodes have been reported. The working electrode usually used, is tungsten oxide film and as counter electrode nanocrystalline TiO_2, Cr_2O_3 and V_2O_5 is used as well as mixed Ti-Ce/Zr oxide films [1, 2].

An electrochromic cell using CVD MoO_3-WO_3 film as working electrode and sol-gel derived TiO_2 film as counter electrode was tested.

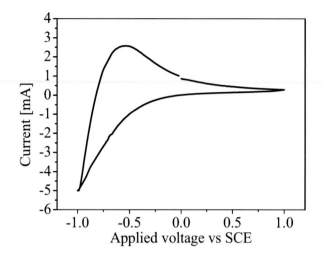

Figure 29. Cyclic voltammetric curve of the EC cell: MoO_3-WO_3/liquid electrolyte ($LiClO_4$+PC)/ TiO_2.

The electrochemical investigation has been performed for that cell: MoO_3-WO_3 (working electrochromic electrode)/liquid electrolyte ($LiClO_4$+PC)/ TiO_2 (counter electrode) and a SCE (standart calomel electrode) as reference electrode. Figures 29 and 30 present the cyclic change of the current and the transmittance as a function of the applied voltage. The found reversibility is very good. At the chosen voltage range (from -1 to +1 V), the color change appears only in the cathodic EC film, in this case CVD MoO_3-WO_3 film, which at the negative voltage colored in deep blue, correspondingly, there is a minimum in the transmittance curve, as shown in figure 30. After going towards the positive applied voltages, the EC film bleached and the cell returns to initial transparent state.

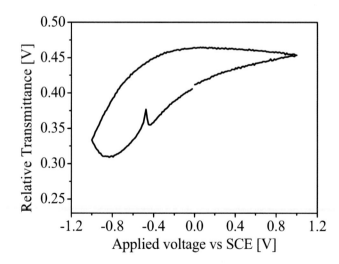

Figure 30. The transmittance dependence on the applied voltage, during the EC cell cycling. The EC cell construction is: conductive glass/MoO_3-WO_3/liquid electrolyte ($LiClO_4$+PC)/ TiO_2/conductive glass.

The stability testing of that EC cell had been performed by applying a multiple cycling. The measurements required a long time and these experiments prolonged within two weeks. The results show that up to 1500 cycles, the EC cell prototype does not change its color/bleached cycles and the cyclic voltammograms have very good reversibility with similar shapes. No peculiarities are observed, except a slight increase of the current. The both films – MoO_3-WO_3 and TiO_2 are not destructed and have no visual surface defects. The EC cell is not optimized. There has need for more research on the film thickness as well as on type of the used electrolyte.

MoO_3-WO_3 films are colored immediately when the negative voltage is applied (-1 V). If the voltage is removed, the film remains in color state for hours and only after applying the positive voltage, it is bleached. When the positive voltage is set up, the transparent state has been restored for about 3-4 minutes. Such switching times are in the range of that required in architectural smart windows.

In conclusion we may note that the research revealed that atmospheric pressure CVD technology is a suitable deposition method for preparation of mixed molybdenum/tungsten oxide films with excellent optical performance. The interesting effect is that the applied thermal annealings influence slightly the structural arrangements of the thin films, meanwhile affect considerably the optical and the electrochromic properties.

The mixed oxide films were found to combine the desirable properties of the molybdenum and tungsten oxides. From one hand, these films show the absorption peak at 2.5 eV (550 nm), where is the human eye sensitivity maximum just like MoO_3. On the other hand, the porous structure of Mo/W oxide films favors the electrochromic effect, resulting in high color efficiency and optical modulation values approaching the electrochromic properties of the most studied WO_3.

The tested electrochromic cells and their working characteristics prove for the applicability of the APCVD films as functional electrodes in electrochromic smart windows.

REFERENCES

[1] C.G. Granqvist, *"Handbook of Inorganic Electrochromic Materials"*, Elsevier Science, 1995.

[2] P.M.S. Monk, R.J. Mortimer, P.R. Rosseinsky, *"Electrochromism: Fundamentals and Applications"*, VCH, 1995.

[3] C.-C. Liao, F.-R.Chen, J.-J. Kai, *Solar Energy Mater. Solar Cells.* 90 (7-8), (2006) 1147

[4] I. Karakurt, J. Boneberg, P. Leiderer, *Appl. Physics A: Mater. Sci. Processing.* 83 (1) (2006) 1.

[5] S.-H. Lee, R. Deshpande, P.A. Parilla, K.M. Jones, B. To, A.H. Mahan, A.C. Dillon, *Advanced Materials.* 18 (6) (2006) 763..

[6] T.-S. Yang, Z.-R. Lin, M.-S. Wong, *Applied Surface Science.* 252 (5) (2005) 2029.

[7] M. Deepa, A.K. Srivastava, T.K. Saxena, S.A. Agnihotry, *Appl. Surface Science.* 252 (5) (2005) 1568

[8] C.G. Granqvist, *J. European Ceramic Society.* 25 (12 SPEC. ISS.) (2005) 2907.

[9] S. Morandi, G. Ghiotti, A. Chiorino, B. Bonelli, E. Comini, G. Sberveglieri, *Sensors and Actuators B: Chemical.* 111-112 (SUPPL.) (2005) 28

[10] Y.-H. Zhang, G.-X. Xiong, W.-S. Yang, X.-Z. Fu, Acta Physico – *Chimica Sinica* 17 (3) (2001) 277

[11] J.R.Sohn, J.S. Han, J.S. Lim, *Materials Chemistry and Physics.* 91 (2-3) (2005) 558.

[12] K. Galatsis, Y.X. Li, W. Wlodarski, K. Kalantar-Zadeh, *Sensors Actuators.* B 77 (2001) 478.

[13] E.E. Khawaja, S.M.A. Durrani, M.A. Daous, *J. Phys.: Condensed Matter.* 9 (1997) 9381

[14] Zh. Cao, J.R. Owen, *Thin Solid Films.* 271 (1995) 69.

[15] S. Papaefthimiou, G. Leftheriotis, P. Yianoulis, *Thin Solid Films.* 343-344 (1-2) (1999) 183.

[16] C. Genin, A. Driouiche, B. Gerard, M. Fizglarz, *Solid State Ionics.* 53-56 (1992) 315.

[17] M. Fizglarz, *Prog. Solid state Chem.* 19 (1989) 1.

[18] T. Ivanova, K. A. Gesheva, A Szekeres, A. Maksimov and S. Zaitzev, *J. Phys.* IV France 11 (2001) Pr3 – 385 – Pr3 – 390.

[19] A. Szekeres, T. Ivanova, K. Gesheva, *J. Solid State Electroch.* 7 (2002) 17 -20.

[20] K. Gesheva, A. Szekeres, T. Ivanova, *Solar Energy Materials and Solar Cells,* 76 (2003) 563 – 576.

[21] K.A.Gesheva, T.Ivanova, A.Kovalchuk, V.Gurtowoi, O.Trofimov, *Electrochemical Society Proceedings Vol.* 2003-08 (1), Eds. M.D. Allendorf, F. Maury, F. Teyssandier, pp 394-400.

[22] K. A. Gesheva, T. Ivanova, A. Iosifova, D. Gogova, R. Porat, *J. de Physique.* IV France 9 (1999) Pr 8-453 – Pr8-459.

[23] D. Gogova, A. Iossifova, T. Ivanova, Zl. Dimitrova, K.A. Gesheva, *J. Crystal Growth.* 198/199 (1999) 1230-1234.

[24] H.A.Wreidt (for Mo-O) and L.Brewer and R.H.Lamoreaux (for O-W)

[25] K. Gesheva, A. Cziraki, T. Ivanova, A. Szekeres, *Thin Solid Films.* 492 (1-2) (2005) 322-326.

[26] L. Sequin, M. Figlarz, R. Cavagnat, J. Lasseques, *Spectroch. Acta.* A 51 (1995) 1323.

[27] G. Nazri, C. julien, *Solid State Ionics.* 80 (1995) 271.

[28] Y. Zhang, S. Kuai, Zh. Wang, X. Hu, *Applied Surface Science* 165 (2000) 56

[29] J. Jang, *J. Phys. Chem Solids.* 61 (2000) 647.

[30] K.A.Gesheva, T.Ivanova, A.Kovalchuk, V.Gurtowoi, O.Trofimov, *Electrochemical Society Proceedings.* Vol. 2003-08 (1), Eds. M.D. Allendorf, F. Maury, F. Teyssandier, pp 394-400.

[31] K.A. Gesheva, T. Ivanova, *Chemical Vapor Deposition.* 12 (2006) 231.

[32] M. Picquart, S. Castro – Garsia, J. Livage, C. Julien, E. Haro-Poniatowski, *J. Sol-Gel Sci. Technol.* 18 (2000) 199.

[33] K. Gesheva, A. Szekeres, T. Ivanova, *Solar Energy Materials and Solar Cells.* 76 (2003) 563 – 576.

[34] T.Ivanova, K.Gesheva G.Popkirov, M.Ganchev, E.Tzvetkova, *Material Science and Engineering.* B 119 (2005) 232-239.

[35] T. Nishide, F. Mizukami, *Thin Solid Films.* 259 (1995) 212.

[36] M. Epifani, P. Imperatori, L. Mirenghi, M. Schioppa, P. Siciliano, *Chemistry Materials.* 16 (25) (2004) 5495.

[37] S. Lee, M.J. Seong, C.E. Tracy, A. Mascarenhas, J. Pitts, S.K. Deb, *Solid State Ionics.* 147 (2002) 29.
[38] B.W. Faugmann, R.S. Randall, *Appl. Phys. Lett.* 31 (1997) 834
[39] N. Ozer, *Thin Solid Films.* 304 (1997) 310.
[40] A. Abdellaoui, G. Leveque, A. Donnadoeui, A. Bath, B. Bouchikhi, *Thin Solid Films.* 304 (1997) 39
[41] Y. Wang, N. Herron, *J. Chem. Phys.* 95 (1991) 525.
[42] T.Ivanova, K.Gesheva G.Popkirov, M.Ganchev,E.Tzvetkova, *Material Science and Engineering.* B 119 (2005) 232-239.
[43] D.Gogova, K.Gesheva, A.Kakanakova-Georgieva and M.Surtchev, *Materials Letters* 11 (2000) 167.
[44] C. Granqvist, "Handbook of Inorganic Electrochromic Materials", Elsevier Science, 1995
[45] O. Zelaya – Angel, *Mat. Sci. eng.* B 86 (2001) 123
[46] C. Julien, A. Khelfa, O. Hussain, G. Nazri, *J. Cryst. Growth.* 156 (1995) 234.
[47] K.A. Gesheva, T. Ivanova, *Chemical Vapor Deposition.* 12 (2006) 231-236.
[48] K.A. Gesheva, T. Ivanova, F. Hamelmann, Solar Energy Mater. *Solar Cells.* (2006) in press
[49] K.A.Gesheva, T.Ivanova, G.Popkirov, F.Hamelmann, *J. Optoelectronics and Advanced Materials.* 7(1) (2005) 169-175.
[50] K. Gesheva, T. Ivanova, E. Steinman, F. Hamelmann, A. Brechling, U. Heinzmann, *Electrochemical Society Proceedings vol.* 2005-09, (2005) 799- 809
[51] G. Nazri, C. Julien, *Solid State Ionics.* 80 (1995) 271
[52] S. Chong, B. Ingham, and J. Tallon, *Curr. Appl. Phys.* 4 (2004) 197.
[53] W. Dong, A. Mansour, B. Dunn, *Solid State Ionics.* 144 (2001) 1.
[54] A. Donnadieu, *SPIE IS.* vol 4 191
[55] Gorenstein, J. scarminio, A. Lourenco, *Solid State Ionics.* 86-88 (1996) 977-981.
[56] S.I. Gordoba de Torresi, A. Gorenstein, R.M. Torresi, M.V. Vasquez, *Electroanal. Chem.* 381 (1991) 131.
[57] M.G. Hutchins, M.A. Kamel, N. El-Kadry, A.A. Ramadan, K. Abdel-Hady, *Phys. Stat. Sol.* (a) 175 (1999) 991.
[58] H. Kaneko, F. Nagao, K. Miyake, *J. Appl. Phys.* 63 (1988) 510.
[59] H. Akram, M. Kitao, S. Yamada, *J. Appl. Phys.* 66 (1989) 4364.
[60] P.R. Patil, P.S. Patil, *Thin Solid Films.* 382 (2001) 13.

Chapter 2D

INVESTIGATION OF CVD OBTAINED THIN FILMS OF CHROMIUM OXIDE

T. M. Ivanova and K. A. Gesheva
Central Laboratory of Solar Energy and New Energy Sources,
Bulgarian Academy of Sciences, Sofia, Bulgaria

2.D.1. APPLICATIONS OF CHROMIUM OXIDE

From the family of transition metal oxides, the less studied one is the chromium oxide system. Its feature of possessing many crystalline modifications is a characteristic for the transition metal oxides and Cr oxide also can form different crystal arrangements. Many crystalline modifications of Cr oxide exist such as Cr_2O_3 (corundum), CrO_2 (rutile), Cr_5O_{12} (three-dimensional framework), Cr_2O_5 and CrO_3 (unconnected strings of CrO_4 tetrahedra). The only stable bulk oxide, Cr_2O_3, is a magnetic dielectric with corundum structure [1]. There, in the Cr_2O_3 corundum structure, the oxygen atoms form a hexagonal close-packing array. The metal atoms occupy two thirds of the octahedral interstices between two layers. The chromium atoms form graphite-like layers parallel to the oxygen layers [2]. Chromium oxide, Cr_2O_3 consists of the rhombohedral primitive cell, which contains two formula units. Chromium atoms are eight-coordinated with two O_{22} layers. They form CrO_6 distorted octahedral that are linked by faces, edges or corners in such a way that each oxygen atom is linked to four Cr atoms.

Thin oxide films based on chromium are investigated due to their interesting optical and electrical properties. As electrochromic material, Cr oxide properties are not fully studied. It has been suggested to manifest anodic type of electrochromism [3], but detailed research has not been carried out. According to ref. [3], the EC effect can be written as follows:

$$Cr_2O_3 + xLi^+ + xe^- \leftrightarrow Li_xCr_2O_3 \tag{1}$$

From the anodic type electrochromic materials, hydrated nickel oxide and iridium oxide are the most famous and studied. Ni oxide possesses proper optical properties, but is unstable in acidic electrolytes, on the other hand, IrO_2 is very expensive. A. Azens et al. [4] associate the anodic electrochromism in nickel oxide with the fact that the upper part of valence band consisting of Ni 3d electrons counters to O2p electrons can reversibly accept to insert or extract small alkali ions in the film without breaking the metal- oxygen bonds during this process. Similar suggestion can be applied to Cr_2O_3. Its valence band consists of unconnected Cr 3d levels, localized roughly at 3 eV above O 2p states [4]. It can be supposed that on the valence band top are situated the metal non-bonding states which can be reversibly filled/emptied upon charge injection/extraction. This leads to conclusion that Cr_2O_3 thin films can be successfully integrated in optical electrochromic devices.

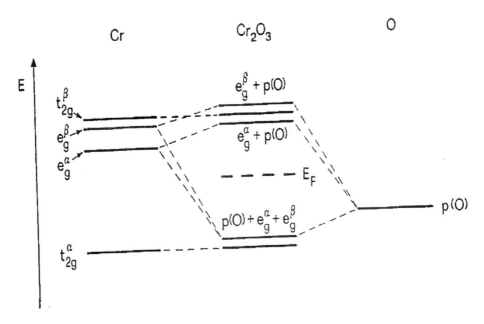

Figure 1. Molecular orbital type diagram of energy states of Cr_2O_3 (M. Catti, G. Sandrone, G. Valerio, R. Dovesi, J. Phys. Chem. Solids 57 (1996) 1735-1741).

Chromium oxide has found wide applications as protective coatings in digital magnetic recorders [6] and as electrodes in Li solid state batteries [7]. Due to its high hardness and resistivity, these oxides in thin film form are studied as resistant coatings on wires and as thermal barrier layers [8]. Other well known applications are usage of chromium oxides as catalysts materials and anticorrosive layers [9].

The optical data and investigations of chromium oxide films are very obscure in the literature nevertheless they are reported to be perspective and promising materials, participating in optical devices such as solar absorbers [10] and as electrochromic cells. They can be also applied as photolithographic masks [11]. Different deposition methods are investigated for the preparation of chromium oxide thin films: such as thermal pulverization [12], evaporation [13], vacuum evaporation [14], ion implantation [15], magnetic sputtering [6, 16, 17], laser ablation [18] and molecular beam epitaxy [19].

CVD technology provides advantages for obtaining thin oxide films at relatively low temperatures. Bermudez et al. [20] had obtained Cr_2O_3 thin films by CVD method using a chromium hexacarbonyl as a precursor. The investigation has been stressed on the film growth and IR study at different growth stages. Chevalier et al. [21] studied thin films of

Cr_2O_3 and Nd_2O_3, derived by metalorganic CVD method. The used precursor has been the chromium acetylacetonate, which inquires temperature of 180°C to sublimate. Then the hot vapors are carried towards the CVD reactor, where the substrate temperature has also been very high – 500°C. The relatively high temperatures are not favorable for the optical applications of the films, when the substrates used are often glass, having its temperature requirements. Other CVD methods found in the literature are KrF laser CVD method [22, 23], where the source vapors are obtained by photodissociation of $Cr(CO)_6$ in an oxidizing atmosphere, using a pulsed UV laser (KrF, λ = 248 nm).

2.D.2. TECHNOLOGICAL DEPOSITION CONDITIONS

The technological process chosen for the fabrication of chromium oxide thin films was the atmospheric pressure CVD method, described in the previous chapters. The precursor used is $Cr(CO)_6$, a white powder that sublimates at temperature range of 60 – 90°C. The chromium carbonyl is stable substance when exposed in air at room temperature and its partial sublimation begins after heating at temperatures above 40°C. That makes this precursor a good and proper vapor source for the low temperature deposition process. The reaction in the vapor phase proceeds in the neighborhood of a heated substrates, where the temperature is kept 200°C. The experiments made at the substrate temperatures of 150 and 300°C revealed a very slow film growth for 40 minutes deposition time, meanwhile powder deposits could be observed on the quartz reactor walls. Similar temperature limitation behavior was found in the case of deposition of MoO_3 films on the same CVD equipment and similar technological conditions. The molybdenum oxide films also exhibit temperature selectivity, growing only at the temperatures of 150 – 200°C and after increasing the deposition temperature, the thin film growth absence was detected.

Technological parameters are varied for improving the optical properties of the films as the basic parameters are the gas flow rate ratio of Ar gas carrying the carbonyl vapors towards the flow rate of oxygen, the sublimator temperatures and the deposition time.

The substrates used were silicon wafers for IR studies, glass for optical measurements and conductive SnO_2:Sb covered glasses with sheet resistance of 8 $\Omega/\square$ (Donelly type). The substrates were cleaned prior deposition by the following procedures: the glass substrates were boiled in ethanol (96%) and additionally washed in distilled water. The silicon wafers were boiled in a mixture of H_2SO_4:H_2O_2 and washed several times in distilled water. After deposition some of the samples were additionally treated at temperatures of 200°C, 300°C, 400°C and 500°C in air for 1 hour.

CVD process allows deposition on various types of substrates with good adhesion. The obtained CrO_x films are visually uniform, smooth and almost transparent. The average growth rate is 16 nm/min, as the film thickness was measured by Profilometry making steps either by shading part of the substrate during deposition or better by plasma etching.

Table 1. CVD technological parameters for the preparation of chromium oxide films. The substrate temperature is 200°C for all cases

$T_{sublimator}$ (°C)	R = $Cr(CO)_6/O_2$	t (min.)
60	1/20	40
70	1/4	40
70	1/12	40
70	1/20	20
70	1/20	40
70	1/32	40
70	1/40	40

The relatively low growth rate allows a uniform ordering of the reacted atoms or molecules over the heated substrates, resulting in uniform and smooth films, without cracking and color changes. CVD thin films of chromium oxide on glass and conductive glass substrates are colored in yellow-brownish.

2.D.3. XRD ANALYSIS OF CR_2O_3 FILMS

X-Ray Diffraction investigation has been performed by means of the URD diffractometer with secondary graphite monochromator using Cu Kα radiation (35 kV, 20 mA) with step 0.05° and detection every 15 sec. Two configurations were used: the standard Bragg-Brentano and grazing-incidence asymmetric X-ray diffraction (GIAXRD), where the incident angle is 3°.

XRD spectra have been recorded for the thin films deposited on the conductive glass substrates at the following technological parameters: the deposition temperature – 200°C, the sublimator was heated at 70°C, and the gas flow ratio of carbonyl vapor to oxygen was 1/40. The deposition time was 40 minutes.

Figure 2 presents the measured XRD spectra for chromium oxide films annealed at 400 and 500°C in air for 1 hour. The spectra analysis reveals that the as-deposited films and CrO_x films annealed at the temperatures lower than 400°C are almost fully amorphous and the detected XRD lines have been related to the crystalline substrate, especially the thin conductive SnO_2:Sb layer (see figure 2, curve 1).

XRD peaks due to the crystal phase of chromium oxide appeared for the thin film annealed at the highest temperature applied – 500°C (figure 2, curve 2). The characteristic peaks attributed to the (012), (104) and (110) directions are found to be related to one specific crystal phase, the tetragonal dichromium trioxide (Cr_2O_3).

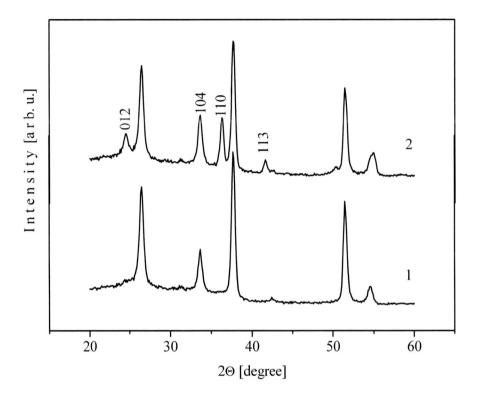

Figure 2. XRD spectra of CVD CrO$_x$ films, annealed at 400°C (curve 1) and at 500°C (curve 2).

Table 2. XRD data for the peak positions of CVD chromium oxide films

Ann. film at 400°C [2Θ]	Ann. film at 500°C [2Θ]	(h k l)	Attributed to
	24.4	012	Cr_2O_3
26.4	26.5		SnO_2
33.8	33.7	104	Cr_2O_3
	36.3	110	Cr_2O_3
37.6	37.7		SnO_2
	41.5	006	Cr_2O_3
	42.7	113	Cr_2O_3
51.6	51.5		SnO_2
54.4	54.9		SnO_2

Table 2 presents the experimental results of XRD measured spectra and the corresponding peaks, which are assigned either to the crystalline Cr_2O_3 or to SnO_2 film from the conductive glass substrate. It is worth noting that for the 400°C annealed film there appears a XRD line, which can be connected with the crystalline Cr oxide. One XRD peak is not sufficient for definite conclusion that the crystallization begins and only after reaching 500°C annealing temperature, the crystallization has been clearly manifested.

The estimated parameters of the tetragonal crystal lattice of Cr_2O_3 from XRD data possess the corresponding values: $a = (4.939 \pm 0.003)$ Å and $c = (13.627 \pm 0.015)$ Å. They are in good agreement with the reference ones according to JSPDC 38 – 1479 card.

2.D.4. TRANSMISSION ELECTRON MICROSCOPY OF CR_2O_3 THIN FILMS

Transmission electron microscopy (TEM) was performed with TEM EM – 400 Philips equipment with a magnification of 14 000 times. The sample preparation was made by using the double stage replica method. The studied Cr_2O_3 films have been obtained on the conductive glass substrates. Technological conditions were: the deposition temperature – 200°C, the sublimator heated at 70°C, and the gas flow ratio (carbonyl vapor to oxygen) was kept 1/32, the process proceeded 40 minutes.

Figure 3a presents a TEM micrograph of the as-deposited CrO_x film. The film structure seems uniform with small grains and a few number of bigger in size grains dispersed randomly on the surface. After the treatment at 300°C, the background decreased and a partial tendency for the crystallization appears (figure 3b). Increasing the annealing temperature to 400°C (figure 3c), the interesting features were manifested. A kind of disappearing of the grain borders has been presented, a formation of a net approaching a two-dimensional growth starts. An interesting effect can be noticed on the micrograph of the CrO_x film morphology after the highest annealing temperature of 500°C (figure 3d). Some inhomogeneities and the formation of separate crystalline phases can be found as the boundaries seem to be washed away. It can be suggested that a structure transformation is proceeding at this temperature.

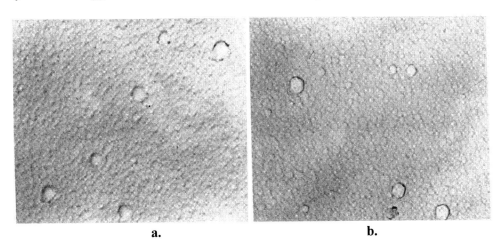

a. b.

Figure 3 (continued)

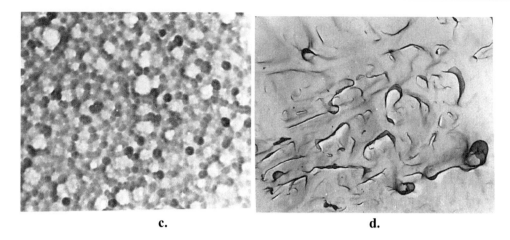

c. d.

Figure 3. TEM micrographs of CVD Cr_2O_3 films on the conductive glass substrates, where a) presents the as-deposited film, b), c) and d) the annealed films at 300, 400 and 500°C (8 mm corresponds to 1 μm).

2.D.5. AFM STUDY OF CVD CR_2O_3 THIN FILMS

The morphology and the surface roughness of CVD Cr_2O_3 film, annealed at temperature of 500°C in air were investigated by Atomic Force Microscopy (AFM). The images presented on figure 4, showed one and the same film surface under different magnifications. As it could be seen, there is a beginning of the film crystallization. The observed film structure consists of grains, almost uniformly distributed. The different contrasts observed in a separate grain can be a sign for inhomogeneous grains in structure, supposing that they are probably agglomerate structures like clusters. The agglomerates consist from very small grains gathered together overlapping, with no clear boundaries. Similar morphologies have been determined for MoO_3, WO_3 and MoO_3-WO_3 films, deposited by the same technology. From these microscopic studies can be assumed that APCVD transition metal oxide films, obtained at the technological conditions considered above have surfaces consisting of nanocluster formations.

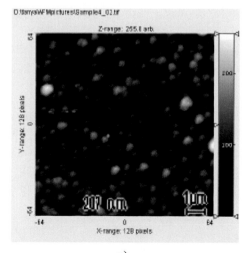

a)

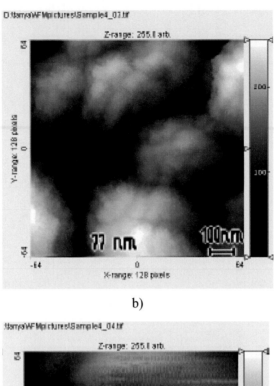

b)

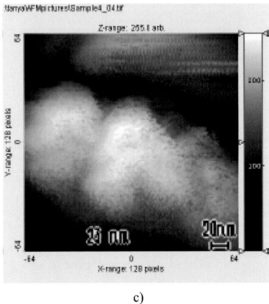

c)

Figure 4. Two dimensional AFM micrographs [at different magnifications as given in a), b), and c) micrographs] of CVD chromium oxide thin film deposited at substrate temperature of 200°C, gas flow ratio 1/20, and deposition time of 40 min. The film was annealed at 500°C for 1 hour in air.

Investigation of CVD Obtained Thin Films of Chromium Oxide 255

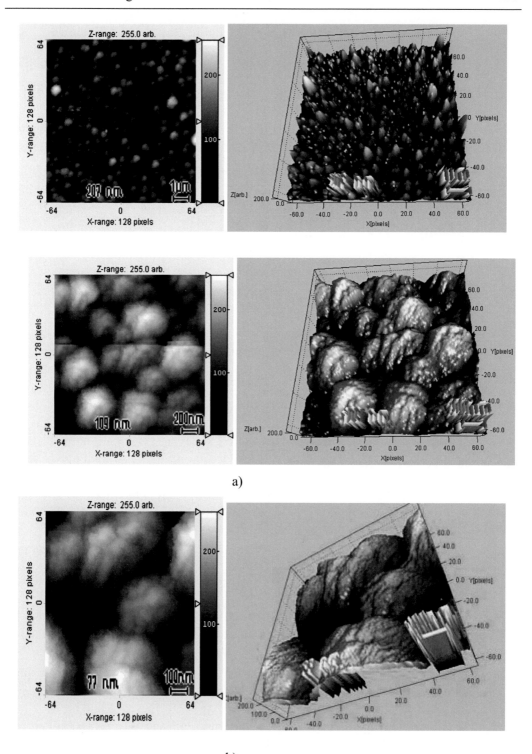

a)

b)

Figure 5 (continued)

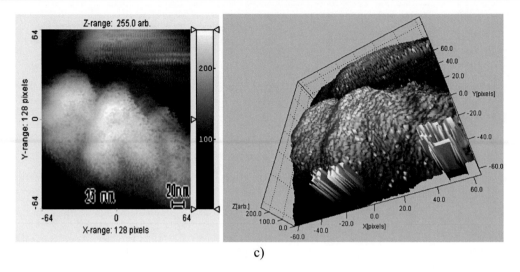

c)

Figure 5. Three-dimensional views of Cr_2O_3 AFM images, together with the two-dimentional ones for the three different magnifications (the films are the same as on figure 4).

The three dimensional view of chromium oxide film clearly reveals the clustering as well as the inhomogeneity of the grains. This interesting feature could not be detected by TEM study.

The average cluster size is 207 nm as shown on the AFM picture (figure 5 c). The value is taken from the picture made at the lowest magnification, where the accuracy of size determination is highest, due to the contribution of the biggest statistical number of grains per unit area. The average Roughness was determined to be S_a = 8.8 nm and the average cluster size – 207 nm.

As a result of the annealing the film grains seem having a complex morphology. There are different contrasts across the grain. These contrasts can be evidence that the grains are inhomogeneous, consisting from very small grains, gathered, overlapped, and with unclear boundaries. There is nanoclusters formation.

2.D.6. IR INVESTIGATION OF CR_2O_3 THIN FILMS

APCVD chromium oxide films deposited on the silicon substrates have been analyzed by Infrared spectroscopy in the range of 200 – 1200 cm^{-1}. The samples were prepared under experimental conditions: the substrate temperature 200°C, the sublimation temperature - 70°C, $Cr(CO)_6$ vapors/O_2 = 1/32 for 40 min deposition time. Some of the samples were additionally annealed at different temperatures (200, 300, 400 and 500°C) in air for 1 hour. The resulted spectra are presented on figure 6.

For the as-deposited film (figure 6, curve 1) the main absorption band is centered at 510 cm^{-1} and is attributed to Cr – O bond [22]. This IR band is broaden and strong. The small shoulder at about 970 cm^{-1} is suggested to be related to the chromyl band vibration (Cr = O). When the film was annealed at the temperature of 200°C, the main absorption peak is not significantly changed (figure 6, curve 2), meanwhile the shoulder at 970 cm^{-1} disappears. There is detected another weak peak located at 300 cm^{-1} and a shoulder at 770 cm^{-1}, which other authors [23] reported to be found in Raman spectrum of Cr_2O_3 film. For the higher annealing temperatures of 300 and 400°C (figure 6, curves 3 and 4) there is found only the broad band around 500 cm^{-1} (a little shifting from the previous analyzed samples can be distinguished). The highest annealing temperature of 500°C resulted in an IR spectrum (figure 6, curve 5) with more specific features, proposing an explanation for the crystallization of the chromium oxide film. The main peak (attributed to Cr – O vibration) is situated at 552 cm^{-1} – broad, intense and shifted from the other spectra with 30 – 40 cm^{-1}. The new weak absorption bands appeared at 408, 447, 618 cm^{-1}.

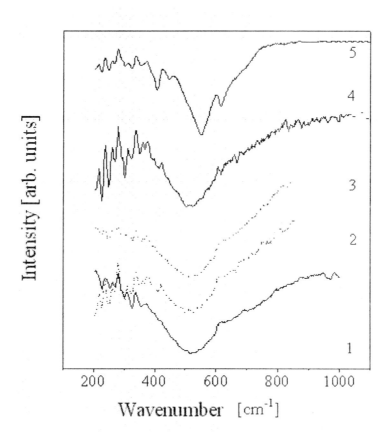

Figure 6. IR spectra of CVD obtained Cr_2O_3 films, where curve 1 presents the as-deposited film, curve 2 – annealed at 200°C, curve 3 – annealed at 300°C, curve 4 – annealed at 400°C and curve 5 – at 500°C in air for 1 hour.

The FTIR measurements were performed using a SPECORD M-82 spectrophotometer in the range 200 - 4000 cm^{-1}. The studied sample has been annealed at the temperature of 500°C. The FTIR spectrum (see figure 7) revealed weak absorption bands at 1008, 881.7, 850.7, 745 and 436.8 cm^{-1} and more pronounced peaks at 612.8 and 542.4 cm^{-1}. The observed band at 745 cm^{-1} is assigned to the longitudinal optic phonon of Cr_2O_3. The absorption bands at 542.4 and 612.8 cm^{-1} and the weak band at 436.8 cm^{-1} are associated with the localized vibrations of CrO_6 octahedra. The IR bands at 612.8 and 436.8 cm^{-1} were also reported by other authors [22]. It must be noted that these authors observed peaks at 413, 537, 619 and 656 cm^{-1} in IR spectra of commercial corundum Cr_2O_3 powder.

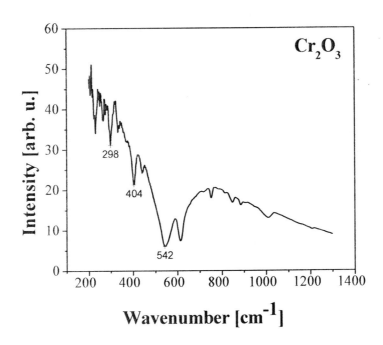

Figure 7. FTIR spectra of CVD chromium oxide film, annealed at 500°C.

FTIR study revealed some absorption bands that are related to crystalline Cr oxide, this is in coincidence with XRD analysis - annealing at 500°C provokes crystallization of Cr_2O_3.

2.D.7. RAMAN SPECTROSCOPY INVESTIGATION OF CHROMIUM OXIDE FILMS

The Raman scattering spectra were measured by a double spectrometer SPEX 403 with photo multiplier detector. During the recording of Raman spectra in the range of 100-1100 cm^{-1} the intensity of laser line (λ=488 nm) excitation was controlled with an accuracy of 1%. The excitation power on the sample surface was 43 mW and the spectral slit width 4 cm^{-1}. Raman spectroscopy has been used to reveal the crystal structure of the annealed Cr_2O_3 film at 500°C. The amorphous structure of the lower temperature treated films would give no efficient characteristic Raman spectra and information.

In the crystal arrangement of Cr_2O_3 structure, Cr atoms are 8-coordinated in two O^{2-} layers in such a way that they create distorted CrO_6 octahedra. The Raman investigations of the vibrational modes of the corundum-like oxides are reported for a wide range of temperatures and pressures to study their magnetic and optical properties [24].

As seen from figure 8, Raman spectrum of Cr_2O_3 film exhibits several bands at 555 cm^{-1} and weaker at 641 and 654 cm^{-1}. Two well-expressed peaks are localized at 852 and 889 cm^{-1}. The Raman peak situated at 555 cm^{-1} is assigned to one of the most intensive modes of Cr_2O_3 as it has been mentioned by many authors [25]. Dolgaev et al. [22] reported this Raman band as a characteristic for Cr_2O_3, expressing the strongest signal. In the same time, these authors found out a broad band at 730 cm^{-1}, which was not detected in our spectrum.

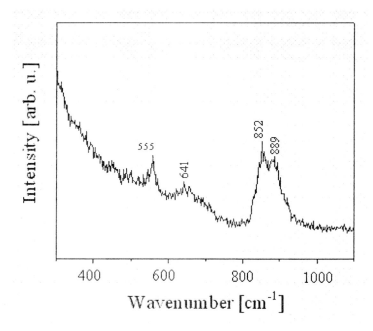

Figure 8. Raman spectrum of Cr oxide thin film, annealed at 500°C.

The Raman band at 615 cm^{-1} has been observed by Mougin et al. [25]. Raman bands in the range of 640 - 654 cm^{-1} are too weak, in this spectral range is supposed to appear Raman lines, connected with the stretching vibrations due to Cr_2O_3. The bonding includes Cr atoms with higher valences [26]. The expressed Raman bands at 852 and 889 cm^{-1} can be related to Cr^{VI} - O terminal stretching vibrations or to the modes of mixed Cr^{III} - Cr^{VI} oxides [26]. Cheng et al. [27] deposited thin films of Cr oxide by laser-initiated organometallic CVD ($Cr(CO)_6$ in oxygen environment) and found out two crystalline phases (Cr_2O_3 and CrO_2) existed in their films.

Raman measurements confirm the crystalline structure of CVD chromium oxide films after annealing at 500°C, a conclusion derived from XRD and IR analysis.

2.D.8. OPTICAL PROPERTIES OF CVD CR_2O_3 THIN FILMS

The optical properties were characterized by UV – VIS Perkin – Elmer Spectrophotometer in the range 250 – 800 nm as a function of annealing temperature. On figure 9 are presented the visible spectra of CVD chromium oxide coatings deposited on the ordinary glass substrates and additionally annealed at 400 and 500°C in air for 1 hour. The as-deposited and the annealed at 200 and 300°C samples show transmittance in the visible region around 35 – 50% (not shown here). The high temperature annealing improved the optical transmittance of CrO_x thin films. The transmittance increased up to 65% (for the annealing temperature of 400°C, figure 9 curve 1) and to 76% (for 500°C, figure 9, curve 2) in the spectral range of 450 – 850 nm.

Absorption coefficient α and the optical band gap E_g values for CVD Cr_2O_3 films were determined from the optical transmission data (figure 10).

$$\alpha d = \ln(I_o/I) \qquad (2)$$

where I_o and I are the intensities of the incident and transmitted radiation, respectively, and d the film thickness. The optical band gap was determined using the relation for band-to-band transition:

$$\alpha(\nu) = A(h\nu - E_g)^n \qquad (3)$$

where A is a constant different for different types of transition, indicated by the value of n. For the allowed indirect gap transition n is equal to 2. From the curve $\alpha^{1/2}$ versus $h\nu$, E_g can be determined by extrapolating the linear part of the plot to $\alpha^{1/2} = 0$.

The calculated value for the optical band gap energy of the chromium oxide thin film annealed at the temperature of 400°C is 2.95 eV for indirect transition. For the crystalline Cr_2O_3 thin film treated at the temperature of 500°C, the optical band gap energy value is higher to 3.2 eV. These values are in good agreement with reported data [28] for CVD chromium oxide films.

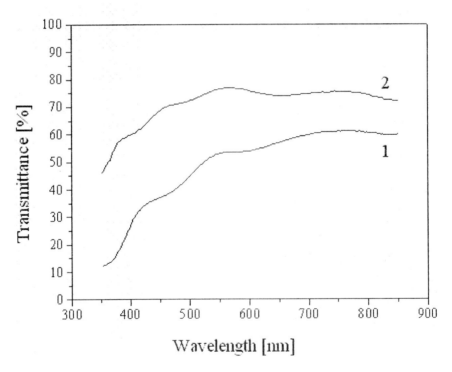

Figure 9. VIS - UV spectra of Cr_2O_3 films, where curve 1 refers to thin film, annealed at 400°C and curve 2 – the film, annealed at 500°C.

Spectroscopic ellipsometry measurements (SE) have been carried out in the spectral range of 300 – 800 nm at the angle of 50°. Ellipsometric data analysis has shown a rather porous film structure, which in combination with the rough film surface is a good basis for a successful reversible ion-intercalation process. SE measurements were performed for the thin films deposited at 200°C, the sublimation temperature of 70°C, the gas ratio (Ar flow passing through the carbonyl vapors chamber towards O_2 flow rate) was kept 1/4 and the deposition time 40 min. The layer thickness was 640 nm as measured after plasma etching by Profilometer and in comparison with the derived thickness values from spectral ellipsometry are 664 nm for the as-deposited chromium oxide film and 652 nm for the annealed film at 500°C. The thickness difference is almost negligible. As approximation theory for analyzing the optical constants, the Bruggeman effective medium was employed.

The spectral distributions of the effective refractive index and extinction coefficient were obtained. The extinction coefficient k is in the range of 0.52 to 0.65 for the as-deposited films and in much more broader range for the annealed at 500°C films: k = 0.2 to 0.85 depending very specifically on the wavelength. The extinction values decrease with increasing the wavelength.

Figure 10 shows the refractive index values in the measured spectral range. It has been revealed an increase of the refractive index after the annealing procedure. The dependence curve for the as-deposited Cr oxide film possesses a minimum at 300 nm and a maximum situated at 750 nm.

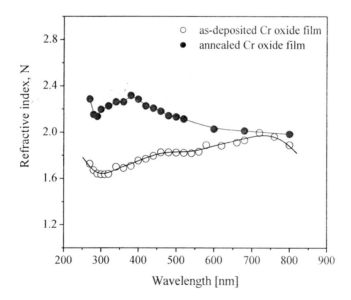

Figure 10. Spectral distribution of the refractive index for Cr_2O_3 films in the as-deposited state and after annealing at 500°C.

Therefore, the values vary in 1.6 to 2.0. After annealing at 500°C, the refractive index has higher values in the range of 2.1 to 2.4. The minimum appears at 250 nm, and the maximum locates at 350 nm. In the visible spectrum, the refractive index curve is flattened with a slight decrease.

The dielectric function of Cr_2O_3 films has been determined by the following equations:

$$\varepsilon = (\varepsilon_1 - j\varepsilon_2) = (N - jk)^2 \text{ and } \varepsilon_1 = N_2 - k_2; \varepsilon_2 = 2N_k \tag{4}$$

where N is the complex refractive index; k – extinction coefficient.

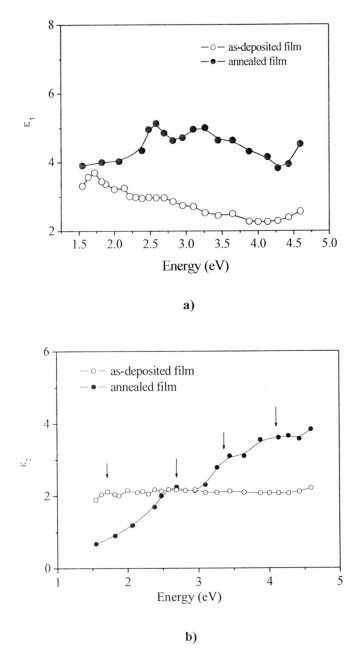

Figure 11. Dielectric function (real and imaginary part) of the chromium oxide films, as it has been determined by spectral ellipsometry.

The observed behavior in the dielectric functions dispersion curves could be connected with some defects existing in the films. The as-deposited Cr oxide film show a less structured dielectric function (ε_2), while for the annealed films several peaks are observed related to defect development at the elevated temperatures, which could be connected with some degree of crystallization taking place in the structure. Independent study of Raman spectra as well as XRD spectra have shown that the annealing above 400°C leads to the crystallization in chromium oxide films [28]. Similar peaks have been reported by Chase L. [29], who studies the optical properties of CrO deposited on (100) TiO_2 substrates by high pressure-decomposition process [30]. He has found a set of three absorption peaks located at E_2=2 eV, E_3=2.7 eV and E_4=3.1 eV. The author emphasizes that this is likely for these peaks correspond to the transitions observed in metallic VO_2 at 2.9, 3.6 and 4.5 eV.

Using Tauc's equation the optical band gap of the chromium oxide films can be estimated The linear relationship is better fitted for the direct than the indirect electron transitions Therefore, it could be concluded that the direct transition model is more appropriate. Similar results were also reported elsewhere [27]. So, in the near-edge absorption region, assuming direct electron transitions, the optical band gap energy E_{og} is determined by:

$$(\alpha E) = C(E-E_{og})^{1/2} \text{ - valid for direct electron transitions} \qquad (5)$$

where α presents the absorption coefficient, $E=h\nu$ - photon energy, h is Planck's constant, ν - photon frequency, C – proportionality constrant and E_{og} – optical bandgap.

Extrapolating the linear part of $(\alpha E)^2$ vs photon energy curve up to its interception with E axis yields the optical band gap value. The obtained values are 3.18 eV for the as-deposited Cr_2O_3 film (with an amorphous structure) and 3.31 eV for 500°C annealed film (crystalline Cr oxide). These values are in good agreement with reported data [32] for CVD chromium oxide films.

2.D.9. ELECTROCHROMIC PROPERTIES OF CHROMIUM OXIDE THIN FILMS

Cyclic voltammetry measurements of CVD Cr_2O_3 thin films were performed either in Li containing electrolyte or in hydrogen type electrolyte. The Li electrolyte consisted of 1 moll $LiClO_4$ dissolved in the propylene carbonate (PC). On the other hand, the proton source was 1 M H_2SO_4 solution with addition of a glycerin. Surprisingly, it was found out that in Li electrolyte the Cr oxide films showed a very fast destruction in a very sharp range of the applied voltages (-1 V; +1V). It was reported that CrO_x films are unstable in organic electrolyte (1) very unlikely compared to other transition metal oxide films. At the same time, the samples studied in the hydrogen type electrolyte showed reversible electrochromic (EC) effect. Of course, this effect was not as strong as it is observed for WO_3, MoO_3 and $Mo_xW_{1-x}O_3$ thin films deposited by similar method [29]. The dependences of current density versus electrode potential are registered at a scanning rate of 20 mV/s.

The studied Cr_2O_3 thin films deposited by atmospheric pressure CVD exhibited cathodic type electrocromism, the color becomes deeper at the negative voltages and the film bleached at the positive values of the applied voltage. Other authors (1) claimed that Cr oxide based films possessed anodic electrochromism. Nevertheless, our experiments clearly revealed that the investigated films both in the amorphous state and after the crystallization showed cathodic EC effect (see figure 12). The obtained charge density depends on the film structure.

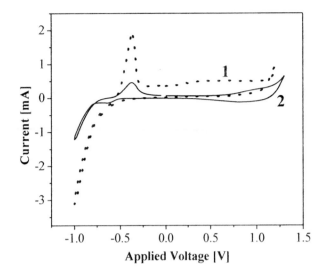

Figure 12. Cyclic voltammetric curves of Cr_2O_3 thin films in as deposited state (curve 1) and after annealing at 500°C (curve 2).

The CE values of the as-deposited and the annealed at 500°C Cr oxide films are shown on figure 13. The obtained CE values are lower as compared to the other electrochromic materials. For example, MoO_3 films show color efficiency up to 40 mC/cm^2 while the most investigated WO_3 films - up to 100 mC/cm^2 [29]. It must be noted here, that after the first cycles the film transparency improves considerably and in bleached state it is higher than the initial transparency. The results of the electrochromic investigation showed that these films could be used as counter electrodes in EC cell. This can be also proved by the results obtained for the diffusion coefficient of H^+ ions into the film structure.

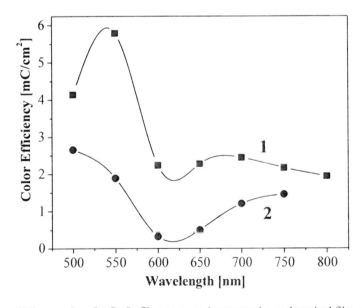

Figure 13. Color efficiency values for Cr_2O_3 films as curve 1 presents the as-deposited film, curve 2 – annealed at 500°C.

The diffusion coefficient D of the diffusing species is of particular interest. The diffusion coefficient can be determined from the formula given in Ref. (24). The estimated values for H insertion in Cr_2O_3 films are in the range of 1.6 to 8.3×10^{-7} cm^2/s. The diffusion coefficients are bigger for the as-deposited film but the difference is not considerable.

MoO_3 and mixed Mo-W oxide films prepared by similar CVD method and technological conditions exhibit some values of D_{H+} in the range of $0.9 - 3 \times 10^{-9}$ cm^2/s. The obtained diffusion coefficients are closer to those received for CVD chromium oxide films. Also, it is well known that these materials based on molybdenum and tungsten oxide are not stable in the acid electrolytes even when glycerin is added to prevent them from destruction.

From the obtained experimental data for inserted and extracted charge, color efficiencies and cyclic voltammetric curves can be concluded that these parameters are very strongly dependent on the preparation conditions including temperature of sublimation, oxygen content in the CVD reactor during film growth and the annealing temperatures. The amorphous and disordered structure favors the electrochromic effect and makes easier the ion injections into film structure, and this could be a reason for the decreased values of the color efficiency for annealed films. The diffusion coefficients have also interesting behaviour. Almost all of their values lie in the range of $1 - 6 \times 10^{-9}$ cm^2/s, only for lower sublimation temperature (60°C) the value is 6.95×10^{-10} cm^2/s. At the same time, this film shows increased CE values especially in the range of 400 – 600 nm (figure 14). There must be noted that for further investigation, film structural improvements are necessary to be done in order to fully examine the EC effect in APCVD Cr_2O_3 films.

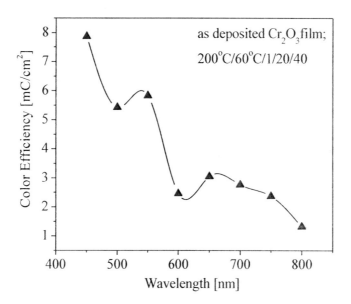

Figure 14. Color efficiency (CE) values of CVD Cr_2O_3 films deposited at the sublimation temperature 60°C in the as-deposited state.

The shape of the recorded CV curves depends on the gas ratio as well as from the additional temperature annealing as can been clearly seen from figures 15 and 16.

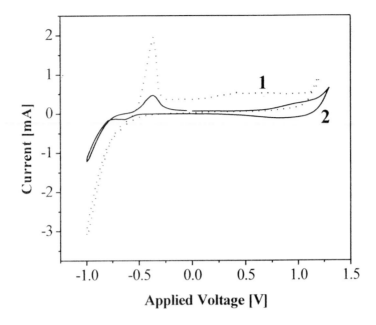

Figure 15. Voltamogramm curves of Cr_2O_3 thin 1/12 films in as deposited state (curve 1) and after annealing at 500°C (curve 2).

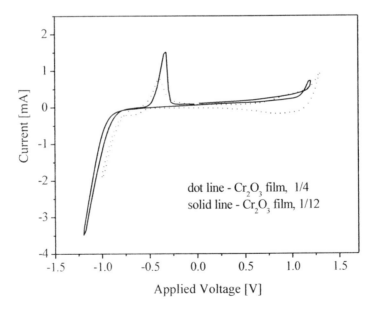

Figure 16. Voltamogramm curves of Cr_2O_3 thin films, deposited at the similar technological conditions, except the gas flow ratios carbonyl vapor towards oxygen is 1/4 and 1/12, as shown on the figure.

Table 3. Electrochromic characteristics of CVD chromium oxide films in dependence on the technological conditions at $T_{sublimator}$ - 70°C. The EC properties are determined at λ= 550 nm

Gas ratio $Cr(CO)_6/O_2$	Annealing Temperature [°C]	CE [cm^2/C]	ΔT [%]	ΔQ [C/cm^2]
1/12	As deposited	4	4	37.3
1/12	500°C	3.2	6	27.8
1/4	As deposited	3.5	11	30.3
1/4	500°C	1.9	7	10.5
1/20*	As deposited	5.8	25	13.5
1/20	500°c	3.4	11	14.9

*$T_{sublimator}$=60°C.

The color efficiency values of the studied Cr oxide films are low enough to be compared f.i. (see table 3) with those of tungsten oxide. The optical modulation is also not acceptable for effective electrochromic applications as working electrodes. These results for the electrochromic properties of chromium oxide films showed that they can be considered as optically passive during ion insertion and extraction, so these films although cathodic type, could be used as counter electrodes in electrochromic device. We note that we have developed Cr_2O_3 phase, only for the CrO_3 is known [3] to be anodic type electrochromic material. By optimization of the technological process parameters APCVD CrO_3 phase could be obtained.

Generally, the Cr oxide films are not enough studied with respect to electrochromism. Poor literature data are available for diffusion constants. The reported values [33] are mainly for Li intercalation for crystalline and heavily disordered oxide films and the values varied in the range of 10^{-8} up to 10^{-9} cm^2/s.

In conclusion, the atmospheric pressure CVD process is suitable method for obtaining thin films of Cr_2O_3 with interesting properties. The Cr oxide films are prepared at relatively low substrate and sublimation temperatures [28, 34]. It was proved that the films remained amorphous even after post-deposition annealing and crystallization appeared at 500°C. IR and Raman studies revealed that the crystallized film possessed corundum Cr_2O_3 structure.

The films show good optical quality with optical transmittance of about 70% in the visible spectral range.

REFERENCES

[1] T. Bredow, *Surface Science.* 401 (1998) 82.
[2] J. E. Maslar, W. S. Hurst, W. J. Bowers Jr., J. H. Hendricks, M. I. Aquino, I. Levin, *Appl. Surf. Sci.*, 180, 102 (2001)
[3] C. G. Granqvist, *Handbook of Inorganic Electrochromic Materials*, Elsevier, Amsterdam, (1995).
[4] A.Azens, G. Vaivars, L. Kullman, C. G. Granqvist, *Electrochimica Acta,* 44, 3059 (1999)
[5] M. Catti, G. Sandrone, G. Valerio, R. Dovesi, *J. Phys. Chem. Solids.* 57 (1996) 1735.

[6] B. Brushan, G. Theunissen, X. Li, *Thin Solid Films.* 311 (1997) 67
[7] P. Norby, A. Norlund Christensen, H. Fjellvac, M. Nielsen, *J. Solid State Chem.*, 94, 281 (1991)
[8] G. Contoux, F. Cosset, A. Celerier, J. Machet, *Thin Solid Films*, 292, 75 (1997)
[9] K. Asami, K. Hashimoto, *Corrosion Sci.*, 17, 559 (1977)
[10] M.G. Hutchins, *Surface Technology.* 20 (1983) 301.
[11] K. Sakurai, T. Kojima, H. Itoh, US patent 821777
[12] B. Brushan, *Thin Solid Films.* 64 (1979) 151.
[13] J.M. Seo, K.H. Lee, *Surface Science.* 369 (1996) 108
[14] S. C. York, M. W. Abee, D. F. Cox, *Surface Science,* 437, 386 (1999)
[15] K. S. Walter, J. T. Scheuer, P.C. McIntyre, P. Kodali, N. Yu, M. Nastasi, *Surf. Coat. Techn.,* 85, 1 (1996)
[16] D.-Y. Wang, J.-H. Lin, W.-Y. Ho, *Thin Solid Films*, 332, 295 (1998)
[17] D. Gall, R. Gampp, H. P. Lang, P. Oelhafen. *J. Vacuum Sci. Technol.* A 14 (2) (1996) 374.-
[18] J. E. Maslar, W. S. Hurst, W. J. Bowers Jr., J. H. Hendricks, M. I. Aquino, I. Levin, *Appl. Surf. Sci.*, 180, 102 (2001)
[19] J. Seo, K. H. Lee, *Surface Science*, 369, 108 (1996)
[20] V. M. Bermudez, W.J. DeSisto, *J. Vac. Sci. Techn.*, 19, 576 (2001)
[21] S. Chevalier, G. Bonnet, J. P. Larpin, *Appl. Surf. Sci.*, 167, 125 (2000)
[22] P.M. Sousa, A.J. Silvestre, N. Popovici, M.L. Paramês, O. Conde, *Material Science Forum.* 455-456 (2004) 20.
[23] P.M. Sousa, A.J. Silvestre, N. Popovici, O. Conde, *Applied Surface Science.* 247 (2005) 423.
[24] O.Seiferth, K. Wolter, B. Dillmann, G. Klyvenyl, H.-F. Freud, D. Scarano, A. Zecchina, *Surface Sci,* 421, 176 (1999)
[25] J. Mougin, T. Le Bihan, G. Lucazeau, *Journal of Physics and Chemistry of Solids.* 62 (2001) 553±563
[26] J. E. Maslar, W. S. Hurst, W. J. Bowers Jr., J. H. Hendricks, M. I. Aquino, I. Levin, *Appl. Surf. Sci.*, 180, 102 (2001)
[27] R. Cheng, C.N. Borca, P.A. Dowben, Sh. Stadler, Y.U. Ildzerda, *Appl. Phys. Lett.,* 78, 521 (2001)
[28] T.Ivanova, M.Surtchev, K.Gesheva, *Phys. Stat. Solidi. (a),* 184, 507 (2001)
[29] L.L. Chase L.L., *Phys. Rev. B*, 10, 2226, (1974).
[30] R. Srivastava and L.L. Chase, *Solid State Commun.*, 11, 349 (1972).
[31] P. Hones, M. Diserens and F. Lévy, *Surface and Coatings Technol.*, 120-121, 277 (1999).
[32] K. Gesheva, A. Szekeres, T. Ivanova, *Solar Energy Materials and Solar Cells.* 76, 563 (2003)
[33] C. G. Granqvist, *Handbook of Inorganic Electrochromic Materials*, Elsevier, Amsterdam, (1995).
[34] T. Ivanova, K. A. Gesheva, E. Steinman, M. Abrashev, *Electrochemical Society Proceedings.* Vol. 2005-09 (2005) 928.

GENERAL CONCLUSIONS

Electrochromic metal oxide films have been studied as very perspective materials for energy control in efficient architectural buildings. The atmospheric pressure CVD technology is developed for preparation of, WO_3, MoO_3, Cr_2O_3 and Mo-W oxide mixed films suitable for usage as active working electrodes or as counter electrodes in electrochromic devices.

On the basis of the performed investigations and the obtained results, the following conclusions can be stated:

- The CVD (chemical vapor deposition) technology for preparation of the thin transparent MoO_3 and WO_3 films is optimized by using the carbonyl precursor at the atmospheric pressure. For the first time are produced MoO_3 films at the low deposition temperatures as low as 150 and 200°C.
- The structural, optical and vibrational properties of MoO_3 films are investigated in details. The MoO_3 films show about 80% transmittance for samples, deposited on the ordinary glass and the transparency is about 60% for the films on the conductive glass substrates. It is determined that with an appropriate selection of the technological parameters we can obtain thin MoO_3 films crystallized either in the layered orthorhombic or in the monoclinic structure. The unique orthorhombic layered structure allows easy intercalation/deintercalation of small ions into the intermediate layer spaces in case of application of the MoO_3 films in electrochromic cells.
- By means of the developed technology, the mixed oxide films on the basis of Mo and W are prepared at the deposition temperature of 200°C. The $Mo_xW_{1-x}O_3$ films are produced by two experimental approaches: through common precursor - a physical mixture of $Mo(CO)_6$ and $W(CO)_6$ and by two separate vapor sources. The obtained mixed $Mo_xW_{1-x}O_3$ films are transparent, uniform and amorphous with the predominant presence of the tungsten oxide.
- It is determined that the absorption peak of the optical absorption for CVD mixed $Mo_xW_{1-x}O_3$ films in comparison with pure metal oxides is a broad band centered at 1.85 eV, lying closer to the absorption maximum of the eye sensitivity. It is established that the mixed $Mo_xW_{1-x}O_3$ film properties are not changing considerable with annealing. This fact leads to the conclusion that the additional thermal treatments can be excluded from the technological process.

- The possibility for the CVD preparation of the mixed $Mo_xW_{1-x}O_3$ films with good optical quality at relatively low deposition temperature without necessary additional annealing as well as the ability for the shifting of the optical absorption peak towards the eye sensitivity maximum make the CVD mixed Mo/W oxide films very perspective for usage as electrochromic electrodes.
- The APCVD method was successfully applied for deposition of chromium oxide films at very low substrate and sublimator temperatures. The CVD Cr – O system is found to be amorphous in structure and crystallization begins after high temperature annealings above 400°C. The optical properties are reaching the requirements for electrochromic applications. Although the electrochromic characteristics are not high enough for working electrodes, the good electrochemical behavior (reversible CV curves, stability in acidic electrolytes) presumes their suitability as counter electrodes for electrochromic cells.
- The chosen CVD technology at atmospheric pressure gives very promising possibilities for preparation of transition metal oxide films with qualities for applicability in real electrochromic devices.

APPENDIX I

In addition to the presented in the main text thin film optical coatings, by the APCVD technology employing either the chloride or carbonyl processes some electronic and protective thin film materials were developed on contract or collaboration basis. Some of the more important publications are given below:

K.Gesheva and E.Vlakhov, "Deposition and Study of CVD-Tantalum Thin Films, *Materials Letters,* v. 5, Ns7-8 1987 pp276-279.

K.A.Gesheva, V.Abrosimova, and G.D.Beshkov, "CVD carbonyl thin films of tungsten and molybdenum and their silicides – a good alternative to CVD fluoride tungsten technology, Journal de Physique IV, colloque C2, suppl. *Au Journal de Physique II,* vol.1, C2, 1991 pp 865-869.

K.A.Gesheva, V.Abrosimova, Rapid Thermal Annealing of CVD molybdenum thin films", *Bulgarian Journal of Physics.* 19, 1-2, (1992) pp.78-81.

K.A.Gesheva, T.A.Krisov, U.I.Simkov and G.D.Beshkov, Deposition and Study of CVD-tungsten and molybdenum thin films and their impact on microelectronics technology", *Appl. Surf. Science.* 73(1993) pp.86-89.

K.Gesheva, G.I.Stoyanov, D.S. Gogova, "Chemical Vapor Deposition of Tungsten Carbide Thin Films and Their Structural Characterization, *Proc. of Eight ISCMP,* Varna, 94, pp.444-447.

K.Gesheva, G.I.Stoyanov, R.Stefanov E.S.Vlakhov, "Microstructure and Electrical Characterization of CVD – W and Mo Films as Contacts in Photocells, *MRS,* 95, vol. 363, pp.119-124.

Kostadinka A.Gesheva, E.S.Vlakhov, and G.I.Stoyanov, "Differences in the microstructure and Electrophysical Behavior of W films deposited by chemical vapor deposition using $W(CO)_6$ and WCl_6 precursors, MRS Conf October 4-6, 1994 Austin Texas, Proc.of Advanced Metallization for ULSI Applications in 1994, Eds. Roc Blumenthal, Guido Janssen), pp.457-461.

K.A.Gesheva, E.S.Vlakhov, G.I.Stoyanov, G.D.Beshkov and M.Marinov. "Deposition and Characterization of CVD –Tungsten and Tungsten Carbonitrides on (100) Si", Ceramics International, v.22, N2, pp.91, (1996).

D.Gogova and K.Gesheva. CVD-WC and WC_xN_y Diffusion Barrier Coatings on WC/Co Metalloceramics", *Materials Letters,* v.35, pp.351-356, 1998.

K.Gesheva, D.Gogova, G.Beshkov and V.Popov, "Preparation of WSi$_2$ by RTA Annealing of CVD-W Thin Films", *Vacuum,* v.51, N2, p.181-184, 1998.

K.A.Gesheva, G.I.Stoyanov, D.S.Gogova and G.D.Beshkov, "Preparation of WSi$_2$ by RTA Annealing of CVD *–W Thin Films, MRS, Proc. Vol.* 402, 1996, p.637-642.

K.A.Gesheva, G.Stoyanov, A.Harizanova and R.Stefanov, "Structure and Properties of W and Mo-subcontact layers influenced by thermal annealing", *MRS. Proc.,* Vol. 403, 1996, p. 477.

Appendix II

Application of CVD Coatings of Transition Metal Oxides in X-Ray Mirrors

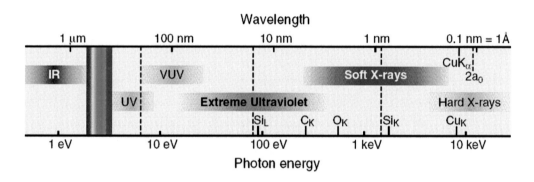

Figure 1. Eelectromagnetic spectrum.

Information about the state-of-the art of X-ray mirrors topic and more could be found at the web site of Lawrence Berkeley National Laboratory - www.ibl.gov (I herein want to thank Eleanor Lee from the LBNL, who emphasized on the increadable work on X-ray mirrors in their Laboratory).

X-rays are a form of electromagnetic radiation (figure 1) with a wavelength in the range of 10 to 0.1 nanometers, corresponding to frequencies in the range 30 to 3000 PHz. Electromagnetic waves with a wavelength approximately longer than 0.1 nm are called soft X-rays. At wavelengths shorter than this, they are called hard X-rays. In contrast to visible light, X-rays cannot be effectively deflected by ordinary mirrors. Either natural crystals have to be used, or (especially for soft X-radiation, where crystals with suitable lattice spacings are not available) a particular preparation of crystal-like structures at the reflecting surface is required, to realize the so-called Bragg reflection in the near-surface region of the mirror.

An actual artificial reflector consists of a stack of typically 100-200 alternating nanometer-thick layers made from two different chemical compounds. They are produced by coating each layer to a precise thickness, with accuracy in the order of picometers.

Layered synthetic microstructures (LSMs) are used in soft X-ray optics as Bragg reflectors for various applications, e.g. collimators, monochromators, mirrors, and focusing optics. In contrast to conventional Langmuir-Blodgett-films, these layer stacks show an outstanding reflectivity of typically 10-80%, even for near-normal incidence of X-radiation. X-rays can be used in a wide spectrum of analytical technologies. Appropriate mirrors will find a widespread use in spectroscopy, e.g. total reflection X-ray fluorescence (TXRF), and in reflectometry (XRR), diffractometry (XRD), microscopy and astronomy.

However, due to the high absorption (in the X-ray reagions the reflectivity of each material drops down with a extend of 10^{-4}) and low refractive index of materials in this wavelength range, neither lenses, nor classical (single reflection) mirrors can be applied. By using positive interference of reflected waves, it is still possible to achieve an acceptable reflectivity.

A multilayer mirror is based on the well-known principle, shown down in figure 2. At each transition from one material to another partial reflection occurs. By carefully selecting the thicknesses of these layers, positive interference occurs, and all individual reflections add up.

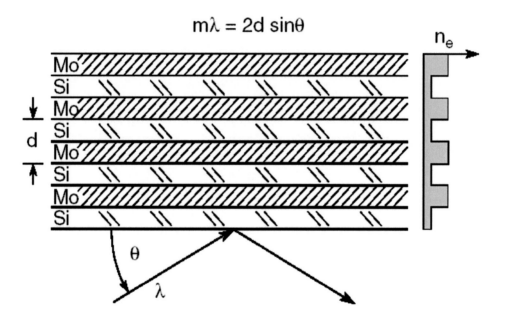

Figure 2. The principle of Bragg diffraction in a multilayer mirror.

The basic relation between the incident wavelength, l, the multilayer periodicity, d, and the grazing angle of incidence, a_i

$$nl = 2d \sin a_i \text{ (Bragg equation)}$$

By adjusting the optical refractive indices within each period of a multilayer system, the multilayer selectivity can be increased. This effort resulted in a highest achieved value of 69.5% reflectivity for a standard Mo/Si multilayer mirror. For the non-standard multilayer mirrors, the best obtained value is currently 69.8%, measured on a $Mo/B_4C/Si/C$ multilayer.

Although these values are already very close to the theoretical maximum for standard Mo/Si systems (74%), the extreme ultraviolet lithography (EUVL) application still demands higher reflectivities.

Applications of the X-Ray Mirrors

X-Ray Microscopy

X-ray microscopes are the short-wavelength analogous of optical microscopes and equally consist of optical elements to collimate the light of a light source onto a sample, and a set of optics for object magnification and imaging. According to Rayleigh's resolution criterion the smallest center-to-center distance Δx between two resolvable objects is:

$$\Delta x = k \frac{1.22 \lambda}{NA},$$

where NA is the numerical aperture of the optical system and k an application dependent parameter. This shows the major advantage of x-ray microscopy above visible light microscopy: the achievable resolution is much higher due to the smaller wavelength of x-rays.

However, additional requirements needed to form an image of the sample are a high optical contrast between different parts of the sample and, obviously, a sufficient transmission of the radiation through the sample. For example for biological samples, consisting of primarily water and carbon, there exist two regions with sufficient optical contrast between the two materials: in the visible region and in a very small region between the oxygen and the carbon absorption edges (2.4 and 4.4 nm), the so-called water window. The penetration depth of the radiation through water is about 10 µm in the water window, just enough to transmit through a complete cell. This is in contrast for example to electron microscopy, which allows very high resolution images, but also requires the cell to be cut into small slices. X-ray microscopy thus enables high-resolution images of complete and (at the moment of exposure) living cells.

Extreme Ultraviolet Lithography (EUVL)

To be able to produce denser computer memory and faster processor chips, it is required that the critical dimensions of the transistors in these chips are as small as possible. This critical dimension is again determined by Rayleigh's criterion applied to the image forming optical system in lithographic equipment. By consequence, lithography companies continuously reduce the wavelength to create smaller transistors. Currently the first demonstration systems for EUV (13.5 nm) lithography are being producing. Since a lithography tool is basically an inversely used microscope, similar issues as for the x-ray microscopy have to be solved: a compact, bright source has to be found and the multilayer mirrors should have maximized reflectivity.

The wavelength of 13.5 nm has been selected because of its proximity to the Si-L absorption edge. Similar to the material selection for the mirrors in the microscope, this enables the use of Mo/Si multilayer mirrors with a theoretical reflectivity of 74%. A value of 69.5% has been demonstrated in our group.5 Although this reflectivity seems very high, the total optics transmission is still low (<3%) due to the ten multilayer optical elements that need to be used. Any increase in reflectivity is, commercially, very desirable. Similar to the situation in x-ray microscopy, the multilayer reflectivity is mainly limited due to the non-zero interface widths.

RECENT DEVELOPMENTS ON X-RAY MULTILAYER SYSTEM L, AS DEPICTED FROM THE WEBSITE WWW.IBL.GOV:

Multilayer of molybdenum carbide and amorphous Si(Mo_2C/Si) developed by Lawrence Livemore and the National Institute of Standards and Technology (NIST).

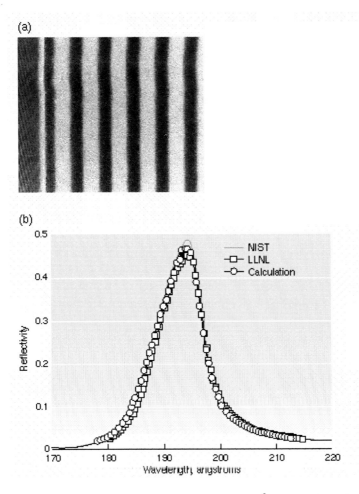

Figure 3. Multilayer of molybdenum carbide and amorphous Si(Mo^2C/Si) are shown in a transmission electron microscopy image at 4 million magnification after annealing at 500 °C. (b) Reflectivity measurements of Mo^2C/Si, taken by Lawrence Livermore and the National Institute of Standards and Technology (NIST), are extremely close to the theoretical calculation of a perfect structure at the wavelengths shown.

At Lawrence Livermore National Laboratory reserchers are pioneering entirely new applications for these materials, which many now believe to constitute an essentially new state of matter.

Multilayers' alternating layers can vary in number from a few to more than 200,000 individual layer in thickness range from a few atoms to a few thousand atoms, corresponding to a maximum structure thickness of about 10 millionths of an inch. The repeat distances in the multilayers, that is, the thickness of two adjacent layers, can be purposely selected to be identical to the interaction lengths characteristic of important physical properties (e.g., magnetic interaction lengths) to yield new properties. Multilayers are part of a larger, established scientific field of so-called designer or "nanostructured" materials that represent the current limits of materials engineering and that are currently impacting numerous Laboratory research programs. Indeed, multilayers are among the first materials to be designed and fabricated at the atomic level, a capacity termed by Barbee as "atomic engineering".To date, Barbee's team of material scientists, engineers, and technicians has synthesized multilayers from 75 of the 92 naturally occurring elements in elemental form or as alloys or compounds. With that wealth of experience, the team has emerged as one of the world leaders in multilayer science and its applications. The team has also partnerships with

Plasma Enhanced Carbonyl CVD process of W and Mo Based Transition Metal Oxides for obtaining of multilayer X-Ray Mirror Optical Systems

An important application for Mo and W based transition metal oxides are the multilayer systems as mirrors for the extreme ultraviolet radiation (EUV, wavelength between 2 and 50 nm). Such multilayer can be obtained by alternating deposition of two materials with different complex refraction index and a double layer thickness in the range of the EUV wavelength. Because of the importance for "next generation" chip technology (the so called extreme ultraviolet lithography EUVL), a lot of studies have been done for the 13.5 nm wavelength. Carbon and oxygen impurities will dramatically decrease the reflectivity at this wavelength. For shorter wavelengths, in the "water window" between the carbon and oxygen absorption edges at 4.4 and 2.4 nm, the water is transparent, while carbon, as a main part of the organic substance, is absorptive, so a beam with such a short wavelength could be used for analysis of water contamination. Furthermore, metal oxides form very stable interfaces and diffusion barriers in multilayered structures, also a very low interface roughness with small roughness correlation lengths. This is very important for deposition of multilayer with a high number of periods and ultra thin layers, what is another advantage of oxide multilayer.

In the Faculty of Physics, Bielefeld University, a plasma enhanced CVD method (PECVD) is applied, employing as precursor Mo and W hexacarbonyl ($Mo(CO)_6$ and $W(CO)_6$ for the crystal planes films and organic precursor - TEOS $Si(OC_2H_5)_4$ for the spacer films. Thus, a multilayer system of 40 double layers has been fabricated with interface roughness of less than 0.2 nm.

Oxygen Plasma

In combination with SiO_2 (deposited from TEOS $Si(OC_2H_5)_4$) multilayer with WO_3 single layer thickness of less than 1 nm and 40 periods have been produced with the PECVD equipment. Multilayer optical systems were obtained in a plasma reactor, where the plasma source is situated above the substrate. The substrate is heated at 160^0C by a high frequency generator, working at 13.5 MHz, and the plasma is driven in oxygen flow of 4 sccm.

Using N_2 as carrier gas the precursor is introduced near the Si substrate through a special nuzzle. A pressure of 1 Pa was maintained by a molecular drag pump. The plasma power during the deposition of MoO_3 and SiO_2 films was 2 W, sufficient for a deposition rate of 4 nm/min for MoO_3 (from $Mo(CO)_6$) and 5 nm/min for SiO_2 (from $Me_5C_5Si_2H_5$).

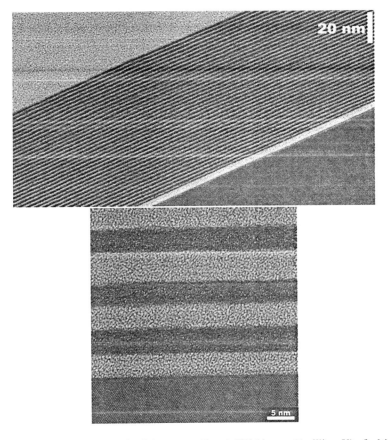

Figure 4. WO_3/SiO_2 multilayer system, d = 2.1 nm, n = 40 and TEM image (4millionsX) of a MoO_x/SiO_y multilayer system, the two an axample for the achievment of very smooth interfaces between the reflecting transition metal oxides surfaces and the surface of the conventional oxide layer, playing the role of "spacer" film (deposited by Frank Hamelmann using PECVD, Faculty of Physics, Bielefeld University, Germany).

Resulted multilayered structure is presented on figure 4. Oxide multilayer systems can provide generally very stable interfaces with no inter diffusion. Since the interfaces are smooth, because of the necessary small period thickness (5 nm) and high number of double layers (24 periods), such system could increase the reflection in the short wavelength region. We see from the figure that the fabricated by PECVD techniques WO_3/SiO_2 multilayer systems has an increased reflectivity not only of monochromatic wavelength, but for wavelengths in the whole water window range. The 2θ values between 50 and 64 correspond exactly to the wavelengths in this range (2.2-4.4 nm).

The in-situ soft X-ray measurement of the reflectivity, applied during the PECVD deposition, consists of a XPS X-ray source coated with carbon, which is emitting radiation of 4.4 nm The source is mounted at 20° to the substrate. A proportional counter under the same angle of incidence detects the reflected radiation. The resulting reflectivity curves show an oscillating behaviour due to wave interference of reflected at the film/substrate interface and the film surface. The values of wavelength λ and angle of incidence α determine the film thickness d for one period of oscillation, as given in first order by Bragg's law $\lambda = 2d\cos\alpha$. This method gives information about surface roughness and film density, also. Due to the absorption of the film, this method is only useful for smooth films up to a thickness of 10 nm (for high absorbing metals) or 100 nm (for low absorbing materials like SiO_2).

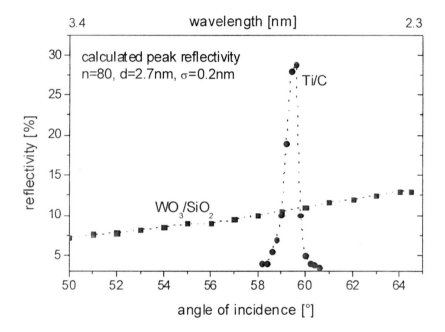

Figure5. Reflectivity of a WO_3/SiO_2 multilayer system as compared to Ti/C based reflector in the water window range.

It is seen from figure 5 the increased reflectivity of the PECVD O3/SiO2 mirror in a large range, corresponding to the water window range(2.2-4.4 nm). Mirrors based on metallic layers are usually reflective in a very short wavelength range as for instance the classical example of Ti/C X-ray mirror, given in the figure.

The Science of X-ray mirrors is a very prospective, very interesting area of material research, recently gathering the interests of prestigious world Laboratories.

REFERENCES

F.Hamelmann, G.Haindl, J.Schmalhorst, A.Aschentrup, E.Majkova, U.Kleineberg, U.Heinzmann, A.Klipp, P.Jutzi, A.Apochenko, M.Jergel, S.Luby, "Metal oxide/silicon oxide multilayer with smooth interfaces produced by in-situ controlled plasma-enhanced MOCVD, *Thin Solid Films.* 358 (2000) 90-93.

F.Hamelmann, S.H.A.Petri, A.Klipp, G.Haindl, J.Hartwich, L.Dreeskornfield, U.Kleineberg, P.Jutzi, U.Heinzmann,"W/Si multilayers deposited by hot-filament MOCVD", *Thin Solid Films.* 338 (1999) 70-74.

F.Hamelmann, A.Aschentrup, J.Schmalhorst, U.Kleineberg, U.Heinzmann, K.Dittmar and P.Jutzi, "Silicon oxide nanolayers for X-ray optics produced by plasma enhanced CVD", *J.dePhys.* IV France 11 (2001), Pr3-431.

K.A.Gesheva, T.Ivanova, F.Hamelmann, "Multilayer systems based on metal oxide thin films for application in electrochromic devices and X-ray mirrors, *Nanoscience and Nanotechnology,* 4 eds. E.Balabanova, I.Dragieva, Heron Press, Sofia, 2004.

K.A.Gesheva, T.Ivanova, F.Hamelmann, O.Trofimov, "CVD Transition Metal Oxide Films as Functional Layers in "Smart Windows" and X-ray mirrors", NATO Science Book Series, *"Functional Properties of Nanostructures Materials",* eds. R.Kassing et.al. (2006), pp.341-349.

Important Publications on the Topic

Eberhard Spiller, "Low-Loss Reflection Coatings Using Absorbing Materials", *Appl. Phys. Lett.,* Vol. 20, N 9, 1 May 1972.

E.Spiller, A.Segmuller, J.Rife, R.-P.Haelbich, *Appl. Phys. Lett.* 37(1980) 1048.

T.W.Barbee, S.Mrowka, M.C. Hettrick, *Appl. Opt.* 24 (1985) 4501.

Hisataka Takenaka, Yasuji Muramatsu, Hisashi Ito, Eric M. Gullikson, and R.C.C. Perera, "Soft X-ray Reflectivity and Structure Evaluation of Ni/Ti and Ni-N/Ti-N Multilayer X-ray Mirrors for Water-Window Region, a work done by the NTT Advanced Technology Co., Musashimo, Tokyo 180-8585, Japan and Lawrence Berkely National Laboratory (LBNL), Berkely, CA 94720, USA

H.Takenaka, T.Kawamura and H.Kinoshita, Fabrication and evaluation of Ni/C Multilayer soft x-ray mirrors, *Thin Solid Films,* 288, 99-102 (1996).

U.Wiesmann, J.Thieme, P.Guttmann, R.Fruke, S.Rehbein, B.Hiemann, D.Rudolph and G.Schmahl, *"First Results of the New Scanning Transmission X-ray Microscope at BESSY – II*

The work is done in: Institute for X-ray Physics, Georg-August-Universitat Gotingen, Geistr. 11, 37073 Gottingen, Germany;

RheinAhrCampus Remagen, *Sudallee.* 2, 53424 Remagen, Germany;

Royal Institute of Technology/SCFAB, BiomedicalandX-ray Physics, SE-106 91 Stockholm, Sweden; e-mail: uweisem@gwdg.de

Jens Birch, Fredrih Eriksson, Goran A.Johansson and Hans M. Hertz, "Recent advances in ion assisted growth of Cr/Sc multilayer X-ray mirrors for the water window", the work done in Thin Film Physics Division, IFM Linkoping University, SE-581 83, Linkoping, Sweden;

Biomedical and X-ray Physics, Royal Institute of Technology, *SCFAB,* SE-106 91, Stockholm, Sweden, Vacuum, Volume 68, Issue 3, November 2002, pp 275-282.

Tadayuki Ohchi, Toshiyuki Fujimoto and Isao Kolma, "Cr/Sc Multilayer Mirror for Soft X-ray", *Analitical Sciences.* (2001), vol.17 Supplement, 2001, p. 159.

H. L. Bai, E. Y. Jiang, C. D. Wang, and R. Y. Tian, " Period expansion of Co/C and CoN/CN soft x-ray multilayers after annealing". *Journal of Applied Physics.* -- September 1, 1997 -- Volume 82, Issue 5, pp. 2270-2276

Other national laboratories, U.S. industries, universities, federal agencies such as the Department of Defense and NASA, and researchers worldwide.

ABBREVIATIONS, ACRONYMS AND SYMBOLS

δ	Bending vibrations
ν	Stretching vibrations
Å	Angstrom
a/e	Figure of merit
AES	Auger Electron Spectroscopy
AFM	Atomic Force Microscopy
a_n	Solar Absorptance
APCVD	Atmospheric pressure chemical vapor deposition
Arb.	Arbitrary
BAS	Bulgarian Academy of Sciences
CBM	Carbonyl black molybdenum
CCVD	Combustion CVD
CE	Color efficiency [C/cm^2]
CL SENES	Central Laboratory of Solar Energy and New Energy Sources
CR	Contrast Ratio
CV	Cyclic Voltammetry
CVD	Chemical Vapor Deposition
D	Macroscopic Displacement field
d	Film thickness
distorted λ	Refers to a display of reflectance against a wavelength abscissa linear in fractions of the total solar flux or of the reradiation loss at a given temperature
D_H, D_{Li}	Diffusion coefficient for hydrogen or lithium ions
E	Macroscopic Electric field
EC	Electrochromic
E_F	Fermi Energy
E_g	Semiconductor optical bandgap
EMD	emissivity modulation devices
e_n	Thermal emittance
ε	The effective dielectric function $\varepsilon(\lambda)$; the dielectric function of the effective medium of a cermet
eV	Electron volt
$F(\lambda)$	Spectral solar irradiance [W/m^2]

f_a, f_b	Filling factors; the volume fraction occupied by phase a or phase b
FTIR	Fourier Transform Infrared Spectroscopy
I	Intensity
Incoloy	Incoloy alloy 800
IPTM	Institute of Microelectronics Technology
IR	Infrared Spectroscopy
ITO	Indium Tin Oxide
JCPDS	Joint Committee of Powder Diffraction Standards
k	Extinction coefficient
LPCVD	low pressure chemical vapor deposition
M	Mole
MBE	Molecular Beam Epitaxy
Me	Metal
MOCVD	metal-organic chemical vapor deposition
n	Refractive index
nm	nanometer
OCMB	Oxychloride black molybdenum; black molybdenum deposited from molybdenum dioxydichloride (MoO_2Cl_2)
OSC	Optical Science Center in Arizona University
P	Polarization
p	Gas pressure
PC	Propylene Carbonate
PECVD	Plasma enhanced CVD
PV	Photovoltaic
PVA	Poly-vinyl alcohol
PVD	Physical Vapor Deposition
Q	Electric charge
R [%]	Reflectance [%]
R	Gas flow ratio
RE	Reference Electrode
rf	Radio frequency
RHEED	Reflection High Energy Electron Diffraction
SCE	Saturated Calomel Electrode
SE	Spectral Ellipsometry
SEM	Scanning Electron Microscopy
Solar absorptance, a	The fraction of energy absorbed from solar flux incident on a surface, calculated for the air mass 2 solar spectrum
T	Transmission [%]
t	time
$T_{deposition}$	deposition temperature
TEM	Transmission Electron Microscopy
$T_{sublimator}$	The sublimator temperature; in the sublimator the precursor is heated
T_s	Substrate temperature
thermal emittance	The fraction of energy emitted by a surface as compared with that of an ideal blackbody at the same temperature (it is given simply as e).
UV	Ultaviolet

Abbreviations, Acronyms and Symbols

V	voltage
VIS	Visible part of the spectrum
WE	Working electrode
XPS	X-Ray Photoelectron Spectroscopy
XRD	X-Ray Diffraction
α	Absorption coefficient
α - MoO_3	orthorhombic α - phase of MoO_3
β - MoO_3	monoclinic phase of MoO_3
ΔOD	optical density
ΔT	Optical modulation [%]
ε	Dielectric function
λ	Wavelength [nm]

INDEX

A

absorption spectra, 106, 113, 230, 238
access, 114
accuracy, 28, 37, 42, 128, 132, 175, 187, 192, 201, 213, 231, 256, 258, 273
achievement, 132, 238
acid, 91, 103, 104, 106, 111, 122, 147, 165, 212, 265
adhesion, 46, 132, 249
adjustment, 26
adsorption, 1, 103, 106
AFM, 145, 147, 148, 150, 151, 152, 153, 154, 171, 175, 176, 217, 218, 219, 253, 254, 256, 283
aggregation, 8
alcohol, 284
alcohols, 98
alloys, 26, 68, 277
alternative, 105, 184, 224, 271
aluminum, 37, 38, 68, 90, 91, 92
aluminum oxide, 91
aluminum surface, 91
ammonium, 68
amorphous phases, 142
amplitude, 192
anatase, 108, 157, 158
aniline, 106
anion, 211
annealing, 4, 50, 51, 70, 83, 100, 132, 134, 136, 139, 141, 142, 147, 148, 149, 150, 151, 152, 155, 159, 170, 171, 172, 173, 174, 175, 177, 178, 179, 180, 181, 183, 184, 185, 189, 190, 191, 193, 195, 197, 202, 203, 204, 214, 218, 219, 220, 222, 224, 226, 229, 230, 231, 232, 237, 238, 240, 251, 252, 256, 257, 258, 259, 261, 262, 264, 265, 266, 267, 269, 270, 272, 281
annihilation, 142
antimony, 111
argon, 27, 46, 127, 128, 129, 130, 142, 213

assessment, 159
assignment, 136
atmospheric pressure, vii, 2, 46, 71, 129, 168, 169, 207, 213, 243, 249, 263, 267, 269, 270
atomic distances, 24, 29
atomic force, 145, 175
atomic force microscope, 175
atoms, 28, 29, 107, 108, 134, 167, 180, 185, 186, 189, 190, 191, 202, 221, 224, 247, 250, 258, 259, 277
attention, 2, 108, 166
automobiles, 112, 115
autonomy, 116
averaging, 6, 11

B

band gap, 145, 167, 168, 192, 197, 198, 201, 202, 204, 231, 232, 259, 260, 263
bandgap, 154, 156, 197, 201, 263, 283
barriers, viii, 1, 277
baths, 214
batteries, 165, 248
behavior, 3, 4, 6, 14, 15, 16, 63, 84, 99, 107, 113, 156, 159, 166, 168, 170, 198, 204, 223, 226, 233, 234, 249, 270
bending, 180, 188, 221, 223, 226, 238
binding, 133, 134
binding energies, 133, 134
binding energy, 133, 134
blackbody radiation, 15
bleaching, 97, 99, 157
blocks, 39
body temperature, 43, 44, 79
bonding, 111, 132, 139, 223, 248, 259
bonds, 136, 137, 180, 183, 186, 189, 190, 203, 221, 222, 224, 226, 238, 240, 248

bounds, 9, 10, 11, 13, 14, 16, 17, 18, 19, 53, 57, 59, 60, 63, 65, 66, 67
building blocks, 158
buildings, 114, 116, 269
Bulgaria, ix, 1, 5, 21, 97
bulk materials, 1

C

candidates, 108, 166
carbon, 31, 34, 40, 41, 49, 67, 91, 108, 129, 133, 134, 165, 214, 275, 277, 279
carbon atoms, 108
carbon film, 31, 34
carbon nanotubes, 165
carrier, 21, 27, 46, 278
catalyst, 99, 214
catalysts, 98, 165, 168, 248
cation, 106
CE, 101, 123, 158, 207, 212, 236, 237, 238, 240, 264, 265, 267, 283
cell, 102, 103, 107, 108, 110, 111, 113, 119, 121, 122, 176, 180, 242, 243, 247, 264, 275
cellulose, 112, 113
ceramic, 5
cerium, 104
changing environment, 116
charge density, 157, 158, 263
chemical bonds, 180, 181, 186
chemical composition, 28, 29, 30, 33, 39, 40, 41, 47, 127, 214
chemical properties, 4, 67, 90
chemical reactions, 118, 128
chemical stability, 165
chemical vapor deposition (CVD), vii, viii, 2, 4, 21, 22, 23, 25, 27, 28, 32, 33, 41, 45, 46, 49, 50, 67, 68, 69, 71, 72, 73, 78, 80, 81, 83, 85, 87, 89, 121, 127, 128, 129, 130, 131, 132, 133, 134, 135, 136, 137, 138, 139, 142, 144, 147, 150, 155, 156, 157, 158, 159, 160, 162, 165, 168, 169, 170, 171, 172, 173, 174, 175, 176, 177, 178, 180, 181, 183, 185, 186, 187, 188, 189, 190, 192, 193, 195, 196, 199, 200, 201, 202, 203, 204, 205, 206, 207, 209, 211, 212, 213, 214, 217, 218, 219, 220, 221, 223, 226, 228, 229, 230, 232, 233, 234, 235, 237, 238, 239, 240, 241, 242, 243, 247, 248, 249, 250, 251, 253, 254, 257, 258, 259, 260, 263, 265, 267, 269, 270, 271, 272, 273, 277, 280, 283, 284
chromium, 26, 90, 91, 247, 248, 249, 250, 251, 254, 256, 257, 258, 259, 260, 261, 262, 263, 265, 267, 270
classes, 97
cleaning, 180

clustering, 219, 256
clusters, 131, 137, 139, 145, 253
CO_2, 115, 128, 129, 170
coatings, vii, viii, 3, 4, 21, 22, 25, 26, 30, 33, 34, 36, 41, 42, 43, 45, 66, 68, 69, 70, 72, 73, 74, 76, 89, 90, 91, 92, 102, 109, 115, 129, 136, 141, 156, 159, 248, 259
cobalt, viii, 26
collaboration, vii, viii, 271
colloids, 114
complexity, 6, 22
components, 6, 8, 12, 22, 69, 72, 106, 109, 142, 212, 221, 223, 230
composites, 92
composition, 1, 2, 3, 6, 10, 11, 12, 13, 16, 17, 19, 21, 22, 25, 26, 27, 28, 29, 30, 31, 41, 44, 46, 50, 51, 52, 53, 57, 63, 64, 67, 68, 72, 127, 128, 131, 132, 139, 141, 142, 144, 145
compounds, 25, 67, 90, 92, 105, 106, 108, 165, 166, 213, 273, 277
comprehension, 13
concentration, 24, 25, 27, 28, 29, 30, 31, 33, 39, 40, 41, 69, 72, 99, 118, 119, 120, 123, 128, 131, 142, 170, 204
conception, 41
condensation, 21, 128
conditioning, 115, 118
conduction, 69, 110, 114, 144, 154
conductivity, 22, 90, 103, 104, 113, 166
conductor, 102, 103, 104, 105, 119
configuration, 16, 41, 73, 98, 107, 109, 115, 132, 198, 211
Congress, iv, 95, 162
constant rate, 52
constituent materials, 2, 11, 68
construction, 105, 115, 169, 242
consumption, 113, 114, 120
contaminants, 22
contamination, 277
continuity, 141
control, 1, 2, 21, 22, 45, 99, 100, 106, 109, 110, 114, 115, 116, 117, 121, 238, 269
conversion, 3, 4, 14, 67, 68, 69, 71, 72, 91, 92, 120
cooling, 3, 24, 74, 76, 89, 114, 115, 117, 159
copper, 92
correlation, 3, 67, 72, 132, 201, 204, 277
corrosion, 1, 26, 72, 90
costs, 115
couples, 98
covering, 46, 64, 88, 181
crystal phases, 167, 177, 181
crystal structure, 26, 50, 107, 176, 189, 222, 258

crystalline, 1, 33, 40, 41, 49, 109, 129, 135, 136, 139, 141, 142, 147, 150, 151, 152, 153, 156, 177, 178, 179, 180, 186, 190, 198, 202, 203, 204, 220, 223, 224, 226, 229, 231, 232, 238, 240, 247, 250, 251, 252, 258, 259, 260, 263, 267
crystallinity, 127
crystallisation, 147, 150, 152
crystallites, 134, 139, 147, 150, 151, 171, 180, 220, 224
crystallization, 134, 142, 145, 151, 156, 175, 177, 179, 185, 189, 191, 203, 219, 223, 224, 226, 229, 231, 238, 240, 251, 252, 253, 257, 258, 262, 263, 267, 270
crystals, 101, 211, 213, 273
current limit, 277
current ratio, 121
customers, 113
cycles, 101, 104, 109, 116, 159, 204, 207, 243, 264
cycling, 74, 90, 99, 104, 159, 211, 212, 242, 243

D

data analysis, 260
database, 198
decay, 120
decomposition, 21, 26, 32, 41, 129, 136, 137, 169, 170, 213, 263
defects, 24, 25, 31, 34, 77, 78, 79, 83, 84, 85, 86, 88, 101, 201, 202, 203, 204, 211, 232, 243, 262
deficiency, 131, 132, 134
deficit, 156
definition, 6
deformation, 24, 139, 180, 181, 183, 189, 190, 222, 224
degradation, 90, 91
degree of crystallinity, 130, 135
demand, 114
density, 25, 26, 64, 122, 123, 129, 130, 131, 139, 141, 142, 144, 156, 165, 169, 206, 214, 232, 234, 240, 263, 279
density values, 131
Department of Defense, 281
Department of Energy, 117
deposition, vii, viii, 1, 2, 3, 4, 21, 22, 24, 25, 26, 27, 39, 40, 41, 42, 43, 45, 46, 47, 49, 53, 67, 70, 72, 73, 74, 75, 78, 80, 81, 83, 85, 87, 89, 91, 127, 128, 129, 130, 131, 132, 133, 135, 136, 137, 139, 142, 144, 145, 147, 150, 152, 154, 155, 156, 157, 160, 165, 166, 168, 169, 170, 171, 172, 180, 181, 183, 185, 193, 195, 197, 198, 199, 200, 201, 206, 213, 214, 220, 230, 231, 238, 243, 248, 249, 250, 252, 254, 256, 261, 267, 269, 270, 277, 278, 279, 284

deposition rate, 26, 46, 169, 170, 278
deposits, 70, 169, 249
derivatives, 180
destruction, 263, 265
detection, 175, 250
deviation, 129, 139, 154
dielectric constant, 154
dielectrics, 1
differentiation, 28
diffraction, 32, 33, 48, 135, 136, 175, 177, 178, 179, 212, 274
diffraction spectrum, 32, 33
diffusion, viii, 1, 22, 45, 49, 70, 111, 112, 116, 118, 119, 120, 122, 128, 193, 206, 235, 264, 265, 267, 277, 278
diodes, 211
dipole moment, 6, 186
direct action, 98
discomfort, 115
disorder, 139, 240
dispersion, 139, 144, 145, 154, 155, 156, 192, 202, 203, 204, 262
displacement, 6
disposition, 178
distilled water, 181, 249
distribution, 12, 28, 29, 49, 110, 171, 261
distribution function, 28
double bonds, 137, 189, 238
drying, 31, 34, 157
durability, 22, 72, 89, 90, 100, 108, 116, 117, 157, 159, 212
duration, 39, 43, 45, 49, 178, 193, 199
dyeing, 108

E

earth, 109
economic efficiency, 105
elastic deformation, 25
electric current, 114
electric field, 6, 7, 10, 98, 119, 204
electrical conductivity, 22, 99
electrical properties, 247
electricity, 3, 112
electrochemical deposition, 127
electrochemical reaction, 118, 119, 121, 235
electrochromic films, 99, 170
electrodeposition, 22, 92, 168, 212
electrodes, 22, 102, 104, 106, 110, 111, 112, 120, 121, 122, 157, 211, 241, 243, 248, 264, 267, 269, 270
electroluminescence, 98
electrolysis, 68

electrolyte, 25, 102, 108, 111, 112, 113, 118, 157, 159, 206, 207, 242, 243, 263
electromagnetic, 6, 115, 273
electron, 24, 28, 31, 34, 35, 36, 40, 47, 48, 98, 102, 103, 104, 105, 106, 107, 114, 118, 119, 121, 132, 159, 166, 172, 175, 176, 204, 231, 252, 263, 275
electron diffraction, 35, 36, 40, 47, 48
electron microscopy, 34, 172, 252, 275
electronic circuits, 113
electronic structure, 97
electrons, 1, 23, 24, 28, 29, 97, 99, 101, 102, 103, 104, 105, 107, 111, 114, 119, 120, 123, 133, 134, 166, 176, 204, 213, 248
electro-optical properties, 139
electroplating, 168
emission, 15, 25, 70, 98, 109
energy, viii, 3, 14, 15, 16, 22, 23, 28, 29, 67, 68, 69, 91, 92, 98, 100, 102, 107, 109, 114, 115, 116, 117, 118, 127, 129, 132, 133, 144, 146, 147, 154, 155, 156, 159, 165, 167, 169, 176, 186, 192, 197, 198, 201, 202, 204, 213, 230, 231, 232, 248, 260, 263, 269, 284
energy consumption, 114
energy density, 3
energy efficiency, 115
energy transfer, 69
environment, 24, 25, 78, 79, 82, 84, 89, 109, 127, 129, 135, 136, 148, 149, 151, 165, 180, 259
environmental conditions, 3, 114
equilibrium, 21, 22, 121, 169
equipment, 28, 29, 38, 109, 113, 117, 122, 169, 172, 224, 249, 252, 275, 278
etching, 113, 249, 261
ethanol, 157, 249
ethylene, 103, 104
ethylene glycol, 104
evaporation, 113, 145, 156, 168, 212, 248
excitation, 97, 180, 187, 258
experimental condition, 256
exposure, 89, 90, 99, 275
external environment, 92
extinction, 13, 66, 111, 123, 144, 198, 199, 201, 202, 204, 232, 261
extraction, 29, 97, 99, 104, 108, 166, 205, 233, 248, 267
extrapolation, 197, 231
eyes, 116

F

fabrication, 2, 3, 4, 21, 23, 26, 68, 69, 71, 111, 249
failure, 3, 4, 85, 86, 89, 90, 101, 116
family, ix, 105, 106, 247

fiber optics, 21
filament, 280
film thickness, 24, 29, 41, 42, 43, 46, 51, 92, 123, 129, 130, 135, 141, 142, 192, 197, 198, 199, 201, 202, 203, 214, 229, 232, 240, 243, 249, 260, 279
flame, 169
flexibility, 1
fluctuations, 8
fluid, 69, 70, 72
fluorescence, 114, 274
focusing, 274
fractional composition, 11, 57, 60, 61, 62, 63
France, 93, 162, 209, 244, 280
FTIR, 238, 239, 240, 257, 258, 284
fuel, 69

G

gas phase, 22, 27, 213, 215
gas sensors, 165, 212
gases, 21, 99, 130, 131, 132, 142, 144, 152, 165
gel, 92, 100, 110, 157, 158
generalization, 9
generation, 3, 109
Germany, viii, 117, 278, 280
glass, viii, 102, 110, 117, 119, 128, 131, 139, 148, 149, 151, 153, 157, 159, 160, 169, 171, 172, 173, 174, 176, 177, 178, 179, 193, 195, 197, 203, 204, 207, 219, 220, 221, 231, 242, 249, 250, 251, 252, 253, 259, 269
glasses, 21, 108, 110, 172, 195, 249
glycerin, 122, 263, 265
gold, 9
government, 117
grading, 44, 64
grain boundaries, 6, 10, 13, 24, 25, 49
grains, 2, 8, 11, 12, 13, 14, 34, 53, 57, 63, 172, 175, 218, 219, 252, 253, 256
graph, 33
graphite, 105, 107, 128, 179, 247, 250
grazing, 176, 179, 250, 274
groups, 98, 100, 103, 107, 181, 221, 224
growth, 2, 21, 24, 31, 40, 41, 44, 50, 52, 53, 82, 127, 130, 131, 150, 157, 168, 170, 183, 185, 190, 194, 196, 199, 206, 213, 214, 219, 231, 237, 238, 248, 249, 250, 252, 265, 280
growth mechanism, 53
growth rate, 40, 41, 130, 131, 157, 185, 196, 199, 213, 219, 237, 249, 250
growth time, 157

H

halogen, 122
hardness, 107, 248
heat, 3, 14, 68, 69, 70, 72, 98, 109, 114, 115, 117
heat conductivity, 72
heat loss, 114
heat transfer, 69, 70, 72, 109
heating, 3, 22, 26, 37, 45, 74, 76, 78, 79, 82, 87, 89, 99, 100, 114, 115, 157, 159, 249
height, 175
hip, 277
homogeneity, 132
Honda, 208
host, 2, 7, 8, 206, 211, 232
HRTEM, 135
humidity, 99, 103
hydrazine, 46
hydrogen, 2, 4, 13, 26, 27, 28, 29, 30, 31, 32, 34, 36, 39, 40, 41, 42, 45, 49, 72, 98, 99, 207, 263, 283
hydrogen gas, 98, 99

I

identification, 177
identity, 5, 6, 28
IFM, 281
images, 112, 147, 148, 150, 152, 175, 176, 217, 218, 219, 253, 256, 275
imaging, 275
impurities, 1, 24, 25, 193, 277
in situ, 157
incidence, 38, 132, 176, 179, 192, 203, 250, 274, 279
inclusion, 8, 24
indication, 112, 152, 176, 191, 202, 221, 231
indicators, 100
indices, 64, 156, 198
indium, 105
industrial processing, 3
industry, viii, 21, 117, 159
inequality, 156
infinite, 6, 107, 108
inhomogeneity, 202, 203, 212, 256
initial state, 103
injections, 265
input, 14, 15, 16, 69
insertion, 97, 101, 102, 104, 108, 111, 165, 166, 211, 233, 237, 240, 265, 267
insight, 139
instruments, 109
integration, 117

intensity, 13, 28, 29, 32, 33, 48, 101, 110, 117, 135, 137, 186, 187, 192, 202, 203, 221, 226, 258
interaction, 3, 6, 7, 14, 98, 167, 277
interactions, 3, 23, 70
interface, 69, 102, 105, 120, 198, 276, 277, 279
interference, 42, 91, 192, 274, 279
interpretation, 9, 52, 137
interval, 157
ion implantation, 248
ion transport, 157
ionization, 28
ions, 22, 45, 97, 98, 99, 101, 102, 103, 104, 105, 107, 108, 111, 118, 119, 122, 128, 132, 158, 166, 168, 189, 204, 205, 206, 207, 213, 233, 235, 236, 240, 248, 264, 269, 283
IR, 60, 61, 62, 90, 99, 102, 109, 114, 122, 180, 181, 183, 184, 185, 186, 189, 190, 202, 206, 221, 222, 223, 232, 237, 238, 248, 249, 256, 257, 259, 267, 284
IR spectra, 181, 183, 221, 222, 223, 239, 257
IR spectroscopy, 185, 186, 221, 237
iridium, 247
iron, 68
Israel, viii, 95

J

Japan, 280

K

K^+, 108, 204
kinetics, 30, 40, 118, 132, 157, 201, 211

L

laser ablation, 248
lattice parameters, 178, 180
lattices, 222
leaching, 165
leakage, 108
LED, 122
lifespan, 69
lifetime, 69, 73, 89, 90, 108, 116, 142, 212
ligands, 98
light transmittance, 115, 122
limitation, 128, 193, 249
linear dependence, 66, 144
liquid phase, 168
liquids, 169, 175
lithium, 104, 159, 165, 166, 207, 233, 235, 283
lithography, 275, 277

localization, 102
location, 9, 13, 14, 15, 48, 106, 112
long distance, 120
low temperatures, 26, 27, 169, 248
lying, 269

M

magnesium, 192
management, 117
manufacturing, 110, 113
market, 111
matrix, 2, 5, 23, 26, 91, 111, 112, 113, 114, 232
measurement, 28, 38, 118, 192, 198, 201, 279
measures, 28, 118, 175, 186, 192
mechanical properties, 22
media, 7, 212
melting, 23, 90, 107
melting temperature, 23
memory, 110, 111, 113, 116, 127, 159, 166, 275
metal oxide, viii, 5, 23, 97, 111, 121, 189, 192, 213, 221, 229, 232, 234, 240, 269, 280
metal oxides, 97, 102, 107, 112, 114, 167, 221, 230, 232, 247, 269, 277
metals, 1, 4, 21, 25, 38, 67, 70, 72, 90, 92, 97, 107, 279
microelectronics, 21, 271
micrometer, 113
microscope, 5, 31, 175, 223, 275
microscopy, 145, 175, 274, 275, 276
microstructure, 1, 3, 6, 9, 11, 12, 13, 14, 16, 17, 19, 53, 211, 271
microstructures, 52, 274
migration, 119, 120
mobility, 23, 104, 107, 119, 120, 165
MOCVD, 168, 280, 284
modeling, 66, 142, 152, 198
models, 6, 9, 99, 101, 102, 110, 117
modules, 117
moisture, 100, 105
molecular beam, 168, 248
molecular beam epitaxy, 168, 248
molecules, 30, 98, 99, 111, 131, 137, 138, 139, 141, 142, 170, 180, 186, 250
molybdenum, vii, viii, 2, 3, 4, 5, 13, 16, 17, 19, 23, 25, 26, 27, 29, 30, 31, 32, 33, 34, 35, 36, 37, 38, 39, 40, 41, 43, 45, 46, 47, 49, 50, 51, 52, 53, 57, 59, 63, 64, 65, 66, 67, 68, 69, 70, 71, 72, 73, 74, 75, 76, 79, 82, 83, 84, 88, 89, 90, 166, 167, 168, 169, 170, 171, 178, 180, 181, 183, 185, 190, 191, 193, 196, 199, 201, 202, 205, 206, 207, 212, 213, 214, 220, 221, 222, 224, 226, 229, 230, 240, 243, 249, 265, 271, 276, 283, 284

monomers, 106
morphology, 32, 34, 53, 135, 147, 148, 150, 151, 171, 175, 180, 200, 211, 218, 219, 252, 253, 256
motion, 122
movement, 119, 120, 132
multilayered structure, 277, 278
multiplication, 37
multiplier, 258

N

Na^+, 122, 166
nanocrystals, 114
nanolayers, 280
nanometer, 98, 106, 114, 273, 284
nanometer scale, 114
nanometers, 273
nanoparticles, 91
National Bureau of Standards, 37
NATO, 280
network, 221
next generation, 277
nickel, 31, 34, 91, 92, 104, 108, 247
niobium, 108
NIR, 115, 132, 157
nitrides, 105
nitrogen, 108
noble metals, 107
noise, 37
numerical aperture, 275

O

observations, 63, 135, 172, 232
observed behavior, 262
oil, 128, 168, 213, 214
optical coatings, vii, viii, 22, 42, 141, 165, 271
optical density, 101, 123, 158, 236, 285
optical parameters, 198, 200
optical properties, 1, 2, 3, 4, 5, 6, 9, 15, 17, 19, 22, 23, 25, 26, 29, 41, 46, 49, 52, 53, 54, 67, 69, 72, 89, 97, 101, 103, 105, 109, 114, 123, 127, 129, 132, 139, 160, 166, 167, 170, 192, 193, 195, 196, 201, 204, 212, 229, 230, 240, 247, 249, 258, 259, 263, 270
optical systems, viii, 22, 278
optimization, 38, 43, 211, 267
optoelectronic properties, viii
orbit, 109, 133
organ, viii, 259
organic compounds, 98, 105, 165
organic polymers, 113

orientation, 2, 8, 12, 34, 48, 64, 178, 179, 192
oscillation, 279
oxidation, 45, 90, 106, 110, 114, 128, 131, 137, 141, 142, 167, 170, 178, 193, 199, 202, 203
oxides, 23, 26, 97, 99, 101, 102, 103, 104, 105, 107, 123, 165, 166, 175, 180, 211, 212, 214, 221, 222, 226, 229, 232, 237, 238, 239, 240, 243, 247, 248, 258, 259, 277, 278
oxygen, 2, 4, 26, 29, 30, 31, 33, 40, 41, 45, 46, 47, 49, 50, 52, 69, 73, 77, 78, 79, 82, 84, 89, 99, 107, 127, 128, 129, 130, 131, 132, 133, 134, 136, 137, 139, 141, 142, 144, 145, 148, 149, 150, 151, 152, 154, 156, 158, 167, 168, 169, 170, 171, 178, 179, 180, 183, 185, 189, 190, 191, 193, 194, 195, 196, 197, 201, 203, 204, 206, 211, 212, 213, 214, 215, 216, 221, 224, 232, 238, 239, 240, 247, 248, 249, 250, 252, 259, 265, 266, 275, 277, 278
oxygen absorption, 277

P

paints, 108
parameter, 2, 102, 108, 130, 169, 180, 196, 275
particles, 2, 5, 6, 7, 8, 9, 12, 23, 26, 28, 49, 52, 53, 111
partnerships, 277
passivation, 90
passive, 104, 106, 114, 157, 159, 267
perchlorate, 104
periodicity, 24, 274
permit, 1, 22, 26
personal communication, 94
PET, 103
phase diagram, 215
phase transformation, 22
phenol, 160
phenothiazines, 105, 106
phonons, 23, 102
phosphorous, 103
photocatalysts, 156
photographs, 75, 77, 78, 79, 80, 81, 82, 83, 84, 86, 88
photons, 1, 15
physical properties, 1, 3, 23, 277
physico-chemical properties, 211
physics, 22
pigments, 100, 108
plasma, 113, 168, 169, 249, 261, 277, 278, 280
platinum, 99
poison, 99
polarity, 103, 186, 198, 233
polarization, 6, 7, 38, 102, 192
polarized light, 192

polymer, 103, 108, 111, 112, 113, 159
polymer electrolytes, 159
polymer film, 112
polymer films, 112
polymer matrix, 111
polymerization, 106
polymers, 106, 112
polypropylene, 104
poor, 41, 46
porosity, 98, 127, 129, 130, 132, 156
positron, 132, 142
positron annihilation spectroscopy, 132
potassium, 92
power, 21, 109, 111, 113, 115, 118, 187, 258, 278
prediction, 17, 65, 66
preference, 214
pressure, 26, 40, 46, 72, 90, 128, 131, 137, 168, 169, 170, 198, 213, 214, 263, 278, 283, 284
prevention, 117
privacy, 111, 116
probability, 8
probe, 175
production, 22, 72, 91, 110, 117, 159
program, vii, 37, 66
propane, 106
proportionality, 73, 263
propylene, 104, 121, 263
protective coating, 45, 68, 113, 248
protons, 98, 102, 104
prototype, 117, 243
pumps, 71
purification, 21
PVA, 159, 160, 284
pyrolysis, viii, 46, 128

Q

quantum confinement, 114, 145
quantum dot, 114
quantum dots, 114
quartz, 42, 46, 70, 72, 89, 92, 122, 249

R

radiation, 3, 6, 7, 14, 21, 28, 32, 36, 37, 41, 43, 69, 73, 98, 99, 109, 115, 117, 177, 178, 179, 180, 250, 260, 273, 274, 275, 277, 279
Radiation, 95
radius, 13
Ramadan, 245
Raman scattering measurements, 132

Raman spectra, 136, 137, 139, 185, 187, 188, 189, 190, 191, 203, 224, 226, 228, 229, 231, 239, 240, 258, 262
Raman spectroscopy, 186, 223, 258
range, 2, 3, 9, 10, 22, 27, 28, 30, 34, 36, 37, 38, 40, 41, 42, 44, 45, 53, 60, 62, 63, 70, 72, 73, 74, 76, 77, 79, 84, 87, 90, 99, 100, 104, 105, 107, 109, 114, 115, 116, 127, 129, 131, 133, 134, 139, 141, 144, 157, 168, 169, 170, 180, 181, 185, 187, 189, 190, 192, 193, 195, 196, 198, 200, 201, 202, 203, 204, 206, 211, 212, 213, 215, 221, 229, 230, 232, 236, 238, 239, 240, 242, 243, 249, 256, 257, 258, 259, 260, 261, 263, 265, 267, 273, 274, 277, 279
reactant, 21, 22, 26, 27, 29, 46, 47
reactants, 21, 22
reaction mechanism, 3
reaction temperature, 22
recall, 240
recrystallization, 139
reduction, 2, 26, 27, 30, 32, 33, 40, 41, 49, 50, 52, 64, 72, 106, 107, 110, 113, 114, 120, 129, 134, 204
reflectance spectra, 23, 36, 42, 45, 73, 192, 229
reflection, 24, 35, 36, 40, 41, 46, 69, 73, 91, 106, 178, 192, 198, 273, 274, 279
reflectivity, 99, 274, 275, 276, 277, 279
reflexes, 177
refraction index, 277
refractive index, 13, 64, 139, 141, 144, 155, 192, 198, 199, 200, 202, 203, 204, 232, 261, 274
refractive indices, 73, 156, 167, 198, 204, 274
refractory, 4, 21, 25, 67, 68, 72, 90
regulation, 117
rejection, 115
relationship, 160, 263
relevance, 2
resistance, 22, 24, 72, 90, 104, 116, 157, 169, 249
resolution, 48, 132, 175, 275
resources, 109
response time, 101, 105, 119
returns, 69, 242
rhenium, 107, 180
rice, 126
room temperature, 26, 37, 38, 72, 73, 92, 100, 157, 159, 169, 213, 249
roughness, 48, 53, 66, 143, 147, 150, 152, 153, 154, 175, 198, 218, 253, 277, 279
ruthenium, 108
rutile, 107, 108, 247

S

salt, 106
salts, 2, 104, 111, 119
sample, 24, 25, 28, 32, 33, 36, 37, 38, 40, 43, 44, 46, 72, 75, 77, 79, 82, 83, 84, 85, 88, 89, 91, 132, 133, 134, 135, 138, 139, 172, 175, 176, 177, 178, 180, 185, 186, 187, 192, 198, 252, 257, 258, 275
satellite, 109
satisfaction, 116
saturation, 203
savings, 115, 117
scanning electron microscopy, 32, 145
scatter, 156
scattered light, 186
scattering, 22, 24, 137, 142, 186, 187, 258
security, 100, 111
selecting, 274
selectivity, 14, 15, 16, 19, 23, 26, 53, 54, 57, 67, 68, 69, 72, 90, 249, 274
SEM micrographs, 148, 149, 171, 172, 219
semiconductor, 8, 99, 114, 156, 159, 197, 211
semiconductors, 1, 102, 107, 114, 145
sensing, 165, 168, 212
sensitivity, 29, 166, 192, 202, 212, 243, 269, 270
sensors, 99, 110
separation, 14, 121, 175, 203
series, 11, 97, 102, 110, 176
shape, 6, 64, 65, 66, 120, 192, 200, 203, 221, 223, 265
shares, 4, 67
sharing, 107, 108, 180, 211, 213
shock, 90
Si_3N_4, 31, 34, 41, 45, 46, 69, 71, 72, 73, 75, 78, 79, 80, 81, 82, 83, 84, 85, 86, 87, 88, 90
sign, 101, 105, 123, 159, 169, 181, 187, 224, 253
signals, 110, 203
signs, 224
silane, 46
silica, 46, 91, 92
silicon, 37, 128, 180, 213, 214, 231, 249, 256, 280
silver, 23
similarity, 189, 222
simulation, 92
Singapore, 162
single crystals, 55
SiO_2, 41, 278, 279
SiO_2 films, 278
sites, 49, 102, 107, 108, 166, 213
skin, 7
sodium, 111
solar cells, 156
sol-gel, viii, 92, 127, 157, 158, 159, 165, 212, 242
solid state, 98, 104, 106, 107, 109, 159, 165, 248
solubility, 214
space environment, 109

Index

species, 110, 111, 114, 119, 120, 122, 175, 265
specificity, 176
spectra analysis, 250
spectrophotometry, 132, 230, 231
spectroscopic ellipsometry, 132, 145, 230, 231
spectroscopy, 142, 180, 186, 192, 214, 256, 274
spectrum, 15, 28, 29, 32, 33, 36, 38, 41, 42, 44, 73, 104, 107, 109, 115, 129, 134, 135, 136, 138, 144, 158, 178, 181, 184, 185, 189, 190, 191, 196, 199, 203, 204, 221, 222, 224, 226, 230, 237, 238, 240, 257, 258, 259, 261, 273, 274, 284, 285
speed, 29, 30, 111, 113, 116, 119
spin, 133, 168
sputtering, 29, 92, 134, 145, 156, 165, 168, 248
stability, 22, 26, 89, 99, 101, 105, 113, 115, 116, 159, 166, 243, 270
stages, 49, 50, 51, 128, 248
standards, 114
steel, 22, 70, 73, 84, 89
stoichiometry, 17, 100, 129, 130, 131, 167, 169, 212
storage, 104, 112, 157
strength, 111, 119
stress, 22, 48, 76, 89, 239
stretching, 136, 137, 138, 139, 180, 181, 188, 189, 190, 221, 222, 223, 224, 226, 238, 259
structural changes, 37
structural defects, 25
structural modifications, 239
structural transformations, 145
students, vii
substrates, viii, 22, 28, 42, 46, 70, 71, 72, 73, 78, 80, 81, 83, 85, 87, 89, 92, 102, 103, 113, 128, 131, 139, 142, 150, 157, 159, 169, 171, 172, 173, 174, 176, 177, 178, 180, 193, 195, 198, 203, 204, 207, 220, 231, 249, 250, 252, 253, 256, 259, 263, 269
sulfate, 92, 111
sulphur, 104
summer, 117
Sun, 115, 124
supply, 118
suppression, 15
surface area, 98, 114
surface layer, 141, 142, 144, 145, 147, 150, 152, 153, 154
surface region, 151, 273
surface structure, 64
Sweden, 280, 281
switching, 98, 101, 104, 105, 108, 111, 113, 114, 116, 117, 127, 159, 211, 212, 243
symmetry, 118, 167
synthesis, 2

systems, 8, 9, 11, 22, 26, 67, 71, 90, 92, 98, 109, 111, 112, 117, 118, 127, 129, 175, 211, 275, 277, 278, 280

T

technology, vii, viii, 1, 2, 3, 22, 23, 24, 25, 27, 30, 38, 41, 45, 74, 90, 91, 100, 109, 110, 111, 115, 117, 123, 129, 160, 212, 243, 248, 253, 269, 270, 271, 277
teflon, 128, 213
TEM, 31, 34, 35, 36, 39, 40, 41, 48, 49, 171, 172, 173, 174, 184, 219, 252, 253, 256, 278, 284
temperature, 2, 3, 14, 15, 16, 22, 25, 26, 27, 28, 29, 30, 31, 32, 33, 34, 35, 37, 38, 39, 40, 41, 42, 43, 44, 45, 46, 49, 68, 69, 70, 71, 72, 73, 74, 75, 76, 77, 78, 79, 82, 85, 87, 89, 90, 91, 92, 98, 99, 100, 101, 105, 109, 115, 116, 120, 127, 128, 129, 130, 131, 134, 135, 137, 138, 139, 141, 144, 145, 150, 152, 155, 156, 157, 160, 165, 168, 169, 170, 171, 172, 173, 174, 175, 176, 178, 180, 181, 183, 184, 185, 187, 188, 189, 191, 193, 194, 195, 196, 197, 198, 199, 200, 201, 202, 203, 204, 206, 213, 214, 215, 220, 222, 226, 230, 232, 234, 235, 237, 238, 240, 249, 250, 251, 252, 253, 254, 256, 257, 258, 259, 260, 265, 269, 270, 283, 284
temperature annealing, 223, 232, 237, 259, 265, 270
temperature dependence, 15, 38, 145
TEOS, 277, 278
theory, vii, 6, 7, 8, 9, 11, 13, 14, 16, 17, 18, 19, 25, 52, 53, 57, 59, 60, 62, 63, 64, 65, 66, 67, 102, 133, 141, 261
thermal activation, 21
thermal decomposition, 2, 127, 170
thermal energy, 3, 22
thermal evaporation, 37, 165
thermal expansion, 23, 24
thermal properties, 68
thermal treatment, 142, 172, 180, 181, 195, 229, 237, 240, 269
thermodynamic calculations, 72
thermodynamic cycle, 3
thermodynamics, 118, 130
thin films, viii, 1, 2, 3, 5, 21, 22, 25, 46, 48, 67, 98, 99, 103, 104, 106, 107, 120, 127, 130, 139, 145, 156, 159, 165, 166, 170, 178, 179, 185, 187, 189, 191, 192, 193, 194, 195, 196, 197, 198, 199, 204, 206, 207, 211, 212, 214, 219, 221, 223, 224, 234, 235, 237, 238, 243, 248, 249, 250, 259, 260, 263, 264, 266, 267, 271, 280
threshold, 156
time, vii, 15, 22, 25, 29, 39, 42, 43, 45, 46, 73, 75, 78, 80, 81, 83, 85, 87, 89, 90, 101, 105, 110, 113,

117, 118, 120, 127, 131, 152, 159, 167, 170, 171, 180, 189, 193, 199, 201, 212, 214, 221, 229, 233, 238, 243, 249, 250, 254, 256, 258, 261, 263, 265, 269, 284
tin, 92, 105, 111
tin oxide, 92, 111
titania, 91, 92, 157
titanium, 111, 157
topology, 2, 7, 8, 11, 19, 53, 63, 67
total energy, 109
tracking, 69
trade, 108, 113
trade-off, 113
transformation, 62, 158, 168, 171, 172, 178, 184, 223, 231, 252
transistor, 113
transition, vii, viii, 23, 26, 67, 70, 90, 91, 97, 99, 100, 101, 102, 105, 107, 122, 123, 159, 166, 167, 169, 175, 180, 186, 197, 211, 212, 213, 229, 239, 247, 253, 260, 263, 270, 274, 277, 278
transition metal, vii, viii, 23, 26, 67, 91, 97, 99, 101, 102, 105, 107, 122, 123, 166, 167, 169, 175, 180, 186, 211, 212, 229, 239, 247, 253, 263, 270, 277, 278
transition metal ions, 167
transition temperature, 100
transitions, 99, 102, 114, 144, 154, 158, 166, 197, 201, 204, 213, 231, 232, 263
transmission, 31, 106, 112, 115, 123, 132, 157, 160, 192, 193, 237, 259, 275, 276
transmission electron microscopy, 31
Transmission Electron Microscopy, 34, 252, 284
transmittance spectra, 181, 197
transparency, 98, 99, 104, 107, 115, 130, 131, 154, 159, 193, 194, 195, 199, 204, 207, 264, 269
transport, 119, 168, 211, 213
trend, 150
tungsten, viii, 2, 4, 26, 67, 90, 99, 103, 107, 108, 111, 118, 127, 128, 129, 131, 132, 134, 136, 139, 141, 142, 143, 144, 145, 146, 156, 166, 168, 211, 213, 214, 221, 222, 224, 230, 231, 233, 237, 238, 241, 243, 265, 267, 269, 271
two-dimensional space, 167

U

uncertainty, 38
uniform, 11, 21, 29, 48, 51, 52, 98, 105, 166, 185, 200, 204, 249, 250, 252, 269
universities, 281
users, 210
UV, 98, 105, 108, 114, 115, 116, 192, 193, 194, 195, 196, 197, 207, 229, 230, 249, 259, 260, 284

UV light, 98, 108
UV radiation, 105

V

vacancies, 1, 101, 158, 178, 193, 201, 211, 232
vacuum, 25, 28, 29, 37, 38, 70, 71, 72, 73, 74, 75, 78, 81, 82, 85, 89, 90, 91, 92, 135, 139, 157, 168, 212, 248
valence, 102, 107, 108, 144, 154, 248
values, vii, 27, 29, 31, 32, 34, 37, 38, 39, 40, 44, 49, 53, 57, 59, 62, 63, 66, 71, 72, 74, 75, 89, 92, 105, 115, 121, 122, 123, 129, 131, 139, 141, 142, 144, 145, 147, 150, 152, 154, 156, 167, 168, 177, 180, 195, 197, 198, 199, 200, 201, 202, 203, 204, 206, 207, 213, 214, 229, 231, 232, 234, 235, 236, 240, 243, 252, 259, 260, 261, 263, 264, 265, 267, 275, 279
vanadium, 92, 99, 107
vapor, 21, 24, 26, 46, 90, 91, 129, 131, 168, 169, 170, 183, 195, 196, 212, 213, 214, 215, 218, 238, 249, 250, 252, 266, 269
variable, 13, 105, 109, 115, 196
variation, 2, 26, 29, 46, 47, 48, 64, 67, 100, 201
vehicles, 100, 110
velocity, 119, 120
versatility, 4, 67
vessels, 214
vibration, 139, 186, 187, 238, 256
vision, 110, 113

W

water vapor, 138
wavelengths, 14, 15, 16, 38, 53, 63, 67, 107, 109, 112, 122, 177, 201, 229, 232, 273, 277, 279
wealth, 277
wear, 21
web, 126, 273
windows, 97, 98, 101, 104, 105, 111, 114, 115, 116, 117, 118, 156, 160, 211, 243
wires, 248
working conditions, 45, 90
writing, 112

X

X-ray diffraction (XRD), 31, 32, 33, 34, 39, 40, 41, 48, 49, 132, 134, 135, 136, 177, 178, 179, 180, 202, 220, 237, 250, 251, 252, 258, 259, 262, 274, 285
X-ray diffraction data, 32, 33

X-ray photoelectron spectroscopy (XPS), 92, 132, 133, 134, 279, 285
x-rays, 275

Y

yield, 14, 65, 89, 106, 277

Z

zirconium, 103
ZnO, 102, 114